AFLATOXIN-CONTROL, ANALYSIS, DETECTION AND HEALTH RISKS

Edited by **Lukman Bola Abdulra'uf**

Aflatoxin-Control, Analysis, Detection and Health Risks
http://dx.doi.org/10.5772/66240
Edited by Lukman Bola Abdulra'uf

Contributors

Asemoloye Michael Dare, Sadaf Rashid, Edward Okoawo, Jonathan Segun Gbolagade, Tunde Sheriffdeen Bello, Habiba Zaffer, Nitin Mahendra Chauhan, Faisal Saeed Awan, Bushra Sadia, Fozia Saleem, Işılay Lavkor, Işıl Var, Fabio Granados, Andrea Molina, Rebeca Zamora, Oluwafemi Ayodeji Adebo, Sefater Gbashi, Ntakadzeni Edwin Madala, Judith Phoku, Lizelle Piater, Patrick Berka Njobeh, Xi-Dai Long, Xiao-Ying Huang, Chun-Ying Luo, Xue-Min Wu, Jin-Guang Yao, Chao Wang, Bing-Chen Huang, Qiang Xia, Jun Lu, Xing-Zhizi Wang, Tian-Qi Zhang, Hyang Sook Chun, Fei Tian, Sevcan Öztemiz, Manaf AlMatar, Seteno Karabo Obed Ntwampe, E.F Itoba-Tombo, John Baptist Nzukizi Mudumbi (Pr.Sci.Nat.), Seval Yılmaz, Emre Kaya, Mehmet Ali Kisacam

Published by InTech

Janeza Trdine 9, 51000 Rijeka, Croatia

Publishing Process Manager Romina Rovan
Technical Editor SPi Global
Cover InTech Design team

First published September, 2017
Printed in Croatia
Legal deposit, Croatia: National and University Library in Zagreb

Additional hard copies can be obtained from orders@intechopen.com

Aflatoxin-Control, Analysis, Detection and Health Risks, Edited by Lukman Bola Abdulra'uf
p. cm.
Print ISBN 978-953-51-3457-2
Online ISBN 978-953-51-3458-9

For EU product safety concerns:
IN TECH d.o.o., Prolaz Marije Krucifikse Kozulić 3, 51000 Rijeka, Croatia,
info@intechopen.com or visit our website at intechopen.com.

Meet the editor

Lukman Bola Abdulra'uf is a lecturer at the Kwara State University, Malete, Ilorin, Nigeria. He started his teaching career at the Kwara State College of Education, Ilorin, in 2006. He had his PhD degree in Analytical Chemistry at the University of Malaya, Kuala Lumpur, Malaysia; his MSc degree at the University of Ilorin, Nigeria; and his BSc degree at the Bayero University, Kano, Nigeria. He is currently a TWAS Postdoctoral Research fellow at the COMSATS Institute of Information Technology, Islamabad, Pakistan. His research focuses on analysis of contaminants such as pesticide residues, mycotoxins, food additives, and veterinary drug residues in food samples using microextraction techniques. His current research interests focus on the synthesis of carbon nanomaterials, ionic liquids, and sol-gel for analytical applications and the use of graphene nanomaterials as electrochemical biosensors.

Contents

Preface

Aflatoxins are a group of highly toxic and carcinogenic substances, which occur naturally, and can be found in food substances. Aflatoxins are secondary metabolites of certain strains of the fungi *Aspergillus flavus* and *A. parasiticus* and the less common *A. nomius*. Aflatoxins B1, B2, G1, and G2 are the most important members, which can be categorized into two groups according to the chemical structure, namely, difurocoumarocyclopentenone series and ifurocoumarolactone. Of the 20 aflatoxins identified so far, only aflatoxins B1, B2, G1, and G2 are known to occur naturally and B and G classes refer to the blue and green fluorescence emitted by their metabolites under ultraviolet (UV) light, and the subtype 1 and subtype 2 imply the major and minor compounds, respectively. Aflatoxins fluoresce strongly under UV radiation (ca. 365 nm). The most common food commodities affected by aflatoxins are cereals (corn, wheat, barley, maize, oats, and rye), nuts (hazelnut, peanut, and pistachio nut), dried fruits (fig), and spices (chili powder). Aflatoxins pose a potential threat to human and animal health through the consumption, contact, or inhalation of foodstuffs and feedstuffs prepared from these commodities. As a result of the adverse health effects of mycotoxins, their levels have been strictly regulated especially in food and feed samples. Therefore, their accurate identification and determination remain a Herculean task due to their presence in complex food matrices. The great public concern and the strict legislation incited the development of reliable, specific, selective, and sensitive analytical methods for mycotoxins monitoring that are discussed in this book.

The book comprises 12 chapters. Chapters 1 to 4 discuss the control and prevention of aflatoxin contaminations in foods, and Chapters 5 to 10 discuss the health risk posed by aflatoxin contaminations in food, while Chapters 11 and 12 discuss the new development in the analysis and detection of aflatoxins in food samples. The book contains up-to-date publications of leading experts, and, therefore, it is hoped that the reference cited by various authors will be a starting point to acquire a deeper knowledge on the prevention, control, identification, and determination of aflatoxins in foods and feedstuffs.

I gratefully acknowledge the efforts and expertise of the contributing authors for their time and efforts in preparing the chapters and for their interest in the book project.

I am indebted to the vice chancellor of Kwara State University, Malete, Ilorin, Nigeria, Prof. Abdul Rasheed Na'Allah and all the academic staff of the Department of Chemistry for the support and encouragement. I also acknowledge the support of my wife (Mrs. Rihanat Abdulra'uf), my children, and my colleagues at the School of Basic and Remedial Studies, Kwara State College of Education, Ilorin, Nigeria, for their unwavering support and encouragement during the chapter review process.

My special appreciation and thanks go to the editorial team and publishing manager of InTechOpen Publisher for their promptness, encouragement, and patience during the review and publication process.

Dr. Lukman Bola Abdulra'uf
Department of Chemistry,
College of Pure and Applied Sciences,
Kwara State University,
Malete, Nigeria

Chapter 1

Control of Aflatoxin Production Using Herbal Plant Extract

Fozia Saleem, Bushra Sadia and Faisal Saeed Awan

Additional information is available at the end of the chapter

http://dx.doi.org/10.5772/intechopen.69867

Abstract

The aflatoxins are a group of chemically similar poisonous, carcinogenic fungal secondary metabolites produced by *Aspergillus flavus*, *A. parasiticus*, and *A. nomius*, which are abundant in warm and humid regions of the world. They are probably the most intensively researched toxins in the world due to their carcinogenic and mutagenic effects. Aflatoxins have also been identified as a potential biological weapon for food and water contamination. The four major aflatoxins commonly isolated from different foods and feed stuffs are AFB1, AF B2, AFG1, and AFG2. Aflatoxin contamination of food and feed has gained global significance as a result of its deleterious effects on human as well as livestock health including gastrointestinal dysfunction, reduced feed utilization, anemia, jaundice, liver damage and immunity suppression. The profitability and marketing of various agricultural products are adversely affected by either contamination of aflatoxins or aflatoxin-producing fungi. The foods at highest risk of aflatoxin contamination are maize, chilies, peanuts, and cotton seeds. There are various physical, chemical, and natural methods investigated to prevent aflatoxin production and the growth of aflatoxin-producing fungus in various agricultural products. Here, we describe various natural plant extracts that would be potential source of controlling aflatoxin production in agricultural products.

Keywords: aflatoxin, *Aspergillus flavus*, *Aspergillus parasiticus*, plant extract, agricultural products

1. Introduction

Aflatoxins are poisonous, carcinogenic, mutagenic, immunosuppressive, and teratogenic secondary metabolites formed by *Aspergillus flavus, A. parasiticus* [1], and *A. nomius* [2]. These fungi are ubiquitous species and generally contaminate agricultural products such as rice, wheat, maize, barley, sorghum, black pepper, chili, ginger, coriander, turmeric, pistachios, almonds, walnuts, Brazil nuts, peanuts, oilseeds (cotton, sunflower, sesame, and soybean), milk, cheese, and animal feed [3–9]. The Food and Agriculture Organization (FAO) estimated that around 25% of the world's cereals are contaminated by mycotoxins, including aflatoxins [10]. Aflatoxins were first identified as causative agent of "Turkey X disease" in 1961. Due to this disease, about more than 100,000 young turkeys, ducks, and poultry birds died in England by eating contaminated Brazilian groundnut meal [11–15].

The aflatoxin was a combination of three words: first letter "A" from genus *Aspergillus*, next three letters "FLA" from species *flavus*, and the noun "TOXIN" [16]. Aflatoxins are quite stable and found resistant to degradation. Among the 18 different groups, aflatoxins B1, B2, G1, G2, M1, and M2 are the major classes and derivative of bifuranocoumarins. The aflatoxins B1 and B2 give blue color, while G1 and G2 give a yellowish green color under UV light. Aflatoxins M are hydroxylated derivatives of aflatoxins B and first isolated from milk. *A. flavus* produces only AFB1 and AFB2, but it is also able to synthesize cyclopiazonic acid. However, *A. parasiticus* produces AFBI, AFB2, AFG1, and AFG2 [17, 18].

The International Agency for Research on Cancer (IARC) classified AFB1 as class I human carcinogens [19] and have a positive association between dietary aflatoxins and liver cell cancer (LCC). This was the third leading reason of cancer death around the world [20]. The cytochrome p450 metabolized AFB1 in their epoxide form. Depurination occurs, when epoxide reacts with DNA or RNA. That will obstruct DNA and protein synthesis in active tissues of bone marrow, intestine, and liver. The order of toxicity of aflatoxins is AFB1 > AFB2 > AFG1 > AFG2 [21], and the critical point, which determined the biological activity of this group of mycotoxins, is terminal furan moiety of aflatoxin [22]. In cereal and their derivatives, maximum residual limits (MRLs) of aflatoxins are 2 $\mu g\ kg^{1}$ for AFB1 and 4 $\mu g\ kg^{-1}$ for the sum of four aflatoxins. In processed cereal-based foods and baby foods for infants and young children, the level of AFB1 is 0.1 $\mu g\ kg^{-1}$. These values were recommended by the European Union Commission Regulation (EC) [21]. According to the Food and Drug Administration (FDA), the safe limit of aflatoxins is 20 ppb (**Figure 1**) [23].

In developing countries, about 4.5 billion people are chronically exposed to uncontrolled amounts of aflatoxins [24]. Consumption of contaminated products causes aflatoxicosis in humans and animals. Aflatoxicosis may be acute and chronic. Acute condition caused death, while chronic condition results in immune suppression and cancer. In human, it is characterized by vomiting, abdominal pain, pulmonary edema, convulsions, coma, and death with cerebral edema and fatty involvement of the liver, kidneys, and heart [25]. Due to aflatoxicosis, in Kenya about 215 people died in 2004 [26–28]. In animal, aflatoxicosis is characterized by gastrointestinal dysfunction, reduced feed utilization, anemia, jaundice, liver damage, decreased milk and egg production, and immunity suppression [29]. In plants, AFs retarded

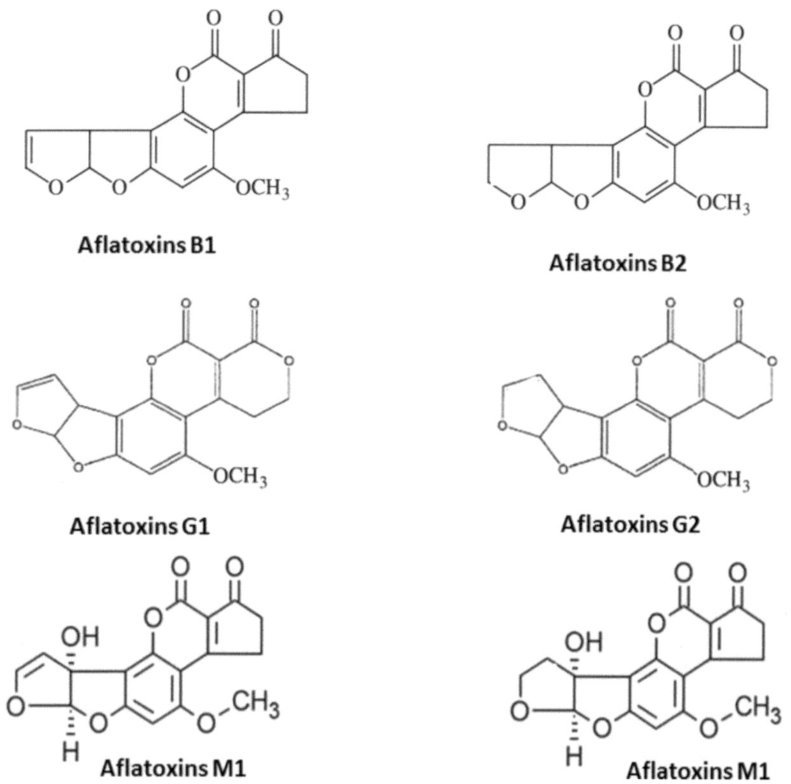

Figure 1. Major classes of aflatoxins.

seed germination, seedling growth, and root elongation. It also inhibits chlorophyll, carotenoid, and some enzymes synthesis [30].

Although *A. parasiticus* and *A. flavus* are related fungi, they are different from each other on the basis of their color and length of conidiophore. Sterigmata were the main characteristics, which differentiate the two *Aspergillus* species. The sterigmata of *A. flavus* were biseriate as compared to *A. parasiticus*, which has uniseriate sterigmata [31]. In 2006, Cary and Ehrlich reported that about 12 *A. flavus* groups are 96% similar to *A. parasiticus*. Another character that distinguishes the two fungi species is their adapted environment. The *A. flavus* acclimated to aerial and foliar environment, mostly prominent in tree nuts, corn, and cottonseed, while the *A. parasiticus* adapted to soil environment and dominated in peanuts [32]. *A. flavus* exists in two forms: one is the S type, while the other is the L type on the basis of morphological, physiological, and genetic characteristics [33]. On average, S-strain isolates produce much more aflatoxins than L-strain isolates [34]. S strain synthesized frequently small sclerotia that are less than 400 μm and processes lesser conidia as compared to L-strain isolates whose sclerotia sizes are greater than 400 μm [34, 35].The members of genus *Aspergillus* mostly contaminate agriculture commodities in tropic and sub-tropic region. Contamination may occur at different stages such as in pre-harvesting stage, harvesting stage, post-harvesting stage, or in storage and transportation stage. In pre-harvesting, the field fungi attack on growing crop because of different reasons. It may be the environmental stress (hot and dry condition and soil moisture), mechanical damage (by arthropods, birds, rodents, and nematodes), or delayed harvesting. While in post-harvesting, contamination occurred due to improper drying, storage in polythene bags, damage during shelling, or storage in poorly ventilated warm environment.

Contamination rate of aflatoxin depends upon humidity, temperature, storage, and soil conditions [36]. Optimum condition for fungal growth in cereal is moisture content about 18% (equal to 85% relative humidity) and temperature about 12–42°C with an optimum at 27–30°C in tropical and sub-tropical areas [37]. An important point to be considered was the time of incubation that effects the production of toxin by *Aspergillus* species [38]. Optimum duration for the production of aflatoxins was 14 days of incubation at 30°C. When the length of incubation time increased, there will be reduction in aflatoxin level because of re-adsorption or degradation by fungus [39]. The fungal growth is effected by 20% CO_2 and 10% O_2 level [40]. The metals such as manganese and zinc are crucial for aflatoxin production. But the mixture of cadmium and iron mixture reduces the mold growth and aflatoxin synthesis [41].

The infectious cycle of *Aspergillus* species is mostly dependent upon host species. Overwinter fungus developed either mycelium or sclerotia (resistant structure) that have the ability to grow on soil surface [42, 43]. Under favorable condition (high temperature and moisture level) in summer, it either produced hyphae or conidia (asexual spores). Through air or insects, conidia spread in soil and on silk and kernels and contaminate agriculture commodities (**Figure 2**) [44].

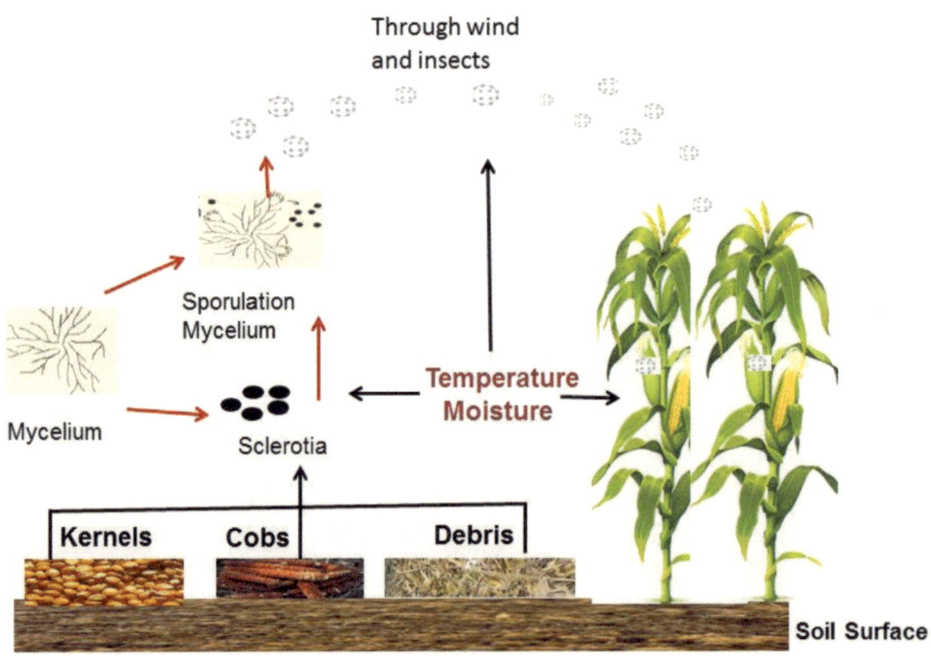

Figure 2. Life cycle of *A. flavus* in field.

Aflatoxin contamination is inescapable due to health hazards in human and animal, crops deterioration, and economical loses. In the past, many strategies (physical, chemical, and biological) are used to avoid aflatoxin contamination. Physical strategies usually used are rodent-proof room, cold storage of feeds with less than 100-g/kg moisture level, use rapid drying and gamma radiation, and so on. In chemical strategies, propionic acid, acetic acid, benzoic acid, citric acid, hydrogen peroxide, copper sulfate, and ammonium hydroxide are used to inhibit the growth of fungi and aflatoxin production. But the formation of toxic residue by chemical treatment was the main concern that causes potent health problems. As compared to chemical, physical practice is a healthier option but it is slow processes. Other strategies used were the biological control in which different microorganisms such as bacteria, yeast, and non-toxic stain of *A. flavus* and *A. parasiticus* were used to detoxify aflatoxins by microbial binding and biotransformation [45–48]. This is a laborious and costly process. Therefore, to avoid potential risk, the use of safe, renewable, and biodegradable natural plant extracts to remove aflatoxin contamination [49] is required.

2. Effect of active ingredients of medicinal plants on aflatoxins producing fungus

Modern research found that phytophenols as plant secondary metabolite existed above 8000 structures. These structures resemble with tannin and phenolic acid [50]. Phytophenols showed antiallergenic, antioxidant, anti-inflammatory, antimicrobial, antiorthrogenic, and antithrombotic activity [51]. These plant compounds exhibited key biological activity in the degradation of many microorganisms [52]. Plants, herbs, essential oils, and spices in powder or extracts form are used to detoxify microbes due to the presence of flavonoids, betalain, phenolics, phytoalexins, and thiosulfonates. But mostly antimicrobial and antioxidant activities of plant extracts were due to their phenolic alignments [53].

A recent study exposed the antifungal and antiaflatoxigenic nature of phenolic components of plant extracts [54–56]. The syringaldehyde, sinapic acid, and acetosyringone were the plant phenolic compounds that inhibited the production of aflatoxin B1 [57]. However, salicylic acid, thymol, vanillyl acetone, cinnamic acid, and vanillin were phenolic compounds that ceased *A. flavus* growth by targeting oxidative mitochondrial stress as defense system [58].

Medicinal plants have been used from centuries for the treatment of various diseases. There are about 53,000 medicinal plants around the world [59]. In developing countries, according to World Health Organization, about 70–95% people used medicinal plants as primary health care for the treatment of diseases [20]. In current scenario, 70% of synthetic medicines are derived from plants [60]. Medicinal plants have antifungal, antimicrobial, anthelmintic, antibiotic, antiviral, anti-inflammatory, antiarthritic, antirheumatic, and antihemorrhoidal properties.

The various medicinal plants native to Southeast Asia including bitter cucumber (*Momordica charantia*), Asiatic pennywort (*Centella asiatica*), betel nut (*Areca catechu*), betel vine (*Piper betle)*, Chaa Phluu (*Piper sarmentosum*), false coriander (*Eryngium foetidum*), Chinese radish (*Raphanus sativus*), clove (*Syzygium aromaticum*), Eucalyptus (*Eucalyptus globules*), Indian mulberry (*Morinda citrifolia*), Madagascar periwinkle (*Catharanthus roseus*),bmangosteen (*Garcinia mangostana*), mandarin (*Citrus reticulate*), onion (*Allium cepa*), pepper (*P. nigrum*), pomegranate

(*Punica granatum*), tomato (*Lycopersicon esculentum*), hedge flower (*Lantanacamara*), roselle (*Hibiscus sabdariffa*), Non Taai Yaak (*Stemona tuberosa*), Raang Chuet (*Thunbergia laurifolia*), Saab Sue (*Chromolaena odorata*), turmeric (*Curcuma longa*), water primrose (*Jussiaeda repens*), and wishing tree (*Cassia bakeriana*) were tested for their ability to control aflatoxins producing fungus [61]. The above study found that ethanolic extracts of some medicinal plant showed the inhibition of aflatoxins producing fungus.

The highest activity was showed by betel vine, a traditional Thai medicine, followed by false coriander, Indian mulberry, Chaa Phluu, Chinese radish, and clove. The leaf of betel vine is used topically for urticaria, contains eugenol and chavicol, and mostly chewed by mouth as antiflatulent, antimicrobial, and antipruritic [62].

Crude ethanolic extract of olive callus in different ratios was used to inhibit the aflatoxins synthesis [63] by the addition of appropriate amounts of extracts onto potato dextrose agar (PDA) to obtain the final concentration of 0.5 and 1%, and *Aspergillus* was then point-inoculated into PDA. The results showed that ethanolic extract of olive callus had no inhibitory effect on fungal growth but it reduced 90% of aflatoxin synthesis. The main compounds in olive callus are reported as caffeic acid, coumarin; o-, p-, or m-coumaric acid and catechin which facilitate the reduction of aflatoxin. Only o-coumaric acid and caffeic acid showed antifungal and antibacterial activity.

Various concentrations (0, 2, 4, 6, 8, and 10% (w/v)) of clove, garlic, and carrot's crude aqueous extracts were tested for their possible inhibitory effect on *Aspergillus* growth and aflatoxin production in 50 g of rice. The results showed that garlic and clove at 10% (w/v) and carrot at 2% inhibited the *Aspergillus* growth and also reduced the level of aflatoxin production in rice [64]. Crude extracts of garlic, eugenol, and onion were used to reduce *A. flavus* growth as well as aflatoxin synthesis in maize and SKMY liquid medium [65]. The study showed that garlic extract inhibited 61.94% fungus growth. However, onion extract ceased about 60.44% aflatoxin synthesis. While on maize grain, eugenol extract reduced 60.35% aflatoxins synthesis. Hussain and Ali [48] compared the antifungal activity of some herbal spices, chemicals, and plants to inhibit the growth of aflatoxins producing fungus like *A. flavus and A. parasiticus*. They found that benzoic and propionic acid showed complete inhibition of *A. flavus* at (0.1–0.5%) and *A. parasiticus* at (0.2–0.5%), while clove (0.5%), garlic (0.5%), and onion (0.5%) showed complete inhibition of both *Aspergillus*.

The aqueous and phenolic extracts of several other natural and medicinal plants have been tested against *Aspergillus* [66]. Aqueous extracts of *Lupinus albus* (Leguminosae), *Ammi visnaga* (Umbelliferae), and *Xanthium pungens* (Compositae) were found to cease the growth of *A. flavus* and also the production of aflatoxin [67]. It was also found that the inhibitory effect was proportional to the applied concentration.

3. Role of essential oils on the inhibition of aflatoxins producing fungus and its production

The search for naturally occurring compounds or metabolites having bioactivity against aflatoxins producing fungi has been the target of interest in the search for ecologically friendly products [68]. There are many essential oils produced by medicinal plants that have been tested for their inhibiting ability of aflatoxin production [69, 70].

Essential oils were extracted from 16 aromatic plants, that is, safflower (*Carthamus tinctorius*), marigold (*Tagetes erecta*), coriander (*Coriandrum sativum*), pomelo (*C. maxima*), mangosteen (*G. mangostana*), *Kaempferia parviflora*, ginger (*Zingiber officinale*), pepper (*P. nigrum*), Boraphet (*Tinospora crispa*), aloe (*Aloe vera*), lavender (*Lavendula officinalis*), rosemary (*Rosemarinus officinalis*), cinnamon (*Cinnamomum cassia*), eucalyptus (*E. globules*), thyme (*Thymus vulgaris*), and white wood (*Melaleuca cajuputi*), and their ability to inhibit the *Aspergillus* on PDA by agar diffusion test [71].

Different ratios (50, 25, 12.5, 6.25%) of each essential oil were placed onto a cylinder cup (6 mm dia) on agar plate streaked with *A. flavus*. It was observed that the essential oil extracted from white wood showed the highest inhibition followed by the essential oils of cinnamon and lavender, respectively. Sindhu et al. [72] used *Curcuma longa* leaves essential oil of 0.01, 0.05, 0.1, 0.5, 0.75, 1, and 1.5% concentration in YES broth that was inoculated with *A. flavus* spores. *C. longa* oil of 1 and 1.5% concentration reduced 95.3 and 100% aflatoxin (AFB1, AFG1) synthesis, respectively. They analyzed α-phellandrene, terpinolene, and p-cymene as an active compound in turmeric leave oil extract by gas chromatography-mass spectrometry (GC-MS). Mahmoud [73] also used 0.01% of five essential oils namely geraniol, nerol and citronellol (aliphatic oils), cinnamaldehyde (aromatic aldehyde), and thymol (phenolic ketone) to suppress the *Aspergillus* growth. The result showed the complete inhibition of *A. flavus* growth.

4. Conclusions

Despite all efforts, it has been very difficult to control the exposure of man and animals to aflatoxins, because of their natural occurrence in the environment. Although the prevention of aflatoxin contamination by inhibiting the fungal growth in food and feeds is the best practice, other measures are also necessary. The advantage of using active compound based on natural plant is that they are safer, ecologically friendly than any chemical compounds, and synthetically produced antimicrobial agents. Other procedures such as the removal or decomposition of aflatoxins are also necessary as the prevention of contamination alone may not always be successful.

Author details

Fozia Saleem[1], Bushra Sadia[2] and Faisal Saeed Awan[1]*

*Address all correspondence to: awanfaisal@yahoo.com

1 Centre of Agricultural Biochemistry and Biotechnology, University of Agriculture Faisalabad, Faisalabad, Pakistan

2 U.S.-Pakistan Center for Advanced Studies in Agriculture and Food Security (USPCAS-AFS), Faisalabad, Pakistan

References

[1] Bhat R, Rai RV, Karim A. Mycotoxins in food and feed: Present status and future concerns. Comprehensive Reviews in Food Science and Food Safety. 2010;**9**:57-81

[2] Murphy PA, Hendrich S, Landgren C, Bryant CM. Food mycotoxins: An update. Journal of Food Science. 2006;**71**:51-65

[3] Adzahan N, Jalili M, Jinap S. Survey of aflatoxins in retail samples of whole and ground black and white peppercorns. Food Additives and Contaminants Part B. 2009;**2**:178-182

[4] Reddy KRN, Reddy CS, Muralidharan K. Potential of botanicals and biocontrol agents on growth and aflatoxin production by *Aspergillus flavus* infecting rice grains. Food Control. 2009;**20**:173-178

[5] Fernandez-Cruz ML, Mansilla ML, Tadeo JL. Mycotoxins in fruits and their processed products: Analysis, occurrence and health implications. Journal of Advance Research. 2010;**1**:113-122

[6] Saleem NM, Ahmad R. Mycotoxins in food from Jordan: Preliminary survey. Food Control. 2010;**21**:1099-1103

[7] Afsah-Hejri L, Arzandeh JS, Mirhosseini H. Optimization of HPLC conditions for quantitative analysis of aflatoxins in contaminated peanut. Food Control. 2011;**22**:381-388

[8] Iqbal SZ, Paterson RRM, Bhatti IA, Asi MR. Aflatoxin concentrations in chilies vary depending on variety. Mycoscience. 2011;**52**:296-299

[9] Firdous S, Ejaz N, Aman T, Khan N. Occurrence of aflatoxins in export-quality Pakistani rice. Food Additives and Contaminants Part B. 2012;**5**:121-125

[10] Dowling TS. Fumonisin and its toxic effects. Cereal Foods World. 1997;**42**:13-15

[11] Hussein HZ. Activity of pomegranate peels and clove powders in detoxification of aflatoxin B_1 and ochratoxin A from contaminated poultry diet. Journal of Plant Pathology and Microbiology. 2015;**6**:1

[12] Howes AD, Newman KE. Compositions and methods for removal of mycotoxins from animal feed. US 6045834. 2000

[13] Marvin HJP, Kleter GA, Frewer LJ, Cope S, Wentholt MTA, Rowe G. A working procedure for identifying emerging food safety issues at an early stage: Implications for European and international risk management practices. Food Control. 2009;**20**(2):345-356. ISSN 0956-7135

[14] Lutfullah G, Hussain A. Studies on contamination level of aflatoxins in some cereals and beans of Pakistan. Food Control. 2012;**23**:32-36

[15] Jouany JP. Methods for preventing, decontaminating and minimizing the toxicity of mycotoxins in feeds. Animal Feed Science and Technology. 2007;**137**:342-362

[16] Ellis WO, Smith JP, Simpson JP, Oldham JH. Aflatoxins in food: Occurrence, biosynthesis, effects on organism's detection and methods of control. Critical Reviews in Food Science and Nutrition. 1991;**30**:404-439

[17] Diener UL, Cole RJ, Sanders TH, Payne GA, Lee LS, Klich MA. Epidemiology of aflatoxin formation by *Aspergillus flavus*. Annual Review of Phytopathology. 1987;**25**:249-270

[18] D'Mello JPF, Macdonald AMC. Mycotoxins. Animal Feed Science and Technology. 1997;**69**:155-166

[19] World Health Organization. IARC, Monograph on the Evaluation of Carcinogenic Risk to Humans. Vol. 82. Lyon: World Health Organization; 2002. p. 171

[20] World Health Organization (WHO). The world medicines situation 2011. In: Traditional Medicines: Global Situation, Issues and Challenges. Geneva: WHO Press; 2011

[21] Quinto M, Spadaccino G, Palermo C, Centonze D. Determination of aflatoxins in cereal flours by solid-phase microextraction coupled with liquid chromatography and post-column photochemical derivatization-fluorescence detection. Journal of Chromatography. 2009;**1216**:8636-8641

[22] World Health Organization. IARC, Monograph on the Evaluation of Carcinogenic Risk to Humans. Vol. 1. Lyon: World Health Organization; 1987. p. 82

[23] Achakzai AKK, Bazai ZA. Occurrence and health hazard status of aflatoxin in human food and animal feed of wheat from Pakistan: A review paper. Pure and Applied Biology. 2015;**4**:611

[24] Shukla R, Kumar A, Prasad CS, Srivastava B, Dubey NK. Antimycotic and anti-aflatoxigenic potency of *Adenocalymma alliaceum* Miers. on fungi causing biodeterioration of food commodities and raw herbal drugs. International Biodeterioration and Biodegradation. 2008;**62**:348-351

[25] United States Department of Agriculture (USDA). Grain Fungal Diseases & Mycotoxin Reference. Washington, DC: United States Department of Agriculture; 2004

[26] Baumgartner T, Heinrichs M, Vonlanthen A, Fischbacher U, Fehr E. Oxytocin shapes the neural circuitry of trust and trust adaptation in humans. Neuron. 2008;**58**(4):639-650

[27] Lewis BP, Shih I, Jones-Rhoades MW, Bartel DP, Burge CB. Prediction of mammalian microRNA targets. Cell. 2003;**115**:787-798

[28] Lewis L, Onsongo M, Njapau H, Schurz-Rogers H, Luber G, Kieszak S, Nyamongo J, Backer L, Dahiye AM, Misore A, DeCock K. Aflatoxin contamination of commercial maize products during an outbreak of acute aflatoxicosis in eastern and central Kenya. Environmental Health Perspectives, 2005;**113**(12):1763

[29] Logrieco A, Bottalico A, Mule G, Moretti A, Perrone G. Epidemiology of toxigenic fungi and their associated mycotoxins for some Mediterranean crops. European Journal of Plant Pathology. 2003;**109**:645-667

[30] Jonathan SG, Adeniyi MA, Asemoloye MD. Fungal biodeterioration, aflatoxin contamination, and nutrient value of "suya spices". Scientifica. 2016;**2016**

[31] Raper, K.B. and Fennell, D.I., 1965. The genus Aspergillus. The genus Aspergillus.

[32] Cary JW, Ehrlich KC. Aflatoxigenicity in *Aspergillus*: Molecular genetics, phylogenetic relationships and evolutionary implications. Mycopathologia. 2006;**162**:167-177

[33] Bayman P, Cotty PJ. Genetic diversity in *Aspergillus flavus*. Association with aflatoxin production and morphology. Canadian Journal of Biotechnology. 1993;**71**:23-31

[34] Bennett JW, Klich M. Mycotoxins. Clinical Microbiology Reviews. 2004;**16**:497-516

[35] Cotty PJ. Virulence and cultural characteristics of two *Aspergillus flavus* strains pathogenic on Conon. Phytopathology. 1989;**79**:808-814

[36] Mojtahedi H, Danesh D, Haghighi B, Barnett R. Postharvest pathology and mycotoxin contamination of Iranian pistachio nuts. Phytopathology. 1978;**68**:1800-1804

[37] Hesseltine C. Recent research for the control of mycotoxins in cereal. Pure and Applied Chemistry. 1973;**35**:251-258

[38] Park K, Bullerman L. Effects of substrate and temperature on aflatoxin production by *Aspergillus parasiticus* and *Aspergillus flavus*. Journal of Food Protection. 1983;**46**:178-184

[39] Kheiralla Z, Hassanin N, Amra H. Effect of incubation time, temperature and substrate on growth and aflatoxin production. International Biodeterioration and Biodegradation. 1992;**30**:17-27

[40] Bankole SA, Adebanjo A. Mycotoxins in food in West Africa: Current situation and possibilities of controlling it. African Journal of Biotechnology. 2003;**2**:254-263

[41] Gilbert J, Anklam E. Validation of analytical methods for determining mycotoxins in foodstuffs. TrAC Trends in Chemistry 2002;**21**(6):468-486

[42] Angle JS, Lindgren RL, Gilbert-Effiong D. Survival of *Aspergillus flavus* conidia in soil. Biodeterioration Research. 1989;**2**:245

[43] Wilson DM, Widstrom N, McMillian WW, Beaver RW. Aflatoxins in corn. In: Proceedings of the 44th Annual Corn and Sorghum Research Conference, American Seed Association. 1989

[44] Payne GA. Process of contamination by aflatoxin-producing fungi and their impact on crops. Mycotoxins in Agriculture and Food Safety. 1998;**9**:279-306

[45] Selvi AT, Joseph G, Jayaprakasha G. Inhibition of growth and aflatoxin production in *Aspergillus flavus* by *Garcinia indica* extract and its antioxidant activity. Food Microbiology. 2003;**20**:455-460

[46] Gowda N, Malathi V, Suganthi R. Effect of some chemical and herbal compounds on growth of *Aspergillus parasiticus* and aflatoxin production. Animal Feed Science and Technology. 2004;**116**:281-291

[47] Passone MA, Resnik S, Etcheverry MG. The potential of food grade antioxidants in the control of Aspergillus section Flavi, interrelated mycoflora and aflatoxin B_1 accumulation on peanut grains. Food Control. 2008;**19**:364-371

[48] Hussain A, Ali J. Inhibition of aflatoxin producing fungus growth using chemical, herbal compounds/spices and plants. Pure and Applied Biology. 2012;**1**:8

[49] Thippeswamy S, Mohana DC, Abhishek RU, Manjunath K. Inhibitory activity of plant extracts on aflatoxin B_1 biosynthesis by *Aspergillus flavus*. Journal of Agricultural Science and Technology. 2014;**16**:1123-1132

[50] Dai J, Mumper RJ. Plant phenolics: Extraction, analysis and their antioxidant and anticancer properties. Molecules. 2010;**15**(10):7313-7352

[51] Ajila CM, Aalami M, Leelavathi K, Prasada Rao UJS. Mango peel powder: A potential source of antioxidant and dietary fiber in macaroni preparations. Innovative Food Science and Emerging Technologies. 2010;**11**:219-224

[52] Bamoniri A, Haghir Ebrahimabadi A, Mazoochi A, Behpour M, JookarKashi F, Batooli H. Antioxidant and antimicrobial activity evaluation and essential oil analysis of *Semenovia tragioides* Boiss from Iran. Food Chemistry. 2010;**122**:553-558

[53] Jayaprakasha GK, Jaganmohan Rao L, Sakariah KK. An improved HPLC method for the determination of curcumin, demethoxycurcumin and bisdemethoxycurcumin. Journal of Agriculture Food Chemistry. 2002;**50**:3668-3672

[54] Hua Y, Sahashi K, Hung G, Rigo F, Passini MA, Bennett CF, Krainer AR. Antisense correction of SMN2 splicing in the CNS rescues necrosis in a type III SMA mouse model. Genes and Development (this issue). 2010. DOI: 10.1101/gad.1941310

[55] Kim W, et al. Yos9p detects and targets misfolded glycoproteins for ER-associated degradation. Molecular Cell. 2005;**19**(6):753-764

[56] Razzaghi-Abyaneh M, Shams-Ghahfarokhi M, Rezaee MB, Jaimand K, Alinezhad S, Saberi R, Yoshinari T. Chemical composition and anti-aflatoxigenic activity of *Carum carvi* L., *Thymus vulgaris* and *Citrus aurantifolia* essential oils. Food Control. 2009;**20**:1018-1024

[57] Hua SST, Baker JL, Flores-Espiritu M. Interactions of saprophytic yeasts with a *nor* mutant of *Aspergillus flavus*. Applied and Environmental Microbiology. 1999;**65**:2738-2740

[58] Kim M, Lee JH, Koh H, Lee SY, Jang C, Chung CJ, Sung JH, Blenis J, Chung J. Inhibition of ERK-MAP kinase signaling by RSK during Drosophila development. The EMBO Journal .2006;**25**(13):3056-3067

[59] Hamilton AC. Medicinal plants, conservation and livelihoods. Biodiversity and Conservation. 2004;**13**:1477-1517

[60] Waxler NEM. Plural medicine in India and Srilanka: Do ayurvedic and Western medical practices offer? Social Science and Medicine. 1988;**27**:531-544

[61] Thanaboripat D, Prugcharoen P, Ruangrattanametee V. Inhibitory effect of some medicinal plant extracts on the growth and aflatoxin production of *Aspergillus*

flavus. In: Proceedings of the 3rd International Conference on Biological Control and Biotechnology. 2005.

[62] Saralamp P, Chuakul W, Temsiririrkkul R, Clayton T. Medicinal Plants in Thailand. Vol. 1. Bangkok: Department of Pharmaceutical Biology, Faculty of Pharmacy, Mahidol University; 1996

[63] Paster N, Juven BJ, Harshemesh H. Antimicrobial activity and inhibition of aflatoxin B_1 formation by olive plant tissue constituents. Journal of Applied Bacteriology. 1988;**64**:293-297

[64] Thanaboripat D, Nontabenjawan K, Leesin K, Teerapiannont D, Sukcharoen O, Ruangrattanametee V. Inhibitory effect of garlic, clove and carrot on growth of *Aspergillus flavus* and aflatoxin production. Journal of Forestry Research. 1997:**8**:39-42

[65] Bilgrami KS, Sinha KK, Sinha AK. Inhibition of aflatoxin production & growth of *Aspergillus flavus* by eugenol & onion & garlic extracts. Indian Journal of Medical Research. 1992;**96**:171-175

[66] Bilgrami KS, Misra RS, Sinha KK, Singh A. Effect of some wild and medicinal plant extracts on aflatoxin production and growth of *Aspergillus flavus* in liquid culture. Journal of Indian Botany Society. 1980;**59**:123-126

[67] Mahmoud A-LE. Inhibition of growth and aflatoxin biosynthesis of *Aspergillus flavus* by extracts of some Egyptian plants. Letter in Applied Microbiology. 1999;**29**:334-336

[68] Passone MA, Girardi NS, Etcheverry M. Antifungal and anti-aflatoxigenic activity by vapor contact of three essential oils, and effects of environmental factors on their efficacy. Food Science and Technology. 2013;**53**:434-444

[69] Omidbeygi M, Barzegar M, Hamidi Z, Naghdibadi H. Antifungal activity of thyme, summer savory and clove essential oils against *Aspergillus flavus* in liquid medium and tomato paste. Food Control. 2009;**18**:1518-1523

[70] Nogueira JHC, Gonçalez E, Galleti SR, Facanali R, Marques MOM, Felicio JD. *Ageratum conyzoides* essential oil as aflatoxin suppressor of *Aspergillus flavus*. International Journal of Food Microbiology. 2010;**137**:55-60

[71] Thanaboripat D, Suvathi Y, Srilohasin P, Sripakdee S, Patthanawanitchai O, Chareonsettasilp S. Inhibitory effect of essential oils on the growth of *Aspergillus flavus*. KMITL Science and Technology Journal. 2007;**61**:18-24

[72] Sindhu S, Chempakam B, Leela NK, Bhai RS. Chemoprevention by essential oil of turmeric leaves (Curcuma longa L.) on the growth of Aspergillus flavus and aflatoxin production. Food and Chemical toxicology. 2011;**49**(5):1188-1192

[73] Mahmoud A-LE. Antifungal action and anti-aflatoxigenic properties of some essential oil constituents. Letter in Applied Microbiology. 1994;**19**:110-113

Chapter 2

Natural Products for Preventing and Controlling Aflatoxin Contamination of Food

Fei Tian and Hyang Sook Chun

Additional information is available at the end of the chapter

http://dx.doi.org/10.5772/intechopen.68413

Abstract

Aflatoxins are the most potent naturally occurring toxin and liver carcinogens known and their contamination of food is a significant risk factor for human health. Conventional chemical and physical approaches have been insufficient to eliminate aflatoxins from food, and the application of synthetic compounds can give rise to notable drug resistance and serious environmental and health problems. Awareness of these problems has led to an urgent need to identify safer alternative strategies. There are various natural compounds that influence aflatoxin contamination of food in different ways, including by inhibiting the growth of aflatoxigenic fungi, blocking aflatoxin biosynthesis, and removing or degrading aflatoxins. These inhibitors, many of which have shown great potentials for the control of aflatoxin contamination, have great promise for the development of new approaches to combatting aflatoxin contamination, and are capable of replacing or complementing conventional strategies. While more and more natural inhibitors are being identified, the modes of action of most of these are poorly understood. Further studies are necessary to better understand the mechanism of action of these compounds before their widespread commercial use. The objective of this chapter is to present the results of studies of the control of aflatoxin contamination using natural products.

Keywords: aflatoxin, natural products, antifungal, antiaflatoxigenic, detoxification

1. Introduction

Aflatoxins are a group of toxic secondary metabolites synthesized by fungi of the *Aspergillus* species, particularly *A. flavus* and *A. parasiticus* [1]. They are the most widely distributed soil-borne molds on earth and are capable of surviving on many organic nutrient sources, including stored grains and fodders, dead plants, insect and animal carcasses, and even immunocompromised humans and animals [2]. When grown under appropriate conditions, these fungi exist in the form of mycelia or conidia (asexual spores), while under adverse conditions such as a lack of nutrients or water, their fungal mycelium will transform to resistant structures known as sclerotia, which can survive in extremely harsh environmental conditions [3]. Aflatoxin-producing fungi affect many agricultural crops such as rice, corn, wheat, peanuts, and chilies. Pre- and post-harvest contamination of these crops with aflatoxins is common and annually causes great economic loss [4, 5].

Aflatoxins were first identified in 1960 in England as the cause of the Turkey X disease [6]. There are four major aflatoxins produced in nature: B1, B2, G1, and G2. They are named based on their fluorescence under ultraviolet light, and their relative mobility in thin-layer chromatography on silica gel. Most *A. flavus* produce aflatoxins B1 and B2, whereas *A. parasiticus* produce aflatoxins B1, B2, G1, and G2. Aflatoxin M1 is another frequently detected aflatoxin in nature; it is a hydroxylated derivative metabolized by cows from aflatoxin B1 and secreted in milk [7].

Aflatoxins are the most potent naturally occurring toxins and liver carcinogens known, and their contamination of food is a significant risk factor for human health, particularly in developing countries that lack detection, monitoring, and regulating measures to safeguard the food supply. It has been reported that approximately 4.5 billion people living in developing countries are chronically exposed to uncontrolled amounts of aflatoxins [7]. Long-term low-dose dietary exposure to aflatoxins is also a major risk for hepatocellular carcinoma. Aflatoxins have been designated as human liver carcinogens by the International Agency for Research on Cancer [8]. Therefore, the control and elimination of aflatoxigenic fungi and aflatoxins in food have great significance. To minimize potential exposure to aflatoxins, maximum levels of aflatoxins have been established by different countries [9]. The U.S. Food and Drug Administration specified a maximum of 20 ppb total aflatoxins for interstate trading of food and feedstuffs and 0.5 ppb aflatoxin M1 in milk. The European Commission has set the limits on cereals and derived products at 4 ppb for total aflatoxins and 2 ppb for aflatoxin B1, and for nuts and dried fruits subject to further processing at 10 ppb for total aflatoxins and 5 ppb for aflatoxin B1. The Korea Ministry of Food and Drug Safety imposed limits for aflatoxin B1 of 10 ppb and total aflatoxins of 15 ppb.

Chemical and physical approaches are widely used to minimize the risk of aflatoxin contamination of food. These are usually focused on inhibiting the development of spores and mycelia, and/or inactivation of aflatoxins by their transformation to nontoxic compounds. The most common methods include the use of synthetic fungicides, X-radiation, dehulling or cooking processes, and control of environmental factors during harvest and storage [10, 11]. These strategies are usually expensive, time-consuming, and inefficient. Some of them also cause major changes in the physical properties of food and a serious loss of nutritive value and therefore are inappropriate to eliminate aflatoxins from food [12]. Synthetic chemicals are still the most widely used recourse to prevent fungal contamination of food crops. However, there are strict regulations on chemical compound use in food, and there is political pressure to remove hazardous chemicals from the market [13]. In addition to these limitations, the application of synthetic fungicides can also give rise to notable drug resistance and serious environmental and health problems [14]. Awareness of these problems has led to an urgent need to identify safer alternative strategies.

Natural products are chemical compounds or substances produced by a living organism, and their use as biocontrol agents provides an opportunity to avoid synthetic fungicides. Over the years, efforts have been made to identify new antifungal materials from natural sources for controlling aflatoxin contamination of food [15]. Many bacteria, fungi, and plants that share ecological niches with and encounter aflatoxigenic fungi have the ability to synthesize compounds that inhibit aflatoxin synthesis or remove aflatoxins from food without significant losses in nutritive value; they therefore could be used to replace or complement conventional strategies. Basically, there are three possible ways of using natural products to avoid the harmful effects of aflatoxin contamination of food and feed: (1) prevent and control aflatoxigenic fungus contamination (fungal growth inhibition), (2) inhibit aflatoxin biosynthesis (aflatoxin production inhibition), and (3) decontamination of aflatoxin-containing food and feed (aflatoxin detoxification). These microbial metabolites and plant constituents are natural products and therefore are desirable for use in food because they can be easily degraded in nature. A variety of naturally derived compounds have been studied for their antifungal and antiaflatoxigenic activities, many of which have shown great potential for controlling aflatoxin contamination. The objective of this chapter is to present the results of studies of the control of aflatoxin contamination using natural products from bacteria, fungi, and plants.

2. Fungal growth inhibition

Fungi of *Aspergillus* sp. are the only source of aflatoxin contamination in food. The production of aflatoxin is greatly dependent on the growth condition of *Aspergillus* fungi; thus, it can be suppressed effectively through inhibit fungal growth. Many compounds produced by bacteria, fungi, and plants are possessed of abilities to inhibit fungal growth at different levels, such as inhibit conidia production and germination, interrupt membrane formation, or damage cell membrane, and disrupt fungal mitochondria (**Table 1**).

The use of bacteria is a promising solution to alleviate fungal contamination in food. In recent years, the study and application of antifungal bacteria has received strong interest. Significant progress has been reported on the isolation and characterization of antifungal compounds. Various bacterial compounds including organic acids, phenyllactic acids, reuterin, and cyclic dipeptides, proteinaceous compounds, and fatty acids have been reported to be able to inhibit the growth of aflatoxigenic fungi (**Table 2**).

Target of action	Antifungal product	Source	Activity against	Reference
Conidia production and/or germination	*Aegle marmelos* essential oil	*A. marmelos* (leaves)	*Alternaria* sp., *Colletotrichum* sp., *Curvularia* sp., *Ustilago* sp.	[16]
	Aldehydes (C6)	Plants[1]	*Alternaria* sp., *Botrytis* sp.	[17]
	(E)-Anethole, p-anisaldehyde, carvacrol, (−)-carvone, 1,8-cineole, (+)-limonene, myrcene, (±)-α-phellandrene, (±)-α-pinene	Plants	*Botrytis* sp., *Monilinia* sp., *Mucor* sp., *Penicillium* sp., *Rhizopus* sp.	[18]
	Benzaldehyde	Plants	*Monilinia* sp., *Botrytis* sp.	[19]
	Chitosan	Plants	*Botrytis* sp., *Rhizopus* sp.	[20]
	Fusapyrone	*Fusarium semitectum*	*Botrytis* sp.	[21]
	1-Octen-3-ol	*P. paneum*	*Penicillium* sp.	[22]
	Terpenoid	*Nasutitermes* sp.	*Metarhizium* sp.	[23]
Membrane formation and/or integrity	Brefeldin A	*Eupenicillium brefeldianum*	*Pisolithus* sp.	[24]
	Carvacrol, thymol	Plants	*Candida* sp.	[25]
	Clove essential oil	*Syzygium aromaticum*	*Candida* sp., *Aspergillus* sp. dermatophyte fungi	[26]
	Defensins	Plants	*Neurospora* sp., *Saccharomyces* sp.	[27]
	Eugenol, methyl eugenol	Plants	*Candida* sp.	[28]
	Geraniol	Plants	*Candida* sp., *Saccharomyces* sp.	[29]
	Ocimum sanctum essential oil	*O. sanctum*	*Candida* sp.	[30]
	Osmotin	Tobacco	*Aspergillus* sp., *Rhizoctonia* sp., *Macrophomina* sp., *Bipolaris* sp., *Fusarium* sp., *Phytophthora* sp., *Trichoderma* sp.	[31]
	Phytochemicals	*Thymus vulgaris* L.	*Rhizopus* sp.	[32]
	Zeamatin	*Zea mays*	*Candida* sp., *Neurospora* sp., *Trichoderma* sp.	[33]

Target of action	Antifungal product	Source	Activity against	Reference
Cell organelles function	Cruentaren	*Byssovorax cruenta*	*Candida* sp., *Metschnikowia* sp., *Saccharomyces* sp., *Rhodotorula* sp., *Botrytis* sp., *Mucor* sp., *Rhizopus* sp.	[34]
	Defensin	Plants	*Candida* sp.	[35]
	Anethum graveolens essential oil	*A. graveolens L.*	*Aspergillus* sp.	[36]
	Haliangicin	*Haliangium luteum*	*Aspergillus* sp., *Botrytis* sp., *Fusarium* sp., *Mucor* sp., *Pythium* sp., *Saprolegnia* sp.	[37]
	Phytoalexins	*Musa acuminata*	*Cladosporium* sp., *Pyricularia* sp., *Plasmopara* sp., *Sphaeropsis* sp.	[38]
	Plagiochin E	*Marchantia polymorpha* L.	*Candida* sp.	[39]
	Pyrrolnitrin	*Burkholderia cepacia*	*Streptomyces* sp.	[40]
	Tagetes patula essential oil	*T. patula L.*	*Botrytis* sp., *Penicillium* sp.	[41]
	UK-2A, UK-3A	*Streptomyces* sp.	*Saccharomyces* sp.	[42]

[1] Natural product exists in different plants.

Table 1. Natural products against fungal growth.

Strain	Activity against	Inhibitory compound	Target of action	Reference
Amorphophallus campanulatus (Roxb)	*A. flavus*	Amblyone	Unknown	[43]
Bacillus subtilis AU195	*A. flavus*	Bacillomycin D	Conidial germination	[44]
B. pumilus	*A. parasiticus*	Cyclic polypeptide	Unknown	[45]
B. subtilis KS03	*A.* flavus, *A. parasiticus*	Iturin A	Cell surface hydrophobicity	[46]
B. subtilis YM 10-20	*A. flavus*	Iturin-like compound	Conidial germination	[47]
Humicola fuscoatra NRRL 22980	*A. flavus.*	Monorden, monocillin IV, cerebrosides	Unknown	[48]
Lactobadllus casei subsp. *pseudoplantarum*	*A. flavus*	Proteinaceous	Unknown	[49]
Lactococcus lactis ATCC 11454	*A. flavus*	Heat-stable low-molecular weight compounds	Unknown	[50]

Strain	Activity against	Inhibitory compound	Target of action	Reference
Lb. casei subsp. *pseudoplantarum*	*A. flavus*	Proteinaceous	Unknown	[51]
Lb. casei subsp. *pseudoplantarum* 371	*A. parasiticus*	Proteinaceous	Unknown	[52]
Lc. lactis subsp. *diacetylactis* DRC1	*A. flavus*	Proteinaceous	Conidial germination	[53]
Lc. lactis subsp. *lactis* CHD28.3	*A. flavus, A. parasiticus*	Proteinaceous	Unknown	[54]
Pseudomonas aeruginosa K-187	*A flavus, A. parasiticus*	Pafungin	Hyphae lysis	[55]
Streptomyces sp. DPTB16	*A. flavus*	4′-Phenyl-1-napthyl-phenyl acetamide	Unknown	[56]
Streptomyces sp. MRI 142	*A. parasiticus*	Aflastatin	Unknown	[57]
S. albidoflavus ANU 6277	*A. flavus*	3-Phenylpropionic acid	Unknown	[58]

Table 2. Antifungal compounds against aflatoxigenic fungi growth.

Many bacteria produce organic acids such as lactic, acetic, and propionic acids. The production of these weak organic acids results in an acidic environment that generally restricts the growth of both bacteria and fungi [59]. Phenyllactic acid has been widely reported to have antifungal activities, and its broad-spectrum antibacterial and antifungal action makes it one of the most extensively studied antifungal organic acids derived from bacteria. Over the last decade, a number of studies have identified phenyllactic acid as the causative agent of antifungal activity. Its lack of toxicity to both animals and humans body, and its lack of any smell make phenyllactic acid a potential candidate for the control of food spoilage [60]. In addition, phenyllactic acid can also play a synergistic role with other metabolites [61, 62]. Reuterin is another antifungal compound produced by bacteria. This low-molecular-weight compound has also been reported to possess broad-spectrum antimicrobial activity. It has been demonstrated to be capable of inhibiting the growth of a wide range of molds including *A. flavus* [63]. Some fatty acids produced by bacterial strains have also received great attention for their antifungal properties. For example, 2-hydroxy-4-methylpentanoic acid produced by *Lactobacillus plantarum* VE56 and *Weissella paramesenteroides* LC11 is thought to act in synergy with other inhibitory metabolites and was shown to cause growth arrest in *Aspergillus* species [64]. Peptides inhibiting fungal growth have also been isolated from some bacterial strains. For example, Garofalo et al. demonstrated the existence of a series of peptides responsible for the antifungal activity of *Lb. rossiae* LD108. These peptides induced a clear delay in fungal growth on different bakery products, and were shown by Matrix assisted laser desorption/ionization time-of-flight mass spectrometric analysis to cause gluten proteolysis [65].

Fungal metabolites have also been used to reduce aflatoxin contamination in various crops. A recent study showed that culture filtrates of *Trichoderma* spp. at 200 mL/kg showed 72–93% inhibition of mycelial growth of *A. flavus* [66]. Nakaya reported the production of a small basic antifungal protein by the mold *A. giganteus* [67], and thoroughly characterized the structure of this protein as a highly twisted β-barrel stabilized by four internal disulfide bridges, which resembles the structure of some antifungal polypeptides produced by plants, such as defensins and thionins [68]. Similar proteins with high sequence homology have been described in other fungi, such as *Aspergillus niger* and *Penicillium chrysogenum* [69]. It is possible that the production of these antifungal proteins provides the producer with a competitive advantage against other fungal strains in the same environment.

Plants lack an immune system and must depend on other mechanisms to defend themselves against fungal invaders. One such mechanism is the synthesis of bioactive compounds that act specifically to inhibit fungal growth. Many plant extracts, particularly essential oils, have been reported to possess significant antifungal activity. An extract of *Azadirachta indica* was observed to be a good inhibitor of the growth of both *A. flavus* and *A. parasiticus in vitro* [70], and the oil from *Ocimum canum* exhibited activity against a broad range of fungi, including aflatoxin-producing fungi [71]. Several peptides and proteins are also associated with the antifungal activity of plants. Huang et al. reported that the grains of Tex6 wheat contain zeamatin, a thaumatin-like protein belonging to the PR5 group of pathogen related proteins, which inhibits the growth of *A. flavus* [72]. Chen et al. identified a 14-kDa protein that was present in resistant maize genotypes but in only a very low concentration in susceptible genotypes. This protein was identified as a trypsin inhibitor that also inhibited conidial germination and hyphal growth of *A. flavus*. Further studies showed that this protein inhibits the α-amylase from *A. flavus* [73]. Chitin is a common constituent of fungal cell walls. All organisms that contain chitin also contain chitinases (EC 3.2.1.14), which are presumably required for morphogenesis of cell walls [74]. Other organisms that do not contain chitin may produce chitinases to degrade the polymer for food. Plants have also been found to contain chitinase. Because plants do not contain chitin in their cell walls, it has been postulated that they produce chitinase to protect themselves from chitin-containing parasites including fungi [75]. Roberts and Selitrennikoff reported the isolation of a chitinase from barley grain that acting alone could inhibit fungal growth [76]. The antifungal activity of bacterial chitinases was also investigated, because plant and bacterial chitinases differ markedly in their antifungal activity, and this difference in biological activity correlates with differences in their substrate specificities [77]. Seeds of many plants contain high concentrations of chitinases, glucanases, and ribosome-inactivating proteins that may help protect seeds and seedlings from fungal infection [78]. One study showed that a maize chitinase preparation was highly active and caused a 100-fold reduction in the minimum dose of nikkomycin required to inhibit fungal growth [79]. Careful analysis of the maize preparation revealed several proteins and several enzyme activities. Further study revealed that zeamatin, a 22-kDa protein, is responsible for this synergistic activity, and showed that zeamatin exerts its antifungal effects by damaging fungal membranes [33].

Antifungal peptides and proteins have also been found in insects. Cecropins, originally isolated from the immune hemolymph of the *Cecropia moth*, are a key component of the immune response in insects. They have been shown to possess strong inhibitory activity against fungal growth. It has been reported that the structural features of the cecropins include a strongly basic N-terminus, an intermediate hinge region containing glycine and/or proline, and a hydrophobic C-terminus, which are all necessary for its lethal activity. Studies suggested that the fungicidal activity of cecropins is mediated by the formation of pores across cell membranes that lead to leakage of cytoplasmic contents and ultimately to cell death [80]. In addition, Powell et al. found that one kind of cecropin, cecropin B, can inhibit fungal growth by suppressing the germination of fungal conidia [81]. Consequently, these peptides have been studied for engineering fungal disease resistance in plants.

3. Aflatoxin production inhibition

One important side effect of fungal growth inhibition is the rapid spread of resistant strains. Therefore, inhibitors of aflatoxin production may be a better choice for control and prevention of aflatoxin contamination of food. Current methods to control aflatoxin contamination are mainly based on chemical strategies (pesticides and fungicides). However, the excessive use of chemical treatments has many undesirable consequences: (1) marked pollution of the environment, (2) an increase in resistant pathogen populations, and (3) the presence of chemical residues in food. Specific microbial metabolites and plant constituents have been shown to be effective inhibitors of aflatoxin production without significantly affecting fungal growth; in fact, numerous compounds and extracts possessing inhibitory activity for aflatoxin biosynthesis have been reported. However, tools and techniques have only recently become available to investigate the molecular mechanisms by which these inhibitors regulate aflatoxin biosynthesis [82].

Microbially derived inhibitors of aflatoxin production are of practical use because of their strong activity and the possibility of large-scale production. For example, a number of *Lentinula edodes* isolates are able to inhibit aflatoxin production and the isolate CF42 shows significant inhibitory activity. This effect is probably the result of the presence in the extracts of a number of different compounds with different inhibitory strategies. This is supported by the results obtained with fractionation of *L. edodes* CF42 filtrates, which leads to a decrease in their inhibitory effect but not to a complete loss of effect. Reverberi and coworkers also reported that culture filtrates of *L. edodes* isolate CF42 are able to inhibit aflatoxin production. *L. edodes* is able to release and accumulate lentinans and other β-glucans in the culture media. A recent study reported a direct relationship between aflatoxin inhibition and the β-glucan content of lyophilized *L. edodes* filtrates, suggesting that β-glucans could be amongst the factors responsible for their inhibitory effect on aflatoxins [83]. The ability of fungal β-glucans to act as free radical scavengers was recently shown in animal models [84]. *In vivo* research showed that glucans and glycoproteins extracted from fungi protect macrophages from the damage caused by lipoperoxide accumulation, mainly by activating the transcription of genes related to the macrophage antioxidant system [85]. Because the molecular analyses carried out on *A. parasiticus* mycelia treated with CF42 filtrates showed a significant activation of hsf2-like transcription factors of the fungal antioxidant system, a similar effect on macrophages could be hypothesized. It could be suggested that culture filtrates of *L. edodes* interfere with the cascade of signals that allows aflatoxin biosynthesis. It has also been hypothesized that accumulated β-glucans in the culture filtrates of *L. edodes* are able to inhibit aflatoxin production by *A. parasiticus* through the enhancement of the internal antioxidant system.

Lyophilized filtrates of *L. edodes* could be applied alone or in association with other food-grade compounds, to prevent aflatoxin production in food and feed [86]. In addition, the polysaccharides of this basidiomycete have low cytotoxicity for animal cells and could contribute to the nutritive value of the food or feed supplemented with these extracts [87].

Plant-derived inhibitors of aflatoxin production have great potential because not only are they highly effective but also the genes responsible for their biosynthesis could be transferred into susceptible host plants to create transgenic plants that resist aflatoxin contamination by *in situ* production of aflatoxin production inhibitors. For example, gallic acid is an effective plant-derived compound that inhibits aflatoxin production by *A. flavus* and disrupts expression of early and late aflatoxin biosynthesis genes. Evidence suggests that its aflatoxin production inhibitory activity may correlate with its strong antioxidant activity [82]. Transgenic plants with elevated levels of gallic acid that suppress aflatoxin production have been created [88]. Many essential oils obtained from parts of higher plants have also been shown to possess antiaflatoxigenic properties [89]. Various individual and combined plant extracts have been evaluated for their efficacy against aflatoxin production *in vitro*. For example, *Satureja hortensis* L. essential oil was found to inhibit production of aflatoxin B1 and G1 by *A. parasiticus*. The aflatoxin-production inhibitors were separated using reverse-phase high-performance liquid chromatography and finally identified as carvacrol and thymol. Further testing revealed that both carvacrol and thymol were able to effectively inhibit production of aflatoxin B1 and G1 in a dose-dependent manner [90].

In principle, there are three possible ways to inhibit aflatoxin biosynthesis (**Figure 1**). First, there can be alteration of the physiological environment or disturbance of the signaling inputs perceived by the fungus. For example, eugenol is a major phenolic component of essential oils extracted from cloves, cinnamon, and nutmeg. It has been shown in multiple experiments to inhibit aflatoxin biosynthesis. Evidence suggests that eugenol inhibits aflatoxin biosynthesis by lowering the physiological requirement for the enzymes activities involved in responding to oxidative stress. Eugenol treatment of fungi growing on Potato Dextrose Agar plates has been shown to result in the reduction of enzyme activities (glutathione peroxidase, microsomal reductases, superoxide dismutase, and xanthine oxidase) involved in responding to oxidative stress, concomitant with the inhibition of aflatoxin production by up to 50% [91]. Zingerone is another plant-derived aflatoxin inhibitor isolated from certain parts of *Zingiber officinale* or *Amomum melegueta*. Zingerone has a greater effect on aflatoxin biosynthesis than on fungal growth. Kim et al. found that zingerone, at a concentration of 5 mM, reduced aflatoxin production to 11% of the control with little reduction in fungal growth. They also found that yeast mutants with increased sensitivity to mitochondrial oxidative stress were more susceptible to combined H_2O_2 and zingerone treatment than a wild-type strain [92]. This result indicates that the antiaflatoxigenic activity of zingerone may be attributed to its alteration of the mitochondrial function in aflatoxin-producing fungi. Flavonoids and isoflavonoids are also inhibitory to aflatoxin production, but most are active only at high concentrations. In an early study, flavonoids (eriodictyol and luteolin) isolated from peanut shells were tested for their effects on aflatoxin production by *A. flavus* and *A. parasiticus*. Eriodictyol showed considerable inhibition of aflatoxin production with minimal influence on fungal growth, while luteolin was much more potent against *A. parasiticus* (IC_{50}< 0.35 mM) than against *A. flavus*

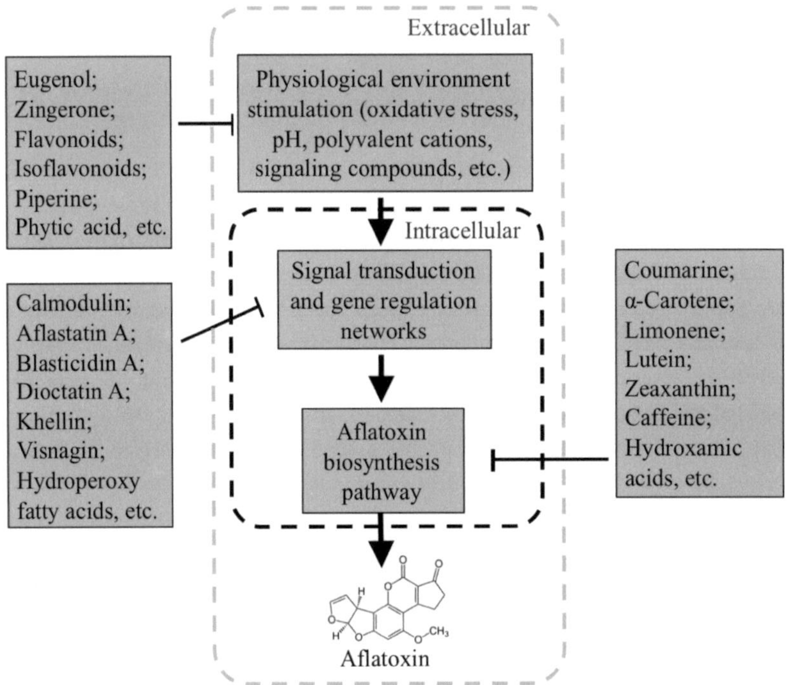

Figure 1. Schematic representation of aflatoxin production inhibition by natural products at different levels.

(IC_{50}~6 mM). The authors suggested that the differences in the responses of *A. flavus* and *A. parasiticus* to luteolin may be caused either by differences between the two fungi or by the culture conditions [93]. In fact, *A. flavus* and *A. parasiticus* are known to respond differently to oxidative stress. It has been reported that some oxidizing agents such as cumene hydroperoxide stimulate aflatoxin biosynthesis in *A. parasiticus* but not in *A. flavus* [94]. Glyceollin is a soybean isoflavonoid that has promise for engineering aflatoxin resistance in plants. It has been reported that 20 μM glyceollin strongly inhibited aflatoxin production in a high-glucose liquid medium, and the authors also speculated that glyceollin contributes to the resistance of soybean to aflatoxin contamination [95], although the mechanism by which glyceollin inhibits aflatoxin biosynthesis is uncertain. However, as a natural plant defense compound with a known biosynthetic pathway, glyceollin is particularly promising for the construction of aflatoxin resistant plants. Other biflavonoids have also been tested for their antiaflatoxigenic activity. Gonçalez et al. found that some of the major biflavonoids isolated from *Ouratea* species had excellent inhibitory activity at micromolar concentrations, reducing aflatoxin B1 production to <30% of the control at approximately 9 μM [96]. The antiaflatoxigenic effects of flavonoids and isoflavonoids might result from their antioxidant activity, because recent studies showed that aflatoxin production was closely related to the peroxidation of the fungal cell and several antioxidants have been reported to strongly inhibit aflatoxin production [97]. More recently, a study further underlined the importance of the role played by oxidative stress in the fungal cell in aflatoxin biosynthesis [98]. However, it has not been definitively demonstrated whether antioxidants work by direct interaction with reactive species or by the stimulation of the fungal cell antioxidant system. Furthermore, the molecular basis of the relationship between cell antioxidant defenses and aflatoxin formation is not yet fully understood. Information about the intracellular mechanism that leads to aflatoxin synthesis could be useful to achieve control over aflatoxin production. Piperine, a natural constituent found in many pepper species, inhibited aflatoxin production by *A. parasiticus* without obvious reduction in fungal growth. Piperine possesses direct antioxidant activity against various free radicals which may be related to its antiaflatoxigenic activity [99]. Phytic acid is an abundant component of seeds that can act as a chelator of polyvalent cations, especially zinc. The effect of phytic acid on aflatoxin production by *A. parasiticus* strongly depends on the pH of the medium: it was reported that 14.3 mM did not inhibit aflatoxin production at pH ≤ 4.5, but when the pH was about 6.6, phytic acid strongly inhibited aflatoxin production [100]. Regulation of aflatoxin production by phytic acid is attributed to its chelation of zinc and other polyvalent cations. It is also a natural antioxidant, and this antioxidant activity may also contribute to its antiaflatoxigenic properties by inhibiting iron-catalyzed free radical production and lipid peroxidation [101].

The second way to inhibit aflatoxin biosynthesis is to interfere with the signal transduction networks or gene expression regulation of aflatoxin biosynthesis by, for example, using calmodulin inhibitors, most of which are alkaloid and peptide compounds that have been isolated from a wide variety of natural sources, including many plant species [102]. Multiple lines of evidence support the idea that calcium-dependent signaling plays an important role in the regulation of aflatoxin biosynthesis [103]. Calmodulin-binding domains have been identified in the primary sequences of aflatoxin pathway transcriptional regulators (AflR and AflJ) and biosynthetic enzymes, presenting the possibility that calmodulin may influence transcriptional regulation of the aflatoxin biosynthesis gene cluster [104]. Aflastatin A and blasticidin A are well-known microbial-derived aflatoxin inhibitors. They are structurally related compounds produced by *Streptomyces* sp. that strongly inhibit aflatoxin production in *A. parasiticus* [57]. It has been reported that aflastatin A and blasticidin A inhibit the biosynthesis of important intermediates of aflatoxins (e.g., norsolorinic acid) and the transcription of aflatoxin biosynthetic genes [105]. Even though their mode of action is unknown, it was suggested that this inhibition is probably a result of perturbations in primary metabolism [106]. It has been reported that aflastatin A significantly enhances glucose utilization and the accumulation of ethanol in fungal cells. The level of transcription of genes for aldehyde dehydrogenase and acetyl-CoA synthetase, which are involved in ethanol utilization, was also suppressed by aflastatin A [105]. Dioctatin A is another antibiotic isolated from *Streptomyces* sp. that inhibits both conidiation and aflatoxin biosynthesis in *A. parasiticus*. Dioctatin A treatment also reduced the expression of AflR and aflatoxin biosynthesis genes. The molecular target of dioctatin A has not yet been identified, although it was suggested that dioctatin A inhibits aflatoxin biosynthesis through the FadA heterotrimeric G-protein signaling cascade [107]. Khellin and visnagin, products of the plant *Ammi visnaga,* were tested on *A. flavus*. Both showed potent inhibitory activity (IC_{50}< 0.1 mM) for aflatoxin production [108]. Khellin and visnagin are pharmacological agents that can inhibit cyclic adenosine monophosphate (cAMP) phosphodiesterases inhibitory activity, and cAMP has been shown to influence aflatoxin production [109]. Hydroperoxy fatty acids from plants, including methyl jasmonate

(MeJA), 9S-hydroperoxy-trans-10,cis-12-octadecadienoic acid (9S-HPODE), 13S-hydroperoxy-cis-9,trans-11-octadecadienoic acid (13S-HPODE), and 13S-hydroperoxy-cis-9,trans-11,cis-15-octadecatrienoic acid (13S-HPOTE), can mimic fungal signal factors and could potentially interact with G-protein-coupled receptor complexes upstream of the heterotrimeric G-protein complex that has been shown to regulate aflatoxin production [110]. Plant hydroperoxy fatty acids have varied effects on aflatoxin production. For example, MeJA significantly inhibited aflatoxin production by *A. flavus* on agar plates with an IC_{50}< 10 nM, but stimulated aflatoxin production by *A. parasiticus* after 7-day incubation in YES medium [111]. 13S-HPODE and 13S-HPOTE, at 100 μM, eliminated aflatoxin production by *A. parasiticus* in A&M medium, whereas 9S-HPODE slightly increased or decreased aflatoxin production, depending on the concentration tested [111]. Burow and coworkers reported that, in *in vitro* experiments, 13S-hydroperoxy fatty acids at concentrations of 10 and 100 μM repressed aflatoxin pathway gene expression and significantly reduced aflatoxin production in *A. parasiticus*. It has also been reported that treatment with 1 μM 13S-hydroperoxy linoleic acid significantly decreased aflatoxin production when it was repeatedly added to growth media at 24-h intervals. However, the same concentrations of 9S-hydroperoxy linoleic acid did not reduce aflatoxin production. These results suggested that specific seed lipoxygenase activity could provide resistance to mycotoxin contamination by *Aspergillus* sp. [112].

The third way to inhibit aflatoxin biosynthesis is to block the activity of aflatoxin biosynthesis-related enzymes. For example, coumarins have been found to strongly inhibit aflatoxin production without causing significant reductions in fungal growth [108]. It has been suggested that structural similarities between these coumarone and aflatoxins may result in competitive inhibition of biosynthetic enzymes. Terpenoids are a major class of natural products synthesized in plants through the mevalonic acid pathway. There are reports that different terpenoids, including camphene, α-carotene, limonene, lutein, and zeaxanthin, are inhibitory to aflatoxin biosynthesis in solid or liquid media [113]. The inhibition by terpenoids may occur at the level of whole-pathway regulation. For example, α-carotene was found to be able to block the synthesis of norsolorinic acid, the first stable aflatoxin precursor, thereby preventing the accumulation of subsequent pathway intermediates. Caffeine is another well-studied inhibitor of aflatoxin production, with studies showing that decaffeinated coffee beans and powder support higher aflatoxin production than normal beans and powder, and that incorporation of coffee into growth medium at concentrations of 1% (w/v) inhibits total aflatoxin production by 25%, with no significant reduction in fungal growth. The inhibitory effect of caffeine on glucose uptake is considered to be the possible mode of action for its antiaflatoxigenic activity [114]. Hydroxamic acids, such as 6-methoxy-benzoxazolin-2-one (MBOA), are also strong inhibitors of aflatoxin biosynthesis. MBOA significantly inhibits α-amylase induction [115]. It was suggested that the perturbation of sugar utilization by MBOA might be the major reason for its antiaflatoxigenic activity.

4. Aflatoxin detoxification

Aflatoxins are extremely stable under most conditions encountered during food storage, handling, and processing. Therefore, preventing contamination with aflatoxigenic fungi is the most rational and economic approach for controlling aflatoxin contamination of food. However, detoxification of aflatoxin is required for food already contaminated with aflatoxin. Although various methods have been described for detoxification of aflatoxins in foods, the most commonly used physical and chemical approaches are usually high cost or complex processes, and many also result in nutrient loss and food safety issues.

Biological detoxification of aflatoxins by employing natural products has been shown to be very effective in removing aflatoxin from food. In principle, there are four possible biological approaches to avoid the toxic effects of aflatoxins on the human body: (1) remove aflatoxins through surface adsorption to bacterial or fungal cells; (2) transform aflatoxins into nontoxic compounds by enzymatic degradation; (3) introduce aflatoxin adsorbents into contaminated food and feed to bind the toxins and inhibit their absorption from the gastrointestinal tract; and (4) metabolize aflatoxin into relatively nontoxic compounds via different metabolic pathways (**Figure 2**).

Aflatoxin detoxifying microorganisms were first demonstrated in 1996, when Ciegler et al. identified a *Flavobacterium aurantiacum* strain. In their research, they also found that both growing and resting cells of *F. aurantiacum* could remove aflatoxin from contaminated milk, oil, peanut butter, peanuts, corn, and soybeans. The detoxification was found to be irreversible with no new toxic products being formed [116]. Lillehoj et al. found that while both living and dead cells of *F. aurantiacum* were capable of removing aflatoxin from solution, aflatoxins removed by living cells could not be recovered while toxins removed by dead cells were simply adsorbed to the cell walls [117]. Line and Brackett also found that the degradation of aflatoxin B1 by *F. aurantiacum* was independent of the nutrients in the culture medium, suggesting that this organism can be used for detoxification in different fermentation processes [118].

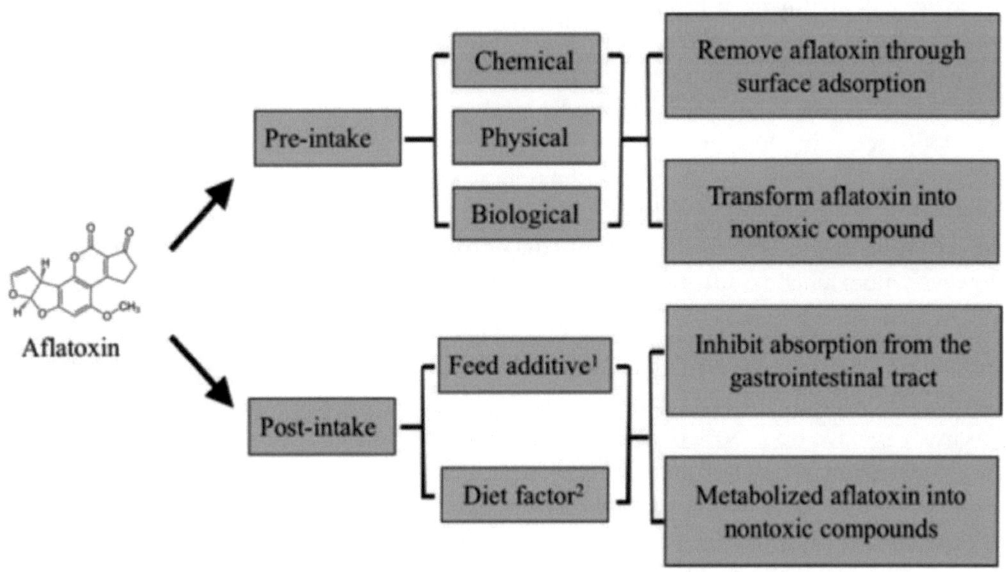

Figure 2. Schematic representation of aflatoxin reduction with different mode of action. [1] Detoxification in animal body; [2] Detoxification in human body.

Over the last few decades, considerable literature has accumulated that describes methods for removing aflatoxins using different microorganisms. Pure cultures of bacteria and fungi that detoxify aflatoxins, which include *Rhizopus* sp., *Aspergillus* sp., *Candida* sp., *Corynebacterium* sp., *Trichoderma* sp., *Mucor* sp., *Neurospora* sp., and *Rhodococcus* sp., have been isolated from complex microbial populations by screening and enrichment culture techniques [119, 120]. Among these, *Rhizopus* sp. was reported to be specifically suitable for large-scale detoxification of food and feed by solid-state fermentation. However, when used in food, viable microorganisms must be controlled to avoid undesired fermentation and undesirable compounds [121]. Among the different potentially decontaminating microorganisms, yeasts and lactic acid bacteria (LAB) have been widely used as starter cultures in the food and beverage industry for thousands of years. Therefore, yeasts and LAB have enormous potential as tools for tackling the problem of aflatoxin contamination of foods and feed [122].

Many reports state that the use of brewer's yeast cells as an animal feed additive resulted in a reduction in the toxic effects of aflatoxins [123]. In an early study, some yeast strains isolated from West African maize were found to be able to bind 15–60% (w/w) of aflatoxin B1 and this toxin binding was highly strain specific [124]. Yeast cells of *Saccharomyces cerevisiae*, which are generally used as performance promoters in poultry feeds, have also been shown to have beneficial effects against aflatoxin B1 exposure [125]. In fact, there have been many reports of yeasts and yeast cell components providing in varying degrees of protection of animals from aflatoxins in feed [126]. Baptista et al. reported that the addition of dried yeast and yeast cell walls to rat rations along with aflatoxin B1 resulted in a significant reduction of aflatoxin toxicity [127]. An *in vitro* study showed that modified mannan-oligosaccharides derived from *S. cerevisiae* showed a dose-dependent binding of aflatoxin as high as 95% (w/w) [128]. A later study confirmed that glucomannans from yeast cells have protective effects against the toxicity of aflatoxins in broiler chickens [129]. However, Baptista et al. found in a feeding experiment that mannan-oligosaccharides did not show significant suppressive effects on aflatoxin-induced damage in rats [127]. Unfortunately, no explanation could be given for this difference. It is well known that yeast cells bind sterols from the medium via cell wall mannan [130]. According to some of the studies reported, dead yeast cells still have this binding ability [127]. Therefore, it is likely that the removal of aflatoxin is not by covalent binding or metabolism, but by means of adhesion to cell wall components. It has been reported that the mannan components of the cell wall play a major role in aflatoxin binding by *S. cerevisiae* [128]. However, more kinetic studies are needed to assess the role of different components of the cell wall in aflatoxin binding.

The use of LAB in food fermentation dates back several centuries. Early studies showed that different LABs, including *Lactobacillus rhamnosus*, *L. acidophilus*, *L. gasseri*, and *L. casei* Shirota, could effectively remove up to 80% of aflatoxin B1 from contaminated culture media [131]. Among these, *L. rhamnosus* strains GG and LC705 showed similar aflatoxin B1 binding, even though they showed differences with respect to other metabolites. Later, more strains of LAB were found to be capable of binding aflatoxins in a strain-specific manner [132]. Several studies also indicated that the aflatoxin binding ability of LAB is highly strain specific [133]. Haskard et al. studied the mechanism of aflatoxin binding to *L. rhamnosus* using enzyme treatments and showed that the binding is predominantly to carbohydrate and some protein components of the cell wall [134].

Urea treatment decreased the binding significantly, indicating that hydrophobic interactions play an important role. Recent studies have shown that peptidoglycan is most likely the carbohydrate involved in the aflatoxin B1 binding process [135]. Haskard et al. found that the binding of aflatoxins to the cell surface of *L. rhamnosus* strains LGG and LC105 is considerably strong [136]. Living cell retained 38 and 50% (w/w) of bound aflatoxin after repeated washings with water. Even stronger binding was found in heat- and acid-treated cells which retained 66–71% (w/w) of the toxin, indicating a higher stability binding complex. This stronger binding was attributed to better access of aflatoxins to the treated cells. In addition, it has also been noted that autoclaving and sonication did not release any detectable toxin from prewashed cells, indicating the high stability of the complex. Binding of aflatoxins was also found to be unaffected by pH but could be easily disrupted with organic solvents, suggesting that hydrophobic interactions rather than cation exchange are the major mechanism of binding [134].

Enzymatic inactivation of aflatoxins is another attractive strategy for food decontamination. Several microorganisms can transform aflatoxin B1 to aflatoxicol and other less toxic or non-toxic compounds. It has been reported that aflatoxin B1 can be detoxified into aflatoxin B2a during yoghurt fermentation [137]; that aflatoxin B1 is detoxified during fermentation of milk by lactic bacteria [138]; and that *Armillariella tabescens* produces detoxification enzymes that show AFB1 detoxification activity [139]. A crude enzyme preparation isolated from *Stenotrophomonas* also showed strong aflatoxin-degrading activity and could degrade 85.7% of aflatoxin B1 [140]. Several other microbes, including *Corynebacterium rubrum, Aspergillus repens, Trichoderma viride, Mucor griseocyanus, Dactylium denroides, Mucor alternans, Rhizopus arrhizus, Rhizopus oryzae,* and *Rhizopus stolonifer,* have been reported to possess aflatoxin-degrading enzymes [141]. However, the degradation of aflatoxins is generally slow and incomplete: it was reported that *D. dendroides, A. repens,* and *M. griseocyanus* take 3–4 days to transform 60% of aflatoxin B1 to aflatoxicol [142].

Mycotoxin-producing fungi are also able to degrade or transform aflatoxins and possibly use them as a source of energy under suitable conditions. Several investigators have observed that aflatoxigenic strains produce large amounts of aflatoxins that usually decrease during continued incubation of the cultures [143]. *A. parasiticus* and *A. flavus* have been demonstrated to be able to degrade aflatoxins in a process that is strongly affected by the mycelia and culture conditions (pH, temperature, and inoculum, etc.), and probably involves peroxidases and P450 monooxygenases [144]. Detoxification of aflatoxin B1 by a cell-free enzyme preparation from *Armillariella tabescens,* an edible fungus used in Chinese traditional medicine, has also been reported [139]. Ames tests revealed a complete loss of mutagenicity and infrared spectroscopy of the product purified by thin-layer chromatography indicated that the difuran ring skeleton had opened; however, the structure of the product remains to be determined.

Another practical approach to reducing the toxicity of aflatoxin to humans and animals is the addition of non-nutritional inert adsorbents to the diet. These adsorbents sequester the aflatoxins in the gastrointestinal tract, thereby minimizing their toxic effects. Various adsorbents have been tested, including activated carbon, bentonite, cholestyramine, hydrated sodium calcium aluminosilicate, and zeolite, and produced promising results with respect to aflatoxin binding [145]. However, their application in food is limited because of their negative impacts such as reducing nutrient utilization. Therefore, the use of microorganisms and other natural products has become increasingly attractive as a reliable alternative to chemical adsorbents in the gastrointestinal tract.

The potential application of natural products as aflatoxin binders in human foods and animal feeds depends on their stability and the residence time of the complex in the gastrointestinal tract. The adhesion of aflatoxin-binding microorganisms to intestinal cells appears to be highly strain specific. Yeast cells generally show very low adhesion to the intestine but are capable of withstanding the harsh environment of the gastrointestinal tract [146]. Animal feeding experiments have shown that the addition of whole cells or cell walls of *S. cerevisiae* to the diet resulted in a significant reduction of aflatoxin toxicity, indicating the possible stability of the yeast--aflatoxin complex during its passage through the gastrointestinal tract [125]. A recent study in mice showed that *S. cerevisiae* over a 6-week period improved weight gain and reduced the genotoxicity produced by aflatoxin B1 [147]. Yeast cell wall components have also been evaluated as aflatoxin adsorbents. An *in vitro* study showed that esterified glucomannan from yeast cells displayed a very high capacity (97%) to adsorb aflatoxin B1 from aqueous solutions [148]. The addition of esterified glucomannan (0.1%) to chicken feed containing aflatoxins (2 mg/kg) significantly reduced the potentially adverse effects of the aflatoxins on hematological parameters, total protein, albumin values, and aspartate amino-transferase activity in broiler chickens. Other *in vivo* studies also showed that esterified glucomannan decreased the number and severity of pathological changes caused by aflatoxin treatment [129].

LAB cells usually show considerably higher adhesion to intestinal cells compared with that of yeast cells. However, it has been reported that aflatoxin binding considerably reduced the adhesive properties of some LAB strains and resulted in the faster excretion of immobilized aflatoxin B1 [149]. Gratz et al. also found that pre-exposure of the cells of a *L. rhamnosus* strain to aflatoxin B1 reduced its binding to intestinal mucus and thus resulted in the faster removal of bound aflatoxin [150]. An *in vitro* study using the chicken duodenum loop technique showed that a *L. rhamnosus* strain removed up to 54% (w/w) of the added aflatoxin B1 and reduced as much as 73% of intestinal adsorption [151]. Some researchers have suggested that the aflatoxin molecules bind to bacterial cell wall components such as polysaccharide and peptidoglycan [152]. Bacterial cell surface hydrophobicity may also play an important role in the binding of aflatoxins.

Aflatoxins absorbed into the bodies of humans or animals may also be metabolized into relatively nontoxic compounds via different metabolic pathways. The process of detoxification of aflatoxins usually involves removing the double bond of the terminal furan ring or opening the lactone ring. Once the lactone ring is opened, further reactions can occur to alter their binding properties to DNA and proteins [153]. The main reactions of aflatoxin metabolism in humans and animals are hydroxylation, oxidation, and demethylation. There are numerous studies concerning the metabolism of aflatoxin *in vitro* and *in vivo*. Salhab and Edwards found that the liver preparations of rabbit and trout were able to metabolize aflatoxin B1 into aflatoxicol by reducing the cyclopentenone carbonyl of aflatoxin B1 in an *in vitro* experiment [154]. An *in vivo* study by Roebuck and Wogan also found that aflatoxicol is the major metabolite of aflatoxin B1 in duck liver, whereas aflatoxin B1 was mainly converted into aflatoxin P1 and aflatoxin Q1 (relatively nontoxic) in human and monkey livers [155]. In fact, there is a great diversity among different animal species in the metabolism of aflatoxins. For example, aflatoxin B1 was able to convert into aflatoxin M1 in ducks, rats, and monkeys but not in humans [156], while in chicken liver, aflatoxin B1 was metabolized into a peptide conjugate of aflatoxin B2a and a glucuronide conjugate of aflatoxin M1 [157]. Donnelly et al. found that lipoxygenase and prostaglandin H synthase were the main enzymes responsible for the biotransformation of aflatoxin B1 in human lung, while in rat liver, aflatoxin B1 is transformed by a mixed-function monooxygenase [158]. In addition, various forms of cytochromes were found to have different biotransformation capacities for aflatoxins. Yoshizawa et al. reported that, in rat liver, transformation of aflatoxin M1 was strictly mediated by cytochrome P448, while transformation of aflatoxin Q1 was catalyzed by both cytochrome P450 and P448 [159]. In human liver, the cytochrome P450-dependent polysubstrate monooxygenase system is the major isoform involved in aflatoxin transformation [158].

5. Application of natural inhibitors

The preferred strategy for reducing the concentrations of aflatoxins in foods is prevention of aflatoxin formation during preharvest and postharvest of the various susceptible crops. In this context, non-aflatoxigenic *Aspergillus* strains have been used to prevent preharvest aflatoxin contamination of crops, such as peanuts, maize, and cottonseed, and have shown great potential. Recent advancements in the use of biocontrol strategies involving microorganisms should soon lead to increased practical applications for the benefit of the food industry. Some microorganisms such as *R. stolonifer* and *A. fumigatus*, which have been used in aflatoxin removal experiments, are not likely to be used in the field because of their potential to cause infection of the plants. However, these strains could still be used to provide natural compounds for prevention of aflatoxin formation. Alternatively, the genes responsible for their antiaflatoxigenic activity could possibly be incorporated into the host plant genome to produce crops resistant to aflatoxin contamination. The use of metabolites from microorganisms and plants as natural agents to control aflatoxin contamination has received much attention in recent years. Although the use of natural metabolites has shown promising results under controlled conditions in *in vitro* experiments, these studies need to be extended *in situ* to systems involving foods or feeds. More work is required to further characterize the antifungal and antiaflatoxigenic mechanisms involved.

To achieve effective control of aflatoxin contamination in food, high concentrations of natural compounds are generally needed. The incorporation of natural compounds into packaging materials can be a useful strategy to solve this problem. In the last decade, there have been plentiful studies of the development of active packaging materials. Because the introduction of protective agents in packaging materials can be used to protect food without direct addition of new chemicals, it has received great interest from both the food industry and academic communities. Many natural extracts, such as essential oils and their constituents, are categorized as flavorings in Europe and are categorized as Generally Recognized as Safe (GRAS) by the U.S. Food and Drug Administration. For this reason, they have often been proposed for and used in active packaging. For the purpose of the design of active packaging, it is necessary to establish which compounds have antiaflatoxigenic properties and what concentration is required to obtain maximum inhibition. In addition, because the volatile nature of some components, the release rate of the encapsulated compounds from their polymer matrix should be controlled, thereby magnifying their antiaflatoxigenic action on the product by both direct contact and through the head space of the packaging. The processes of encapsulating natural aflatoxin inhibitors into the polymer matrix should also be carefully controlled. Previous studies have shown notable losses of the active compounds during the film formation step of the casting technique [160].

Nanotechnology-based systems associated with natural compounds are also a good option. There are many well-known benefits of associating natural compounds with nanotechnological drug-delivery systems [161]. One good example is a nanoemulsion: in an emulsified form, natural compounds may be applied as an aqueous-based treatment. In fact, fine droplets may improve the delivery of inhibitory compounds to cereals because they may be able to penetrate into the cracks and crevices on the cereal surfaces. Nanoemulsions are emulsion droplets with a radius below 100 nm, which can be formed using both high-energy and low-energy methods [162]. High-energy methods require specialized mechanical devices, such as high pressure valve homogenizers, sonicators and microfluidizers. These devices are capable of generating intense mechanical forces that can intermingle and disrupt the oil and water phases. Low-energy methods rely on the spontaneous formation of nano-sized oil droplets, which is a physicochemical process that occurs under appropriate conditions with certain combinations of surfactant, oil, and water. The spontaneous emulsification method has recently been reported to be suitable for application in the food industry for fabricating effective antimicrobial nanoemulsions from essential oils [163].

As mentioned previously, genetic engineering is another way to utilize these compounds. Host crop species can be engineered to gain resistance to aflatoxin contamination by incorporation of the genes for biosynthesis of natural inhibitors. There are likely hundreds, if not thousands, of natural compounds that influence aflatoxin biosynthesis at concentrations ranging from submicromolar to millimolar. Unfortunately, many of these inhibitors are not suitable for genetic engineering. The complexity of altering plant natural product pathways makes it difficult to engineer crop species resistant to aflatoxin contamination. In addition, the majority of aflatoxin inhibitors reported so far were tested *in vitro* in media that do not approximate the conditions on the host plant. The tissue specificity and/or inducible expression of inhibitors are also important considerations [164]. Therefore, it is critical to identify the most promising candidates before attempting to engineer aflatoxin-resistant plants. Compared with the production of exogenous inhibitors, the development of plants that already possess aflatoxin inhibitors might be easier, because the biosynthetic pathways are already present in the host, and an increased inhibitor concentration can be achieved by upregulating endogenous genes. Most genetic and molecular approaches aimed at preventing aflatoxin biosynthesis have not yet reached commercial application in the field and require substantial further development.

6. Conclusions

In summary, there are various natural compounds that influence aflatoxin contamination in food through different ways, including inhibition of the growth of aflatoxigenic fungi, blocking aflatoxin biosynthesis, and removal or degradation of aflatoxin. These inhibitors are highly promising for the development of new approaches to fighting aflatoxin contamination in food and have the capability to replace or complement conventional strategies. A common feature of many inhibitors is their antioxidant activity; yet, the relationship of antiaflatoxigenic activity and antioxidant activity is unknown. Some inhibitors of aflatoxin production are specifically targeted to the biosynthesis of aflatoxin without affecting the development of the fungal cells. However, most inhibitors also inhibit fungal growth at higher concentrations. This may indicate that secondary metabolism (aflatoxin) is sensitive to stress resulting from low concentrations of growth-inhibitory compounds. The production of norsolorinic acid, the first stable intermediate in the aflatoxin biosynthetic pathway, was inhibited in parallel with aflatoxin production at the regulatory level of biosynthesis rather than at specific steps within the pathway, indicating the importance of this intermediate. More and more natural inhibitors are being identified, yet the modes of action of most are poorly understood. Further studies are necessary to better understand the mechanisms of action of those compounds before they can be widely used commercially. Using new biological approaches, researchers are now combining datasets from profiling of transcripts, proteins, and metabolites generated using inhibitory compounds with different modes of action, which will provide useful information for dissecting different facets of aflatoxin regulation.

Author details

Fei Tian and Hyang Sook Chun*

*Address all correspondence to: hschun@cau.ac.kr

Advanced Food Safety Research Group, BK21 Plus, School of Food Science and Technology, Chung-Ang University, Anseong, South Korea

References

[1] Gourama H, Bullerman LB. *Aspergillus flavus* and *Aspergillus parasiticus*: Aflatoxigenic fungi of concern in foods and feeds: A review. Journal of Food Protection. 1995;**58**(12): 1395-1404.

[2] Klich MA. Soil fungi of some low-altitude desert cotton fields and ability of their extracts to inhibit *Aspergillus flavus*. Mycopathologia. 1998;**142**:97-100.

[3] Bennett JW, Leong PM, Kruger S, Keyes D. Sclerotial and low aflatoxigenic morphological variants from haploid and diploid *Aspergillus parasiticus*. Cellular and Molecular Life Sciences. 1986;**42**(7):848-851.

[4] Perrone G, Susca A, Cozzi G, Ehrlich K, Varga J, Frisvad JC, Samson RA. Biodiversity of Aspergillus species in some important agricultural products. Studies in Mycology. 2007;**59**:53-66.

[5] Robens J, Cardwell K. The costs of mycotoxin management to the USA: Management of aflatoxins in the United States. Journal of Toxicology: Toxin Reviews. 2003;**22**(2-3):139-152.

[6] Allcroft R, Carnaghan RBA, Sargeant K, O'kelly J. A toxic factor in Brazilian groundnut meal. Veterinary Record. 1961;**73**:428-429.

[7] Van Egmond HP. Current situation on regulations for mycotoxins. Overview of tolerances and status of standard methods of sampling and analysis. Food Additives & Contaminants. 1989;**6**(2):139-188.

[8] Wogan GN. Aflatoxins as risk factors for hepatocellular carcinoma in humans. Cancer Research. 1992;**52**(7 Supplement):2114s–2118s.

[9] Van Egmond HP, Schothorst RC, Jonker MA. Regulations relating to mycotoxins in food. Analytical and Bioanalytical Chemistry. 2007;**389**(1):147-157.

[10] Bata Á, Lásztity R. Detoxification of mycotoxin-contaminated food and feed by microorganisms. Trends in Food Science & Technology. 1999;**10**(6):223-228.

[11] Piva G, Galvano F, Pietri A, Piva A. Detoxification methods of aflatoxins. A review. Nutrition Research. 1995;**15**(5):767-776.

[12] Line JE, Brackett RE. Factors affecting aflatoxin B1 removal by *Flavobacterium aurantiacum*. Journal of Food Protection. 1995;**58**(1):91-94.

[13] Pal KK, Gardener BM. Biological control of plant pathogens. The Plant Health Instructor. 2006;**2**:1117-1142.

[14] White TC, Marr KA, Bowden RA. Clinical, cellular, and molecular factors that contribute to antifungal drug resistance. Clinical Microbiology Reviews. 1998;**11**(2):382-402.

[15] Klich MA. Environmental and developmental factors influencing aflatoxin production by *Aspergillus flavus* and *Aspergillus parasiticus*. Mycoscience. 2007;**48**(2):71-80.

[16] Rana BK, Singh UP, Taneja V. Antifungal activity and kinetics of inhibition by essential oil isolated from leaves of *Aegle marmelos*. Journal of Ethnopharmacology. 1997;**57**(1):29-34.

[17] Hamilton-Kemp TR, McCracken CT, Loughrin JH, Andersen RA, Hildebrand DF. Effects of some natural volatile compounds on the pathogenic fungi *Alternaria alternata* and *Botrytis cinerea*. Journal of Chemical Ecology. 1992;**18**(7):1083-1091.

[18] Caccioni DR, Guizzardi M. Inhibition of germination and growth of fruit and vegetable postharvest pathogenic fungi by essential oil components. Journal of Essential Oil Research. 1994;**6**(2):173-179.

[19] Wilson CL, Franklin JD, Otto BE. Fruit volatiles inhibitory to *Monilinia fructicola* and *Botrytis cinerea*. Plant Disease. 1987;**71**(4):316-319.

[20] El Ghaouth A, Arul J, Grenier J, Asselin A. Antifungal activity of chitosan on two postharvest pathogens of strawberry fruits. Phytopathology. 1992;**82**(4):398-402.

[21] Altomare C, Perrone G, Stornelli C, Bottalico A. Quaderni della Scuola di specializzazione in Viticoltura ed Enologia. University of Torino, Italy. 1998;**22**:59-66.

[22] Chitarra GS, Abee T, Rombouts FM, Posthumus MA, Dijksterhuis J. Germination of *Penicillium paneum* conidia is regulated by 1-octen-3-ol, a volatile self-inhibitor. Applied and Environmental Microbiology. 2004;**70**(5):2823-2829.

[23] Rosengaus RB, Lefebvre ML,Traniello JF. Inhibition of fungal spore germination by Nasutitermes: evidence for a possible antiseptic role of soldier defensive secretions. Journal of Chemical Ecology. 2000;**26**(1):21-39.

[24] Cole L, Davies D, Hyde GJ, Ashford AE. Brefeldin A affects growth, endoplasmic reticulum, golgi bodies, tubular vacuole system, and secretory pathway in *Pisolithus tinctorius*. Fungal Genetics and Biology. 2000;**29**(2):95-106.

[25] Ahmad A, Khan A, Akhtar F, Yousuf S, Xess I, Khan LA, Manzoor N. Fungicidal activity of thymol and carvacrol by disrupting ergosterol biosynthesis and membrane integrity against Candida. European Journal of Clinical Microbiology & Infectious Diseases. 2011;**30**(1):41-50.

[26] Pinto E, Vale-Silva L, Cavaleiro C, Salgueiro L. Antifungal activity of the clove essential oil from Syzygium aromaticum on Candida, Aspergillus and dermatophyte species. Journal of Medical Microbiology. 2009;**58**(11):1454-1462.

[27] Thevissen K, Terras FR, Broekaert WF. Permeabilization of fungal membranes by plant defensins inhibits fungal growth. Applied and Environmental Microbiology. 1999;**65**(12): 5451-5545.

[28] Ahmad A, Khan A, Manzoor N, Khan LA. Evolution of ergosterol biosynthesis inhibitors as fungicidal against Candida. Microbial Pathogenesis. 2010;**48**(1):35-41.

[29] Bard M, Albrecht MR, Gupta N, Guynn CJ, Stillwell W. Geraniol interferes with membrane functions in strains of Candida and Saccharomyces. Lipids. 1988;**23**(6):534-538.

[30] Khan A, Ahmad A, Akhtar F, Yousuf S, Xess I, Khan LA, Manzoor N. *Ocimum sanctum* essential oil and its active principles exert their antifungal activity by disrupting ergosterol biosynthesis and membrane integrity. Research in Microbiology. 2010;**161**(10): 816-823.

[31] Abad LR, D'Urzo MP, Liu D, Narasimhan ML, Reuveni M, Zhu JK, Niu X, Singh NK, Hasegawa PM, Bressan RA. Antifungal activity of tobacco osmotin has specificity and involves plasma membrane permeabilization. Plant Science. 1996;**118**(1):11-23.

[32] de Lira Mota KS, de Oliveira Pereira F, de Oliveira WA, Lima IO, de Oliveira Lima E. Antifungal activity of *Thymus vulgaris* L. essential oil and its constituent phytochemicals against *Rhizopus oryzae*: interaction with ergosterol. Molecules. 2012;**17**(12):14418-14433.

[33] Roberts WK, Selitrennikoff CP. Zeamatin, an antifungal protein from maize with membrane-permeabilizing activity. Microbiology. 1990;**136**(9):1771-1778.

[34] Kunze B, Steinmetz H, Höfle G, Huss M, Wieczorek H, Reichenbach H. Cruentaren, a new antifungal salicylate-type macrolide from *Byssovorax cruenta* (Myxobacteria) with inhibitory effect on mitochondrial ATPase activity. Journal of Antibiotics. 2006;**59**(10):664.

[35] Aerts AM, Bammens L, Govaert G, Carmona-Gutierrez D, Madeo F, Cammue B, Thevissen K. The antifungal plant defensin HsAFP1 from *Heuchera sanguinea* induces apoptosis in Candida albicans. Frontiers in Microbiology. 2011;**2**:47.

[36] Tian J, Ban X, Zeng H, He J, Chen Y, Wang Y. The mechanism of antifungal action of essential oil from dill (*Anethum graveolens* L.) on Aspergillus flavus. PLoS One. 2012;**7**(1):e30147.

[37] Fudou R, Iizuka T, Yamanaka S. Haliangicin, a novel antifungal metabolite produced by a marine myxobacterium. 1. Fermentation and biological characteristics. The Journal of Antibiotics. 2001;**54**(2):149-152.

[38] Jeandet P, Douillet-Breuil AC, Bessis R, Debord S, Sbaghi M, Adrian M. Phytoalexins from the Vitaceae: biosynthesis, phytoalexin gene expression in transgenic plants, antifungal activity, and metabolism. Journal of Agricultural and Food Chemistry. 2002;**50**(10):2731-2741.

[39] Wu XZ, Cheng AX, Sun LM, Sun SJ, Lou HX. Plagiochin E, an antifungal bis (bibenzyl), exerts its antifungal activity through mitochondrial dysfunction-induced reactive oxygen species accumulation in *Candida albicans*. Biochimica et Biophysica Acta (BBA)-General Subjects. 2009;**1790**(8):770-777.

[40] El-Banna N, Winkelmann G. Pyrrolnitrin from *Burkholderia cepacia*: antibiotic activity against fungi and novel activities against streptomycetes. Journal of Applied Microbiology. 1998;**85**(1):69-78.

[41] Romagnoli C, Bruni R, Andreotti E, Rai MK, Vicentini CB, Mares D. Chemical characterization and antifungal activity of essential oil of capitula from wild Indian *Tagetes patula* L. Protoplasma. 2005;**225**(1-2):57-65.

[42] Ueki M, Taniguchi M. The mode of action of UK-2A and UK-3A, novel antifungal antibiotics from Streptomyces sp. 517-02. The Journal of Antibiotics. 1997;**50**(12):1052-1057.

[43] Khan A, Rahman M, Islam MS. Antibacterial, antifungal and cytotoxic activities of amblyone isolated from *Amorphophallus campanulatus*. Indian Journal of Pharmacology. 2008;**40**(1):41.

[44] Moyne AL, Shelby R, Cleveland TE, Tuzun S. Bacillomycin D: an iturin with antifungal activity against *Aspergillus flavus*. Journal of Applied Microbiology. 2001;**90**(4):622-629.

[45] Munimbazi C, Bullerman LB. Inhibition of aflatoxin production of *Aspergillus parasiticus* NRRL 2999 by *Bacillus pumilus*. Mycopathologia. 1997;**140**(3):163-169.

[46] Cho SJ, Lee SK, Cha BJ, Kim YH, Shin KS. Detection and characterization of the *Gloeosporium gloeosporioides* growth inhibitory compound iturin A from *Bacillus subtilis* strain KS03. FEMS Microbiology Letters. 2003;**223**(1):47-51.

[47] Chitarra GS, Breeuwer P, Nout MJR, Van Aelst AC, Rombouts FM, Abee T. An antifungal compound produced by *Bacillus subtilis* YM 10-20 inhibits germination of *Penicillium roqueforti* conidiospores. Journal of Applied Microbiology. 2003;**94**(2):159-166.

[48] Wicklow DT, Joshi BK, Gamble WR, Gloer JB, Dowd PF. Antifungal metabolites (monorden, monocillin IV, and cerebrosides) from Humicola fuscoatra Traaen NRRL 22980, a mycoparasite of *Aspergillus flavus* sclerotia. Applied and Environmental Microbiology. 1998;**64**(11):4482-4484.

[49] Gourama H, Bullerman LB. Antimycotic and antiaflatoxigenic effect of lactic acid bacteria: a review. Journal of Food Protection. 1995;**58**(11):1275-1280.

[50] Coallier-Ascah J, Idziak ES. Interaction between Streptococcus lactis and *Aspergillus flavus* on production of aflatoxin. Applied and Environmental Microbiology. 1985;**49**(1):163-167.

[51] Gourama H, Bullerman LB. Inhibition of growth and aflatoxin production of *Aspergillus flavus* by Lactobacillus species. Journal of Food Protection. 1995;**58**(11):1249-1256.

[52] Gourama H, Bullerman LB. Anti-aflatoxigenic activity of *Lactobacillus casei* pseudoplantarum. International Journal of Food Microbiology. 1997;**34**(2):131-143.

[53] Batish VK, Grover S, Lal R. Screening lactic starter cultures for antifungal activity. Cultured Dairy Products Journal. 1989;**24**:23-25.

[54] Roy U, Batish VK, Grover S, Neelakantan S. Production of antifungal substance by *Lactococcus lactis* subsp. lactis CHD-28.3. International Journal of Food Microbiology. 1996;**32**(1-2):27-34.

[55] Wanga SL, Yieh TC, Shih L. Purification and characterization of a new antifungal compound produced by *Pseudomonas aeruginosa* K-187 in a shrimp and crab shell powder medium. Enzyme and Microbial Technology. 1999;**25**(3):439-446.

[56] Dhanasekaran D, Thajuddin N, Panneerselvam A. An antifungal compound: 4′ phenyl-1-napthyl–phenyl acetamide from Streptomyces sp. DPTB16. Facta Universitatis Ser Med Biol. 2008;**15**:7-12.

[57] Ono M, Sakuda S, Suzuki A, Isogai A. Aflastatin A, a novel inhibitor of aflatoxin production by aflatoxigenic fungi. The Journal of Antibiotics. 1997;**50**(2):111-118.

[58] Narayana KJ, Prabhakar P, Vijayalakshmi MUVVA, Venkateswarlu Y, Krishna PS. Study on bioactive compounds from Streptomyces sp. ANU 6277. Polish Journal of Microbiology. 2008;**57**(1):35.

[59] Ross RP, Morgan S, Hill C. Preservation and fermentation: past, present and future. International Journal of Food Microbiology. 2002;**79**(1):3-16.

[60] Lavermicocca P, Valerio F, Visconti A. Antifungal activity of phenyllactic acid against molds isolated from bakery products. Applied and Environmental Microbiology. 2003;**69**(1):634-640.

[61] Rizzello C, Cassone A, Coda R, Gobbetti M. Antifungal activity of sourdough fermented wheat germ used as an ingredient for bread making. Food Chemistry. 2011;**127**(3): 952-959.

[62] Ryan LA, Zannini E, Dal Bello F, Pawlowska A, Koehler P, Arendt EK. *Lactobacillus amylovorus* DSM 19280 as a novel food-grade antifungal agent for bakery products. International Journal of Food Microbiology. 2011;**146**(3):276-283.

[63] Axelsson LT, Chung TC, Dobrogosz WJ, Lindgren SE. Production of a broad spectrum antimicrobial substance by *Lactobacillus reuteri*. Microbial Ecology in Health and Disease. 1989;**2**(2):131-136.

[64] Ndagano D, Lamoureux T, Dortu C, Vandermoten S, Thonart P. Antifungal activity of 2 lactic acid bacteria of the Weissella genus isolated from food. Journal of Food Science. 2011;**76**(6):305-311.

[65] Garofalo C, Zannini E, Aquilanti L, Silvestri G, Fierro O, Picariello G, Clementi F. Selection of sourdough lactobacilli with antifungal activity for use as biopreservatives in bakery products. Journal of Agricultural and Food Chemistry. 2012;**60**(31):7719-7728.

[66] Reddy KRN, Reddy CS, Muralidharan K. Potential of botanicals and biocontrol agents on growth and aflatoxin production by *Aspergillus flavus* infecting rice grains. Food Control. 2009;**20**(2):173-178.

[67] Nakaya K, Omata K, Okahashi I, Nakamura Y, Kolkenbrock H, Ulbrich N. Amino acid sequence and disulfide bridges of an antifungal protein isolated from *Aspergillus giganteus*. The FEBS Journal. 1990;**193**(1):31-38.

[68] García-Olmedo F, Molina A, Alamillo JM, Rodríguez-Palenzuéla P. Plant defense peptides. Peptide Science. 1998;**47**(6):479-491.

[69] Lee DG, Shin SY, Maeng CY, Jin ZZ, Kim KL, Hahm KS. Isolation and characterization of a novel antifungal peptide from Aspergillus niger. Biochemical and Biophysical Research Communications. 1999;**263**(3):646-651.

[70] Bhatnagar D. The inhibitory effect of neem (*Azadirachta indica*) leaf extracts on aflatoxin synthesis in *Aspergillus parasiticus*. Journal of the American Oil Chemists' Society. 1988;**65**(7):1166-1168.

[71] Bhargava KS, Dixit SN, Dubey NK, Tripathi RD. Fungitoxic properties of *Ocimum canum*. Journal of Indian Botanical Society. 1981;**60**:24-27.

[72] Huang Z, White DG, Payne GA. Corn seed proteins inhibitory to *Aspergillus flavus* and aflatoxin biosynthesis. Phytopathology. 1997;**87**(6):622-627.

[73] Chen ZY, Brown RL, Russin JS, Lax AR, Cleveland TE. A corn trypsin inhibitor with antifungal activity inhibits *Aspergillus flavus* α-amylase. Phytopathology. 1999;**89**(10):902-907.

[74] Gooday GW. Biosynthesis of the fungal wall-mechanisms and implications the first Fleming Lecture. Microbiology. 1977;**99**(1):1-11.

[75] Bell AA. Biochemical mechanisms of disease resistance. Annual Reviews of Plant Physiology. 1981;**32**:21-81.

[76] Roberts WK, Selitrennikoff CP. Isolation and partial characterization of two antifungal proteins from barley. Biochimica et Biophysica Acta (BBA)-General Subjects. 1986;**880**(2-3):161-170.

[77] Boller T, Gehri A, Mauch F, Vögeli U. Chitinase in bean leaves: induction by ethylene, purification, properties, and possible function. Planta. 1983;**157**(1):22-31.

[78] Mauch F, Mauch-Mani B, Boller T. Antifungal hydrolases in pea tissue. 11. Inhibition of fungal growth by combinations of chitinase and p-1,3-glucanase. Plant Physiology. 1988;**88**(3):936-942.

[79] Roberts WK, Selitrennikoff CP. Plant and bacterial chitinases differ in antifungal activity. Journal of General Microbiology. 1988;**134**(1):169-176.

[80] Lockey TD, Ourth DD. Formation of pores in *Escherichia coli* cell membranes by a cecropin isolated from hemolymph of *Heliothis virescens* larvae. European Journal of Biochemistry. 1996;**236**(1):263-271.

[81] Powell WA, Catranis CM, Maynard CA. Synthetic antimicrobial peptide design. MPMI-Molecular Plant Microbe Interactions. 1995;**8**(5):792.

[82] Holmes RA, Boston RS, Payne GA. Diverse inhibitors of aflatoxin biosynthesis. Applied Microbiology and Biotechnology. 2008;**78**(4):559-572.

[83] Wasser SP, Weis AL. Medicinal properties of substances occurring in higher basidiomycetes mushrooms: current perspectives (review). International Journal of medicinal mushrooms. 1999;**1**(1):31-62.

[84] Slameňová D, Lábaj J, Kogan G, Šandula J, Bresgen N, Eckl P. Protective effects of fungal (1→ 3)-β-D-glucan derivatives against oxidative DNA lesions in V79 hamster lung cells. Cancer Letters. 2003;**198**(2):153-160.

[85] Yuan C, Mei Z, Shangxi L, Yi L. PSK protects macrophages from lipoperoxide accumulation and foam cell formation caused by oxidatively modified low-density lipoprotein. Atherosclerosis. 1996;**124**(2):171-181.

[86] Guo FC, Kwakkel RP, Williams BA, Li WK, Li HS, Luo JY, Li XP, Wei YX, Yan ZT, Verstegen MWA. Effects of mushroom and herb polysaccharides, as alternatives for an antibiotic, on growth performance of broilers. British Poultry Science. 2004;**45**(5):684-694.

[87] Xu Y. Perspectives on the 21st century development of functional foods: bridging Chinese medicated diet and functional foods. International Journal of Food Science & Technology. 2001;**36**(3):229-242.

[88] Jermnak U, Yoshinari T, Sugiyama Y, Tsuyuki R, Nagasawa H, Sakuda S. Isolation of methyl syringate as a specific aflatoxin production inhibitor from the essential oil of Betula alba and aflatoxin production inhibitory activities of its related compounds. International Journal of Food Microbiology. 2012;**153**(3):339-344.

[89] Mahmoud AL. Antifungal action and antiaflatoxigenic properties of some essential oil constituents. Letters in Applied Microbiology. 1994;**19**(2):110-113.

[90] Sidhu OP, Chandra H, Behl HM. Occurrence of aflatoxins in mahua (*Madhuca indica* Gmel.) seeds: Synergistic effect of plant extracts on inhibition of *Aspergillus flavus* growth and aflatoxin production. Food and Chemical Toxicology. 2009;**47**(4):774-777.

[91] Jayashree T, Subramanyam C. Antiaflatoxigenic activity of eugenol is due to inhibition of lipid peroxidation. Letters in Applied Microbiology. 1999;**28**(3):179-183.

[92] Kim JH, Campbell BC, Mahoney NE, Chan KL, Molyneux RJ. Identification of phenolics for control of *Aspergillus flavus* using *Saccharomyces cerevisiae* in a model target-gene bioassay. Journal of Agricultural and Food Chemistry. 2004;**52**(26):7814-7821.

[93] DeLucca AJ, Palmgren MS, Daigle DJ. Depression of aflatoxin production by flavonoid-type compounds from peanut shells. Phytopathology. 1987;**77**(11):1560-1563.

[94] De Luca C, Passi S, Fabbri AA, Fanelli C. Ergosterol oxidation may be considered a signal for fungal growth and aflatoxin production in *Aspergillus parasiticus*. Food Additives & Contaminants. 1995;**12**(3):445-450.

[95] Song DK, Karr AL. Soybean phytoalexin, glyceollin, prevents accumulation of aflatoxin B 1 in cultures of *Aspergillus flavus*. Journal of Chemical Ecology. 1993;**19**(6):1183-1194.

[96] Gonçalez E, Felicio JD, Pinto MM. Biflavonoids inhibit the production of aflatoxin by *Aspergillus flavus*. Brazilian Journal of Medical and Biological Research. 2001;**34**(11): 1453-1456.

[97] Paster N, Juven BJ, Harshemesh H. Antimicrobial activity and inhibition of aflatoxin B1 formation by olive plant tissue constituents. Journal of Applied Bacteriology. 1988;**64**(4):293-297.

[98] Jayashree T, Subramanyam C. Oxidative stress as a prerequisite for aflatoxin production by *Aspergillus parasiticus*. Free Radical Biology and Medicine. 2000;**29**(10):981-985.

[99] Lee SE, Mahoney NE, Campbell BC. Inhibition of aflatoxin B1 biosynthesis by piperlongumine isolated from *Piper longum* L. Journal of Microbiology and Biotechnology. 2002;**12**(4):679-682.

[100] Ehrlich K, Ciegler A. Effect of phytate on aflatoxin formation by *Aspergillus parasiticus* and *Aspergillus flavus* in synthetic media. Mycopathologia. 1984;**87**(1-2):99-103.

[101] Graf E, Empson KL, Eaton, JW. Phytic acid. A natural antioxidant. Journal of Biological Chemistry. 1987;**262**(24):11647-11650.

[102] Martínez-Luis S, Pérez-Vásquez A, Mata R. Natural products with calmodulin inhibitor properties. Phytochemistry. 2007;**68**(14):1882-1903.

[103] Praveen Rao, J, Subramanyam C. Calmodulin mediated activation of acetyl-CoA carboxylase during aflatoxin production by *Aspergillus parasiticus*. Letters in Applied Microbiology. 2000;**30**(4):277-281.

[104] Juvvadi PR, Chivukula S. Putative calmodulin-binding domains in aflatoxin biosynthesis–regulatory proteins. Current Microbiology. 2006;**52**(6):493-496.

[105] Kondo T, Sakurada M, Okamoto S, Ono M, Tsukigi H, Suzuki A, Sakuda S. Effects of aflastatin A, an inhibitor of aflatoxin production, on aflatoxin biosynthetic pathway and glucose metabolism in *Aspergillus parasiticus*. The Journal of Antibiotics. 2001;**54**(8):650-657.

[106] Yoshinari T, Akiyama T, Nakamura K, Kondo T, Takahashi Y, Muraoka Y, Nonomura Y, Nagasawa H, Sakuda, S. Dioctatin A is a strong inhibitor of aflatoxin production by *Aspergillus parasiticus*. Microbiology. 2007;**153**(8):2774-2780.

[107] Yoshinari T, Noda Y, Yoda K, Sezaki H, Nagasawa H, Sakuda S. Inhibitory activity of blasticidin A, a strong aflatoxin production inhibitor, on protein synthesis of yeast: Selective inhibition of aflatoxin production by protein synthesis inhibitors. The Journal of Antibiotics. 2010;**63**(6):309-314.

[108] Mabrouk SS, El-Shayeb NMA. Inhibition of aflatoxin production in *Aspergillus flavus* by natural coumarone and chromones. World Journal of Microbiology and Biotechnology. 1992;**8**(1):60-62.

[109] Roze LV, Beaudry RM, Keller NP, Linz JE. Regulation of aflatoxin synthesis by FadA/cAMP/protein kinase A signaling in *Aspergillus parasiticus*. Mycopathologia. 2004;**158**(2):219-232.

[110] Brodhagen M, Keller NP. Signalling pathways connecting mycotoxin production and sporulation. Molecular Plant Pathology. 2006;**7**(4):285-301.

[111] Vergopoulou S, Galanopoulou D, Markaki P. Methyl jasmonate stimulates aflatoxin B1 biosynthesis by *Aspergillus parasiticus*. Journal of Agricultural and Food Chemistry. 2001;**49**(7):3494-3498.

[112] Burow GB, Nesbitt TC, Dunlap J, Keller NP. Seed lipoxygenase products modulate *Aspergillus mycotoxin* biosynthesis. Molecular Plant-Microbe Interactions. 1997;**10**(3): 380-387.

[113] Greene-McDowelle DM, Ingber B, Wright MS, Zeringue HJ, Bhatnagar D, Cleveland TE. The effects of selected cotton-leaf volatiles on growth, development and aflatoxin production of *Aspergillus parasiticus*. Toxicon. 1999;**37**(6):883-893.

[114] Buchanan RL, Hoover DG, Jones SB. Caffeine inhibition of aflatoxin production: mode of action. Applied and Environmental Microbiology. 1983;**46**(5):1193-1200.

[115] Kato-Noguchi H, Macías FA. Possible mechanism of inhibition of 6-methoxy-benzoxazolin-2 (3H)-one on germination of cress (*Lepidium sativum* L.). Journal of Chemical Ecology. 2006;**32**(5):1101-1109.

[116] Ciegler A, Lillehoj EB, Peterson RE, Hall HH. Microbial detoxification of aflatoxin. Applied Microbiology. 1966;**14**(6):934-939.

[117] Lillehoj EB, Ciegler A, Hall, HH. Aflatoxin B1 uptake by *Flavobacterium aurantiacum* and resulting toxic effects. Journal of Bacteriology. 1967;**93**(1):464-471.

[118] Line JE, Brackett RE. Role of toxin concentration and second carbon source in microbial transformation of aflatoxin B1 by *Flavobacterium aurantiacum*. Journal of Food Protection. 1995;**58**(9):1042-1044.

[119] Teniola OD, Addo PA, Brost IM, Färber P, Jany KD, Alberts JF, van Zyl WH, Steyn PS, Holzapfel WH. Degradation of aflatoxin B 1 by cell-free extracts of Rhodococcus erythropolis and *Mycobacterium fluoranthenivorans* sp. nov. DSM44556 T. International Journal of Food Microbiology. 2005;**105**(2):111-117.

[120] Nout MJR. Effect of Rhizopus and Neurospora spp. on growth of *Aspergillus flavus* and *A. parasiticus* and accumulation of aflatoxin B1 in groundnut. Mycological Research. 1989;**93**(4):518-523.

[121] Oliveira CAF, Jager AV, Corassin CH, Bovo F, Reddy KR. Recent trends in microbiological decontamination of aflatoxins in foodstuffs. In: INTECH Open Access Publisher. Croatia, Rijeka; 2013.

[122] Nout MJR, Motarjemi Y. Assessment of fermentation as a household technology for improving food safety: A joint FAO/WHO workshop. Food Control. 1997;**8**(5-6):221-226.

[123] Santin E, Paulillo AC, Maiorka A, Nakaghi LSO, Macari M, Silva AVF, Alessi AC. Evaluation of the efficacy of *Saccharomyces cerevisiae* cell wall to ameliorate the toxic effects of aflatoxin in broilers. International Journal of Poultry Science. 2003;**2**(5):341-344.

[124] Shetty PH, Jespersen L. *Saccharomyces cerevisiae* and lactic acid bacteria as potential mycotoxin decontaminating agents. Trends in Food Science & Technology. 2006;**17**(2):48-55.

[125] Çelýk K, Denlý M, Savas T. Reduction of toxic effects of aflatoxin B1 by using baker yeast (*Saccharomyces cerevisiae*) in growing broiler chicks diets. Revista Brasileira de Zootecnia. 2003;**32**(3):615-619.

[126] Stanley VG, Ojo R, Woldesenbet S, Hutchinson DH, Kubena LF. The use of *Saccharomyces cerevisiae* to suppress the effects of aflatoxicosis in broiler chicks. Poultry science. 1993;**72**(10):1867-1872.

[127] Sampaio Baptista A, Horii J, Antonia Calori-Domingues M, Micotti da Glória E, Mastrodi Salgado J, Roberto Vizioli M. The capacity of manno-oligosaccharides, thermolysed yeast and active yeast to attenuate aflatoxicosis. World Journal of Microbiology and Biotechnology. 2004;**20**(5):475-481.

[128] Devegowda G, Aravind BIR, Morton MG. *Saccharomyces cerevisiae* and mannanoligosaccharides to counteract aflatoxicosis in broilers. In: Proceedings of Australian Poultry Science Symposium Sydney. 1996;**103**:106.

[129] Karaman M, Basmacioglu H, Ortatatli M, Oguz H. Evaluation of the detoxifying effect of yeast glucomannan on aflatoxicosis in broilers as assessed by gross examination and histopathology. British Poultry Science. 2005;**46**(3):394-400.

[130] Thompson ED, Knights BA, Parks LW. Identification and properties of a sterol-binding polysaccharide isolated from *Saccharomyces cerevesiae*. Biochimica et Biophysica Acta (BBA)-General Subjects. 1973;**304**(1):132-141.

[131] El-Nezami H, Kankaanpaa P, Salminen S, Ahokas J. Ability of dairy strains of lactic acid bacteria to bind a common food carcinogen, aflatoxin B 1. Food and Chemical Toxicology. 1998;**36**(4):321-326.

[132] Peltonen K, El-Nezami H, Haskard C, Ahokas J, Salminen, S. Aflatoxin B1 binding by dairy strains of lactic acid bacteria and bifidobacteria. Journal of Dairy Science. 2001;**84**(10):2152-2156.

[133] Turbic A, Ahokas JT, Haskard, CA. Selective in vitro binding of dietary mutagens, individually or in combination, by lactic acid bacteria. Food Additives & Contaminants. 2002;**19**(2):144-152.

[134] Haskard C, Binnion C, Ahokas J. Factors affecting the sequestration of aflatoxin by *Lactobacillus rhamnosus* strain GG. Chemico-Biological Interactions. 2000;**128**(1):39-49.

[135] Lahtinen SJ, Haskard CA, Ouwehandz AC, Salminenz SJ, Ahokasy JT. Binding of aflatoxin B1 to cell wall components of *Lactobacillus rhamnosus* strain GG. Food Additives and Contaminants. 2004;**21**:158-164.

[136] Haskard CA, El-Nezami HS, Kankaanpää PE, Salminen S, Ahokas JT. Surface binding of aflatoxin B1 by lactic acid bacteria. Applied and Environmental Microbiology. 2001;**67**(7):3086-3091.

[137] Megalla SE, Hafez AH. Detoxification of aflatoxin B1 by acidogenous yoghurt. Mycopathologia. 1982;**77**(2):89-91.

[138] Megalla SE, Mohran MA. Fate of a atoxin B-1 in fermented dairy products. Mycopathologia. 1984;**88**:27-29.

[139] Liu DL, Yao DS, Liang R, Ma L, Cheng WQ, Gu LQ. Detoxification of aflatoxin B 1 by enzymes isolated from *Armillariella tabescens*. Food and Chemical Toxicology. 1998;**36**(7):563-574.

[140] Liang ZH, Li JX, He YL, Guan S, Wang N, Ji C, Niu TG. AFB1 bio-degradation by a new strain-Stenotrophomonas sp. Agricultural Sciences in China. 2008;**7**(12):1433-1437.

[141] Teunisson DJ, Robertson JA. Degradation of pure aflatoxins by *Tetrahymena pyriformis*. Applied Microbiology. 1967;**15**(5):1099-1103.

[142] Detroy RW, Hesseltine CW. Transformation of aflatoxin B1 by steroid-hydroxylating fungi. Canadian Journal of Microbiology. 1969;**15**(6):495-500.

[143] Schultz DL, Luedecke LO. Effect of neutral fats and fatty acids on aflatoxin production. Journal of Food Protection. 1977;**40**(5):304-308.

[144] Hamid AB, Smith JE. Degradation of aflatoxin by *Aspergillus flavus*. Microbiology. 1987;**133**(8):2023-2029.

[145] Huwig A, Freimund S, Käppeli O, Dutler H. Mycotoxin detoxication of animal feed by different adsorbents. Toxicology Letters. 2001;**122**(2):179-188.

[146] Lewis SJ, Freedman AR. Review article: The use of biotherapeutic agents in the prevention and treatment of gastrointestinal disease. Alimentary Pharmacology & Therapeutics. 1998;**12**(9):807-822.

[147] Madrigal-Santillán E, Madrigal-Bujaidar E, Márquez-Márquez R, Reyes A. Antigenotoxic effect of *Saccharomyces cerevisiae* on the damage produced in mice fed with aflatoxin B 1 contaminated corn. Food and Chemical Toxicology. 2006;**44**(12):2058-2063.

[148] Diaz DE, Hagler WM, Hopkins BA, Whitlow LW. Aflatoxin binders I: in vitro binding assay for aflatoxin B1 by several potential sequestering agents. Mycopathologia. 2003;**156**(3):223-226.

[149] Kankaanpää P, Tuomola E, El-Nezami H, Ahokas J, Salminen SJ. Binding of aflatoxin B1 alters the adhesion properties of *Lactobacillus rhamnosus* strain GG in a Caco-2 model. Journal of Food Protection. 2000;**63**(3):412-414.

[150] Gratz S, Mykkänen H, Ouwehand AC, Juvonen R, Salminen S, El-Nezami H. Intestinal mucus alters the ability of probiotic bacteria to bind aflatoxin B1 in vitro. Applied and Environmental Microbiology. 2004;**70**(10):6306-6308.

[151] El-Nezami H, Mykkänen H, Kankaanpää P, Salminen S, Ahokas J. Ability of Lactobacillus and Propionibacterium strains to remove aflatoxin B1 from the chicken duodenum. Journal of Food Protection. 2000;**63**(4):549-552.

[152] Bolognani F, Rumney CJ, Rowland IR. Influence of carcinogen binding by lactic acid-producing bacteria on tissue distribution and in vivo mutagenicity of dietary carcinogens. Food and Chemical Toxicology. 1997;**35**(6):535-545.

[153] Basappa SC, Shantha T. Methods for detoxification of aflatoxins in foods and feeds: A critical appraisal. Journal of Food Science and Technology. 1996;**33**(2):95-107.

[154] Salhab AS, Edwards GS. Comparative in vitro metabolism of aflatoxicol by liver preparations from animals and humans. Cancer Research. 1977;**37**(4):1016-1021.

[155] Roebuck BD, Wogan GN. Species comparison of in vitro metabolism of aflatoxin B1. Cancer Research. 1977;**37**(6):1649-1656.

[156] Roebuck BD, Siegel WG, Wogan GN. In vitro metabolism of aflatoxin B2 by animal and human liver. Cancer Research. 1978;**38**(4):999-1002.

[157] Chipley JR, Mabee MS, Applegate KL, Dreyfuss M.S. Further characterization of tissue distribution and metabolism of [14C] aflatoxin B1 in chickens. Applied Microbiology. 1974;**28**(6):1027-1029.

[158] Donnelly PJ, Stewart RK, Ali SL, Conlan AA, Reid KR, Petsikas D, Massey TE. Biotransformation of aflatoxin B1 in human lung. Carcinogenesis. 1996;**17**(11): 2487-2494.

[159] Yoshizawa H, Uchimaru R, Kamataki T, Kato R, Ueno Y. Metabolism and activation of aflatoxin B1 by reconstituted cytochrome P-450 system of rat liver. Cancer Research. 1982;**42**(3):1120-1124.

[160] Sánchez-González L, Cháfer M, González-Martínez C, Chiralt A, Desobry S. Study of the release of limonene present in chitosan films enriched with bergamot oil in food simulants. Journal of Food Engineering. 2011;**105**(1):138-143.

[161] Saraf S. Applications of novel drug delivery system for herbal formulations. Fitoterapia. 2010;**81**(7):680-689.

[162] McClements DJ, Rao J. Food-grade nanoemulsions: formulation, fabrication, properties, performance, biological fate, and potential toxicity. Critical Reviews in Food Science and Nutrition. 2011;**51**(4):285-330.

[163] Chang Y, McLandsborough L, McClements DJ. Physicochemical properties and antimicrobial efficacy of carvacrol nanoemulsions formed by spontaneous emulsification. Journal of Agricultural and Food Chemistry. 2013;**61**(37):8906-8913.

[164] Gurr SJ, Rushton PJ. Engineering plants with increased disease resistance: What are we going to express? Trends in Biotechnology. 2005;**23**(6):275-282.

Chapter 3

The Control of Aflatoxin Contamination at Harvest, Drying, Pre-Storage and Storage Periods in Peanut: The New Approach

Isilay Lavkor and Isil Var

Additional information is available at the end of the chapter

http://dx.doi.org/10.5772/intechopen.68675

Abstract

Aflatoxins produced by *Aspergillus flavus* and *Aspergillus parasiticus* are contaminants of peanut (*Arachis hypogea* L.). Aflatoxin contamination is a serious concern given their hepatotoxic properties and their widespread occurrence during cultivation, harvest, drying and storage. Management of aflatoxin contamination of peanut is very important using cultural practice such as habitat management, soil amendments and pre- and post-harvest managements, using physical control methods, biological control methods and chemical control methods at harvest, drying, pre-storage and storage periods. Some procedures such as upkeep of low temperature and relative humidity (RH) in storage, keeping away the pod- and seed-feeding insects, doing the harvest and post-harvest procedure control, fast post-harvest drying, optimal timing of digging and harvest, providing optimum water to the crop through irrigation, avoiding mechanical damage during cultivation and optimal timing of digging and harvest might prevent the contamination of aflatoxin. In this review, various strategies for control of aflatoxin in peanuts in all periods are discussed.

Keywords: aflatoxin, control, *Aspergillus flavus*, *Aspergillus parasiticus*, peanut

1. Introduction

Peanut (*Arachis hypogaea* L., Family: Fabaceae) is a rich source of fat, proteins and vitamins. Peanuts are a good source of protein and vitamin E and have a good flavour. It is a very popular snack worldwide, and people of all age groups widely consume peanut products [1]. It is widely cultivated in Asia, Africa and the America [2]. Peanuts were found first in Brazil or Peru as early as 950 BC and carried to Africa by early explorers and missionaries. They were brought from

Africa to North America by slave traders in the early colonial days and used as food for slaves on ships [3]. Although there are a few species and kinds which are wild or cultivated, the peanut developed commercially is described as the fruit or pod of A. hypogaea, which belongs to the leguminosae family.

Their bloom grow on the ground, then is fertilized and dried, the stalk extends longitudinally and this ovary is forced underground. The pod holding those seed matures underneath the surface. On maturity, inner lining of the pod darkens, and the seed coat changes from white to reddish-brown. The entire plant, including most of the roots, is removed from the soil during harvesting [3].

As aflatoxin-producing *Aspergillus* species are naturally present in soil, it is difficult to avoid the invasion of these molds. Wounding by insects, mammals, birds and mechanical processes as well as stresses of hot, dry conditions can result in mold infection during the pre-harvesting period. The influence of delayed harvest on contamination is most severe when crops are affected by rain just prior to or during the harvest [4].

Poor agricultural practice and post-harvest treatments of peanuts can lead to an infection by mould fungus *Aspergillus flavus* and *Aspergillus parasiticus* releasing the toxic substance aflatoxins [5]. Infection and contamination of peanuts can occur both in the field (pre-harvest) and during post-harvest drying/curing and in storage facilities (post-harvest). Several species of fungi infect agricultural crops both in the field and during storage. Aflatoxigenic mould growth and aflatoxin contamination may occur in agricultural crops during growth, harvest, transportation and storage [6].

The contamination of food and feed materials with aflatoxins, which have toxic, carcinogenic and mutagenic activity, causes important health problems and economic losses [7]. Among these, aflatoxin B_1 (AFB_1) is the most naturally occurring compound of toxigenic isolates of *Aspergillus* species and was classified by the International Agency for Research on Cancer of WHO as a group 1A (cancerogenic) agent in 1993 [8].

Toxigenic fungi and aflatoxin contamination in peanuts start at farm level, and contamination occurs in both pre- and post-harvest periods. Lavkor et al. [9] reported that the levels of aflatoxins detected in 74.5% raw peanut samples were in the range of 0.3–1333.42 μg/kg [9]. Williams [10] referred to a study close by African markets indicating that more than 40% of the commodities found there exceed reasonable aflatoxin levels and expected more than 4.5 billion individuals in developing nations are at danger because of uncontrolled or ineffectively control input of aflatoxins [10].

It is recognised that high aflatoxin levels in the circulation system discourage the safe framework, accordingly encouraging tumour and HIV and hindering the development of kids. A cross-sectional review led in Ghana and referred by Williams et al. [11] demonstrates that invulnerable frameworks of as of late HIV-contaminated individuals are fundamentally adjusted if they have above-middle levels of regular exposures to aflatoxins [11]. Alluding to another study, Dr. Williams notes, "Individuals with a high aflatoxin biomarker status in the Gambia and Ghana will probably have dynamic jungle fever". In 2014, the Global Forum for

Innovations in Agriculture (GFIA) assembled an abnormal state meeting in Abu Dhabi, UAE, on reforming worldwide agribusiness through developments. Straight to the point Rijsberman, the CEO of the CGIAR Consortium, in his report in view of a Benin study about on the post-weaning introduction to aflatoxin, presumes that aflatoxins have debilitated development in kids and are costing African agriculturists over $450 million USD every year in lost exports [12]. As indicated by Oladale [13], research has shown that aflatoxins can lead to cause infertility, premature births, and postponement of egg production in birds and sudden calamity in egg generation effectively in ovulation birds. Furthermore, loss of flavour, skin staining and even yellowish colouring on skin can be observed in fish [13]. Kooprasertying et al. [14] found that roasted and ground peanuts and raw samples were contaminated with AFs at 100% and 80%, respectively. They determined the high amounts of AFs in peanuts as 362 ng g^{-1}, which means the highest concentration of AFS in peanuts (68. 22 ng g^{-1}). It has been reported and emphasised by researcher that the average intake of AFs was 0.49, 0.40 and 2.13 ng/kg bw/day for raw, roasted and ground peanuts, correspondingly. In addition, the potential risk for cancer was estimated at 0.01–0.12 cancer/year/100,000 persons. According to the results of the research, the researchers suggest that the current situation of aflatoxins contamination in peanuts and peanut products (especially in ground peanuts) has an adverse effect on the health of the Thai population [14].

In order to eliminate aflatoxins from contaminated peanut materials, numerous physical, chemical and biological methods have been developed. In addition, there are some genetic studies for developing peanut cultivars resistant to a broad spectrum of pathogens that pose a recurring threat to peanut health as well [15]. The work to be done in this context should be considered to have a minimum effect on the nutritional value and chemical composition of the nuts. It is known that peanut contents are very valuable and include 7% water, 25.8 g protein, 16.1 g carbs, 4.7 g sugar, 8.5 g fibre and 49.2 g fat (saturate: 6.28 g; monosaturated: 24.43 g; polysaturated:15.56 g, Omega-3:0 g; Omega 6:15.56 g; trans fat) [16].

An expanding amount of logical research has been given to adapt more about aflatoxin development issues and conceivable arrangements, including utilising hereditarily changed or hybridised seeds detailed for mold resistance or through utilisation of items, for example, AflaSafe, now utilised in Africa. AflaSafe's "biological approach" utilises a firmly related, non-aflatoxin-delivering mold to out-compete the aflatoxin-creating molds. In mild atmospheres, aflatoxin issues have been controlled largely with ventilation amid cooler evenings and through lower winter temperatures [17].

During the past decade, there has been increasing interest in the hypothesis that, the absorption of aflatoxin in consumed food is may be inhibited in the gastrointestinal tract. In recent years, some biological control strategies have been used to reduce aflatoxin contamination in various food materials. Aflatoxins contamination may occur in the field before harvest, during harvesting or during storage and processing; thus, methods for the prevention of aflatoxin contamination can conveniently be divided into pre-harvest, harvest and drying of unshelled peanuts, shelling post-harvest storage strategies. In addition, because of the high occurrence of aflatoxins in crops worldwide, fast and cost-effective analytical methods are required for the identification of contaminated agricultural commodities before they are processed into final products. In addition, there have been several reports on AFB_1 outbreaks, especially in many undeveloped countries. Therefore due to its potential threat in every step of the food production, analytical methods have been developed for the determination of AFB_1 in various matrices including liquid chromatography (LC), thin-layer chromatography, TLC), high-performance liquid chromatography (HPLC) immunoaffinity chromatography (IAC), enzyme-linked immunosorbent assay (ELISA), electrochemical immunosensor, etc. [18].

2. Control strategies

The risk of such contamination can be greatly increased because of the poor traditional practices. However, certain treatments have been found to reduce aflatoxin formation in peanuts, and the complete elimination of aflatoxin is not realistically achievable [19].

2.1. Pre-harvest factors influencing aflatoxin contamination of peanuts

2.1.1. Peanut cultivars

In the 1980s, numerous scientists had endeavoured to discover peanut cultivars resistant to *A. flavus* contamination and the production of aflatoxin but they were unsuccesful because of the cultivars exhibiting the complex elements affecting the development and dispersion of the growth and aflatoxin production [20, 21]. As of late, transgenic innovation has been broadly utilised for cultivar change. Transgenic peanuts containing the Bt (*Bacillus thuringiensis*) gene had altogether brought down levels of aflatoxin than non-Bt peanuts in preparatory examination of log-changed information [22]. Guo et al. [23] recognised the resistance-related genes (iso ara h3 and LEA 4) in peanut against *A. parasiticus* disease and resulting aflatoxin contamination, and after that built up a peanut microarray to distinguish hopeful genes that give imperviousness to *A. flavus* contamination [23, 24]. Furthermore, cultivar improvement in expanding the resistance of peanut to ailments can likewise altogether diminish the frequency of fungal contamination contrasted with the unaltered assortments [25].

2.1.2. Soil type

It is outstanding that peanuts can develop in various soil sorts such as light sandy soil and heavier soils. Light sandy soil benefits for the quick multiplication of *A. flavus*, especially under dry conditions in the later development time frame. Despite what might be expected, heavier soil can decrease the level of aflatoxin defilement in peanut grown because of having a higher water-holding limit [26].

Soil preparation is necessary for planting peanut in order to reduce the incidence of aflatoxin contamination. Several chemical control agents have been reported to inhibit aflatoxigenic mold growth and subsequent aflatoxin biosynthesis. Although some studies suggested that pesticides and fungicides might be useful in controlling mycotoxin production under field conditions, other results have found that pesticides were ineffective in controlling mycotoxin production by *Aspergillus* species [26]. Control of pod-feeding insects through the application of recommended insecticides and use of insect-resistant cultivars should be an integral part of the strategy to eliminate pre-harvest aflatoxin contamination [27]. In order to reduce the aflatoxin contamination in peanut soil rehabilitation with gypsum, cereal crop residue and farmyard fertiliser have been applied either singly or in different combinations at various stages of cropping. However, farmyard manure and gypsum at the sowing time were found to be the most effective in reducing aflatoxin contamination [28]. Biological control of toxigenic *A. flavus* strains can be achieved by the application of atoxigenic *A. flavus* strains to maize, groundnut and cotton fields [29]. Probst et al. [30] reported that A. flavus NRRL-21882 is the atoxigenic active ingredient in AflaGuard(Syngenta, Wilmington, DE) which is used for a biocontrol product currently registered for management of aflatoxins in maize in the United States. In addition, the researchers emphasise that isolate mixtures could compete more effectively than individual isolates in a greater diversity of environmental niches. In Argentina, Alaniz Zanon et al. [31] showed significant reductions of aflatoxin levels in peanut kernels harvested in the peanut core area of the country treated with a biocontrol agent based on the native non-aflatoxigenic *A. flavus* AFCHG2 strain [31]. Another study by Alaniz Zanon et al. [32] characterised native non-aflatoxigenic *A. flavus* strains isolated from the main peanut growing region of Argentina based on phenotypic, physiological and genetic characteristics; and to evaluate selected strains as biological control agents as single or mixed inocula to reduce aflatoxin accumulation in peanuts harvested in Northern Argentina. According to the results of [32], they found that an inoculum comprising a mixture of two nontoxigenic *A. flavus* strains proved to be effective in the reduction of aflatoxin accumulation in peanut kernels. In addition, Lavkor et al. [9] reported that *A. flavus* NRRL21882 (Afla-guard) was applied in three different ways in trial experiment, and it reduced aflatoxin amount varying from 98.4% to 99.8% and suppressed aflatoxin contamination of peanuts [9]. In another research, Power et al. [33] used the method of RNA interference (RNAi) as a promising method to reduce or prevent the accumulation of aflatoxin in peanut seed. In this study, they also performed high-throughput sequencing of small RNA populations in a control line and in two transformed peanut lines that expressed an inverted repeat targeting five genes involved in the aflatoxin biosynthesis pathway and that showed up to 100% less aflatoxin B_1 than the control samples. The researchers stated that the research output would increase their understanding of the effectiveness of RNAi and enable the possible improvement of the RNAi technology for the control of aflatoxins and thus probably it can determine the putative involvement of the small RNA populations in aflatoxin reduction [33].

2.1.3. Species of fungi in soil

Soil is a repository of fluctuated microorganisms including organisms, and peanuts are in direct contact with soil populaces of aflatoxigenic growths [34]. Regular fungal contaminants of peanuts involve *Aspergillus, Penicillium, Rhizopus* and *Fusarium* species [35, 36]. Numerous literary works detailed that *A. flavus* and *A. parasiticus* are the two firmly related types of organisms that attack peanuts and in this manner prompt to their defilement with aflatoxins B_1, B_2, G_1 and G_2 [25, 37, 38]. The existence of other fungi, for example, *Penicillium* and

Fusarium species, diminishes the aflatoxin generation because of competitive inhibition [39]. Furthermore, different morphological sorts for similar species likewise influence the aflatoxin contamination of peanuts, for example, S-and L-strains, which was kind of *A. flavus*. Although the occurrence of *A. flavus* S-strain has shown to have a relation with the contamination of peanut with aflatoxin, L-strain was not demonstrated and not definitely associated with the aflatoxin levels in peanuts [25].

2.1.4. Climate

Taylor et al. [40] detailed that aflatoxin defilement occurred in most developing zones; however, the most incidence of aflatoxin was in the hotter, more humid developing locales and took after the same geological example. *A. flavus* can be separated from soil in every single climatic zone, and it is separated moderately more as often as possible in warm temperature zones (latitudes 26–35°) than in tropical or cooler temperature zones. It is very uncommon in latitudes over 45° [41, 42]. In this manner, the aflatoxin defilement of peanuts is frequently found in scopes latitudes 35° [43]. In a study by Wu et al. [44], 2494 peanut samples were been collected from four major peanut producing areas in China and were investigated for the occurrence of aflatoxins. As a result, they found a close relationship can be concluded between the aflatoxin presence and the weather a month before harvest. In this survey in China from 2010 to 2013 in peanuts at harvest,they have emphasised that it is essential for taking preventive measures to alleviate pre-harvest contamination of aflatoxin to peanuts [44].

2.1.5. Weather conditions

Sanders et al. [45] documented that aflatoxin contamination is not generally straightforwardly associated with the rate of attack by *A. flavus*. Cole et al. [46] proposed that after the attack of aflatoxigenic fungi occurred, development of the fungi and aflatoxin creation could not occur until a resistance mechanism separated subsequently of natural anxiety (dry season and high temperature). Dry season and temperature stress are basic variables for aflatoxin contamination of peanuts [45, 46]. Cole and co-workers [46] found that drought stress and soil temperature of 29°C for 85–100 days produced the best number of colonised consumable grade peanuts and great aflatoxin levels [46]. End of season drought stress and lifted soil temperature are more advantage for advancing aflatoxin contamination [47, 48]. The reason is that dry season provokes a huge increase in proline in plants, which can improve aflatoxin occurrence [49]. Along these lines, sufficient rainfall can control or decrease aflatoxin generation of peanuts. Moreover, defilement has been observed to be across the board where peanuts are developed under rain-bolstered conditions compared with those developed under irrigation system [50].

2.1.6. Agricultural practices

Inappropriate agricultural pursuits, such as crop revolution, culturing, planting date, fertilisation and irrigation, can likewise expand the occurrence of *A. flavus* and aflatoxin contamination in peanuts [26]. The proceed with development of peanuts on a similar land may bring about the high disease from fungi and aflatoxin formation [51]. Crop rotation may bring down the rate of between-season survival of various species, particularly if it includes crops that are non-host to *Aspergillus* species [25]. Nevertheless, the impacts of product rotation on aflatoxin rely on upon the planting condition, for instance, in a semi-arid environment, *Aspergillus* occurrence might be high, and crop rotation may have little impact on the fungal action [52].

In non-inoculated, non-insecticide-sprayed territories, thick populace of plants or condensed fertilisation seem to affect the frequency of contamination by aflatoxin [53]. Insects may assume a vital part in the aflatoxin contamination of yields since almost all aflatoxins were found in regions harmed by insects. The insect harms peanut tissue, in this manner making section entrances for the fungus, and after that prompts to the high aflatoxin formation in peanuts [28, 53]. The research of Var and Uçkun [54] showed that the peanuts in the healthy shell have been preserved very well [54]. Numerous authors reported that a few insecticides and fungicides can repress aflatoxigenic fungi development and consequent aflatoxin biosynthesis in field [55, 56]. Bowen and Mack [57] utilised the insecticide to treat peanuts during development and decreased the levels of *A. flavus* disease and aflatoxin contamination. Moreover, early sowing in ideal time, scientific fertilising and solid field administration were required in decreasing the aflatoxin contamination of peanuts [57].

2.1.7. *Phytoalexin production*

Phytoalexins are described as antimicrobial substances combined by plants that collect quickly at territories of pathogen infection. In spite of the fact that the substance nature of the phytoalexins was not determined, it was demonstrated that peanuts created phytoalexins when tested by a few types of fungi, including *A. flavus* [58]. It was proved just as it was in 1972 that he resistance of immature peanut to fungi was expected to phytoalexins yielded in high amounts in light of fungal infection [59]. It was observed that as long as peanuts had phytoalexin generation they were not contaminated with aflatoxins and in immature peanuts the aflatoxin did not form until phytoalexin generation stopped in dry season stressed plants [60]. It has also been found that the water activity (aw) of the peanut kernels plays a crucial role in controlling the capacity of the nucleus to produce phytoalexins. Therefore, peanuts may produce sufficient phytoalexin in high water activity (> 0.97) to prevent the development of A. flavus and aflatoxin contamination.It was observed as kernel aw diminishes, as a result of elongated drought, the capacity of those kernels to create phytoalexins likewise reduces and in the end is lost (aw < 0.95) [60].

2.2. Post-harvest factors influencing aflatoxin contamination of peanuts

Generally, kernel moisture contents of 10% or higher post-harvest peanuts are prone to generate aflatoxins. Timely drying and keeping at safe moisture level can effectively control aflatoxin contamination of peanuts after harvest [26]. Diener and Davis [61] found that aflatoxin generation can be blocked by quickly drying to or beneath an aw of 0.83 for post-harvest peanuts. Before-storage separating to remove contaminated peanuts is the best approach to decrease aflatoxin generation [62, 63]. To keep an expansion in aflatoxin occurring during capacity and transportation, it is essential to control the dampness content, the temperature in the environment and the hygienic conditions [64]. Unsuitable kernel dampness during storage can continue from leaky roofs, reduction because of inappropriate ventilation in the warehouse, high-dampness outside material related with put away peanuts and high-dampness peanuts at first going into storage [65]. Thus, the storage and transportation conditions are the most vital reasons controlling aflatoxin defilement of peanuts.

2.2.1. *Harvest control strategies*

During harvesting, mechanical damage to peanuts must be avoided because it enhances susceptibility to contamination. Moreover, only mature peanuts should be harvested since fungal infection is more likely to occur in shrivelled and cracked kernels [66].

Recently, biosensors based on the use of monoclonal or polyclonal antibodies have seen great development in the field of small molecules analytical determination and specifically in the mycotoxins analyses [67]. Early and reliable precise methods protect health and life by preventing the entry of toxins into food chain. For this reason, it is necessary to transport these fast technologies to commercial products from the research stage using appropriate subsidies [68]. On the other hand, new unthermal preservation methods (Ozone, UV-C, ultrasound and manosound) are used for reducing aflatoxin content on some food and commodities. In addition, some studies try to show that these unthermal preservation methods could be used with hyperspectral imaging methods. Hyperspectral imaging methods could show us about the product or crop composition and distribution of food components [69]. In their research, Kandpal et al. [70] used hyperspectral imaging method for the detection of aflatoxin contamination on corn kernels. They have been reported that corn specimens were inoculated with four different concentrations (10, 100, 500 and 1000 mg/kg) with aflatoxin B1 (AFB1), and control specimens surface was sterilised with a PBs. Both contaminated and control specimens were scanned with an SWIR hyperspectral for the spectral range from 1100 to 1700 nm. The PLS-DA model has been created to arrange control and contaminated kernels and was discovered that most elevated general arrangement exactness yielded of the created model was 96.9% [70]. In their study, Jiang et al. [71] focused to identify the moldy peanuts using near-infrared (NIR) hyperspectral images, and NIR hyperspectral images were obtained at the wavelength ranging from 970 to 2570 nm. In order to select sensitive bands, principle component analysis (PCA) in the spectral dimension was used as well as the spectral vector was employed to identify the moldy information [71]. In another work, utilising a FRET-based method, Sabet et al. [72] have developed a nanobiosensor for detection of AFB_1 in agricultural foods. Aptamer-conjugated quantum dots (QDs) are adsorbed to Au nanoparticles (AuNPs) due to interaction of aptamers with AuNPs leading to quenching effect on QDs fluorescence. Upon the addition of AFB_1, the specific aptamers are attracted to AFB_1, getting distance from AuNPs which result in fluorescence recovery. Under optimised conditions, the detection limit of proposed nanobiosensor was 3.4 nM with linear range of 10–400 nM. Selectivity test demonstrates that the nanobiosensor could be a promising tool for specific evaluation of food stuff. This method was successfully applied for the analysis of AFB_1 in rice and peanut samples [72].

Traditional methods that require intense labour force are currently being used to separate aflatoxinous products. Workers are trying to determine whether there are aflatoxins in the products that pass through the tapes by standing on the UV lamp stands set up in a dark room for

8–12 h a day. Güzel et al. [73] stated that this manual separation technique reduces working efficiency and negatively affects the health of workers exposed to long periods of light. In addition, due to the distraction created by fatigue, the diseased products that need to be separated can escape attention. Therefore, the researchers had developed a UV light-based separator that does not escape toxic products, more rapid sorting and less use of human power. Güzel et al. [73] have believed that many negative conditions will come to an end with their machine.

2.2.2. *Drying of unshelled peanut control strategies*

The drying stage is all important to reduce attack and damage fungi. Lavkor and Bicici [74] reported that peanut kernels aflatoxin analysis was performed at four distinct periods: harvest, post-harvest, drying and pre-storage, and analysis results showed that aflatoxin contamination was not found on 96 samples sundried on drying sheet at experimental area in 2010 and 2011. According to Cole et al., it seems that post-harvest screening is a chance to decrease or eliminate aflatoxin at defiled seed. Probably, there are generally few, but highly contaminated seeds dispersed in the peanut lots when aflatoxin contamination occurred [65]. Practical methods include manual sorting, seed size and density separation, or electronic colour sorting. Electronic colour sorting has proven to be the most effective aflatoxin management strategy available in the processing phase [75]. Guchi [75] reported that electronic colour sorting is another means that can be used. For example, peanut that has been colonised by aflatoxin-producing fungi is often discoloured. Microwave heating shows great potential for the destruction of aflatoxin in contaminated peanut. Aflatoxin B_1 is sensitive to UV radiation and absorbs UV light at 222, 265 and 362 nm with the maximum absorption occurring at 362 nm. One strategy to reduce the entry of aflatoxin into the peanut chain is the use of chemical treatments such as acetosyringone, syringaldehyde and sinapinic acid and ammonia applications during post-harvest to reduce both fungal growth and toxin production [76]. Ozone due to its safety, environment-friendly, low cost and high efficiency in decomposing aflatoxin B_1 has been widely studied and used in the food industry [1]. Proctor et al. [77] achieved the highest level of degradation for aflatoxin B_1 (77±2%) after ozonation of peanut kernels for 10 min at 75°C [77]. In their study, Chen et al. [78] focused on the optimization of aflatoxin reduction by ozone during air drying of peanuts. They have observed that 5% moisture in peanut provided sensitivity of aflatoxins to ozone and reacted with 6.0 mg/l of ozone at the room temperature for 30 min simply degraded. They also found that the diminution of the total aflatoxins and aflatoxin B_1 (AFB_1) was 65.8% and 65.9%, respectively. In this research, they also examined the quality of peanut samples, and it has been observed that no significant differences ($P > 0.05$) were found in the polyphenols, resveratrol, acid value (AV) and peroxide value (PV) between treated and untreated samples. According to the researchers, the results suggested that the ozonation was a promising method for aflatoxin detoxification in peanuts [78].

In another study, Luo et al. [79] examined the ozone treatment effect in reducing aflatoxin B_1 in corn with different moisture content. In this study, the toxicity of the degradation products (DPs) of the ozone-treated aflatoxin B_1 contaminated corn was also evaluated using the human hepatocellular carcinoma cell line (HepG2) as model cells. It was observed that the degradation rate of aflatoxin B_1 in corn increases with ozone concentration and treatment time. It was also observed that aflatoxin B_1 contaminated corn with 13.47% moisture content was easier to be degraded by ozone than with 20.37% moisture content. In this study, when the safety of ozone used on aflatoxin B_1 contaminated corn was evaluated, the results showed that aflatoxin B_1 contaminated corn had high cell toxicity while the toxicity of ozone-treated aflatoxin B_1 contaminated corn had no significant difference with that of the aflatoxin B_1 free culture solution. The researchers suggested that ozonation can quickly and effectively degrade aflatoxin B_1 in corn and diminish aflatoxin B_1 contaminated corn's toxicity, and therefore, ozonation is expected to be an effective, fast and safe method for aflatoxin B_1 degradation in aflatoxin B_1 contaminated corn [79].

Diao et al.'s [1] study aimed to verify the ozonolysis efficiency of AFB_1 and to evaluate the oral safety of ACPs treated by ozone through a short-term subchronic toxicity study with Wistar rats. As a result of the study, they found that 89.40% of aflatoxin B_1 (AFB_1) in peanuts was decomposed by ozone with a concentration of 50 mg/L and flow rate of 5 L/min for 60 h. In their subchronic toxicity experiment, they declined that all rats did not have unusual changes in behaviour, and no signs of intoxication were observed except for several dead rats due to inappropriate gavage or anaesthesia. The researchers suggested that the deleterious effects of AFB_1 could be highly reduced by ozone, and ozone itself did not show any toxic effects on animals in this processing [1].

2.2.3. *Shell extraction*

Mechanical harm to food stuff during shell extraction makes them much more susceptible to attack by moulds such as *A. flavus*. Fungal growth may be much faster in the damaged peanuts compared to intact peanuts in any given environmental conditions. Cracks and breaks in peanut shell are mainly caused during shell extraction by use of machines or trampling. The machines used for this purpose are generally manual or motorized shellers. The latter normally use electricity and can be a simple type that can handle small volumes of peanuts or big type that can handle several bags of peanut per hour [80]. The use of ultraviolet light (UV) is well established for surface decontamination. After the application of UV-C for almond and nuts, it was observed that for AFG_1 and AFB_1 degradation result was found to be 100% and 96.5%, respectively [81]. Furthermore, Sharareh et al. [82] evaluated the effect of ultra-violet irradiation on detoxification of aflatoxin total including aflatoxin B_1 (AFB_1), aflatoxin B_2 (AFB_2), aflatoxin G_1 (AFG_1), aflatoxin G_2 (AFG_2) and aflatoxin total (AFT) content in standard solutions and investigated the structural changes using HPLC, GC/MS and FT-IR techniques. For this purpose, standard vials of aflatoxin solutions with concentrations of 1000 μg/kg AFB_1, 200 μg/kg AFB_2, 1000 μg/kg AFG_1, 200 μg/kg AFG_2 and 2400 μg/kg AFT were treated by UV-irradiation at 366 nm wavelength for 10 min in this study. Aflatoxin contents were analysed by high-pressure liquid chromatography (HPLC) method. As a result, in this research, it was observed that the amount of AFB_1, AFB_2, AFG_1, AFG_2 and AFT reduced approximately 98, 99.5, 99.8, 100 and 99.1%, respectively [82].

2.3. Post-harvest storage control strategies

As evidenced by the storage structures, traditional crop storage is not yet improved. The storage conditions should be cool and dry, should be defended from insects, rodents and birds;

should be easy to clean and should be waterproof and protected from flooding. These conditions are indispensable for modern or traditional storage. These suggestions are so important to prevent *A. flavus* contamination and aflatoxin formation in stored products, especially in peanuts. It has been reported that field application not only reduced aflatoxin contamination in the field but also reduced aflatoxin contamination that occurred in storage [83]. Aflatoxin production could be prevented or at least reduced by modification of atmospheric gases in storage silos such as by using carbon dioxide, nitrogen, carbon monoxide and sulphur dioxide. Previous work on peanuts reported that increases in the concentration of CO_2 in storage silo resulted in significant reductions in aflatoxin production within stored peanuts [84].

Globally, there have been increasing incidences relating to foodborne diseases including aflatoxins in both developed and developing countries. Because of the lack of proper hygiene practices and personal sanitation are not applied for food products, significant public health crisis can result from aflatoxins contamination. Studies conducted in these areas indicate that due to the low consciousness and knowledge of food handlers and workers in these subjects, aflatoxin contamination is seen in food and especially in groundnut products. According to some researchers, raising the level of public knowledge by arranging awareness campaigns can diminish the risk of aflatoxin contamination. The important factors to ensure that food handlers are proficient and knowledgeable on the principles of food safety and personal sanitation are advised trainings, food safety education and the developments of food safety certifications [85]. Therefore, Azaman et al. [85] planned a study that was to identify the differences in terms of knowledge, attitude and practices (KAP) of aflatoxins contamination among stakeholders of peanut-based products and to determine factors that mostly influence stakeholders' hygienic practices in peanut-based products. As a result of the study, they strongly emphasised the need for continuous hygiene improvement and training programmes by the stakeholders of peanut-based products. In addition, they stated that relevant strategies such as promotion and motivational models on health education and food safety campaigns would increase awareness and knowledge on food contaminants [85].

It is known that despite all these there remains aflatoxin contamination in the products. Therefore, in order to minimise aflatoxin exposure among consumers, it is essential to prevent highly contaminated kernels from re-entering food chains, and decontamination of such kernels should complement some sorting practices. Schwartzbord and Brown [86], in their study, focussed on to explore a process to transform oil from contaminated peanuts into a safe edible product. Schwartzbord and Brown [86], in their study, focussed on to explore a process to transform oil from contaminated peanuts into a safe edible product. As a result of the study, the researchers found that in extracted oil included aflatoxin concentration was approximately 10% of that of unextracted oil, which means it had a concentration that was only 5% of the original contaminated peanuts.

Therefore, they displayed that without pre-filtration aflatoxin concentration in the final product was 99.5% less than that found in the original peanuts [86].

Extrusion cooking is an important process widely applied in the food industry. The extrusion of AFT in cereals had been studied by different research groups. In one research, it was investigated the extrusion of AFT contaminated corn grits at 105°C and found that the levels of

AFT were reduced by 50%–80% after processing in the extruded corn grits [87]. In their study, Azaman et al. [85] explored the feasibility of degrading aflatoxin B_1 (AFB_1) in contaminated peanut meal by extrusion cooking. In this study, the effects of barrel temperature, material moisture content, feed rate, and screw speed as well as their interactions on the reduction rate of AFB_1 in peanuts meal were evaluated by response surface methodology (RSM) to optimise the extrusion conditions [85]. Zheng et al. [87] emphasised that the study indicated that extrusion cooking was an effective way to remove total AFB_1 from contaminated peanut meal. Moreover, the researchers stated that extrusion cooking can be used to treat other cereals. Although extrusion cooking has wide application prospects in food processing industry, but the researchers advised that it is required to perform further research to determine whether certain toxic products are generated during the decomposition of AFB_1 [87].

2.4. Analytical methods for the identification of contaminated agricultural commodities

Because of the high occurrence of aflatoxins in crops worldwide, fast and cost-effective analytical methods are required for the identification of contaminated agricultural commodities before they are processed into final products. So far, many aflatoxin detection technologies have been developed for the determination of AFB_1 in various matrices including liquid chromatography (LC), thin-layer chromatography (TLC), high-performance liquid chromatography (HPLC) immunoaffinity chromatography (IAC), enzyme-linked immunosorbent assay (ELISA) and electrochemical immunosensor, LC-MS/MS, Fluorescence polarisation immunoassay, capillary electrophoresis, near infrared spectroscopy, hyperspectral imaging electronic nose [88]. Actually, there are some advantages and disadvantages of these aflatoxin detection technologies, and these are still discussed.

Thiel et al. [89], in their study, had described the application of a technique for the determination of aflatoxins by reverse phase HPLC and fluorescence detection incorporating post-column derivatisation with iodine. They stated that the procedure proved to be extremely sensitive and reproducible [89]. Researchers suggested new tools for screening aflatoxins in food. For this purpose, one is for aflatoxin B1 and the other for total aflatoxin, they developed two prototypes to be used in the ELISA method. For this reason, they highlighted that seven monoclonal antibodies were produced that were with matchless high sensitivity and at the same time good cross-reactivity properties [90]. However due to limitations associated with these methods, including extensive sample preparation, expensive procedure and unavailability for onsite screening, increasing demand has been emerged especially in developing countries for more simple and cost-effective methods [72]. Utilising a FRET-based method, it has developed a nanobiosensor for detection of AFB_1 in agricultural foods. According to Sabet et al. [72] Aptamer-conjugated Quantum dots (QDs) are adsorbed to Au nanoparticles (AuNPs) due to the interaction of aptamers with AuNPs leading to quenching effect on QDs fluorescence. Upon the addition of AFB_1, the specific aptamers are attracted to AFB_1, obtaining distance from AuNPs, which result in fluorescence recovery [72]. Semiconductor quantum dots (QDs), as a new type of fluorescent probes, have unique optical characteristics such as photostability and high quantum yield originated from "quantum size" effect and have been proven to be of many uses in biosensing application. In their research, Sabet et al. found that selectivity test demonstrates that the nanobiosensor could be a promising tool for specific evaluation of food stuff. Moreover, they stated that this method was successfully applied for the analysis of AFB_1 in rice and peanut samples. In recent years, with the rapid development of nanostructured materials and nanotechnology in the fields of biotechnology and contaminant detection, magnetic nanoparticles (MNPs) have been receiving considerable attention. In their research, Sun et al. [91] used artificial antigen-modified MNPs employed as immune sensing probes, and antibody functionalised UCNPs were used as signal probes. Besides in this study, the antibodies-functionalised UCNPs were linked to the surface of the MNPs by antibody-antigen affinity [91]. According to Sun and co-workers, rare earth-doped upconversion nanoparticles were used successfully to assemble for sensing Aflatoxins B_1 in actual food samples (peanut oil) [91]. Ezekiel et al. [92] described a reliable and simple analytical method for the determination of aflatoxins (AFB_1, AFB_2, AFG_1 and AFG_2) in cereals, peanuts, vegetable oils and fermented foods such as beer, soybean sauce and soybean paste based on immunoaffinity column (IAC) cleanup coupled with direct competitive enzymelinked immunosorbent assay (dcELISA) detection and confirmed by ultra-high performance liquid chromatography tandem mass spectrometry (UHPLC-MS/MS) in their research. As a conclusion, they suggested this assay could be used as an effective analytical method for the determination of aflatoxins in complex grain foods [92].

3. Conclusion

As we can see above, there are various methods aimed at minimising the aflatoxins in foods, but there still exists Aflatoxin problem in food. Since it is difficult to achieve zero tolerance with AF contamination in commodities, AFs should be minimised in foods as much as possible to prevent the risk of cancer and the other health problems. Thus, legal tolerance limits based on scientific evidence obtained from risk assessment in different countries have been set for AFB_1 and total aflatoxin (AF) in foods and feeds. The limits vary between 4 and 20 parts per billion (ppb) through different countries [93]. The Codex Alimentarius Commission (CAC) has adopted the maximum permissible limits for AFs in unprocessed peanuts and tree nuts, which is 15 ppb as well as10 ppb in ready-to-eat tree nuts. However, European Union (EU) has adopted the level of 4ppb, which is the strictest limit in the world for AFs [93].

In a study, 60 peanut samples were analysed for aflatoxin B_1 using thin layer chromatography. The Democratic Republic of Congo is among African countries listed with high prevalence of liver cancer. As a result, Kamika and Takoy showed that aflatoxin B1 levels increased from the dry season to the rainy season with values ranging from 1.5 to 390 and 12 to 937, respectively. They reported that 70% of the peanut samples from both seasons exceeded the maximum limit of 5 mg/kg prescribed by the World Health Organization (WHO). Therefore, they emphasised continuous research on aflatoxin B_1 should be sought after [94]. In a study in Zambia, another African country, it showed that the high level of AFs in raw peanuts from both open markets and supermarkets samples are a health hazard for the population of the Lusaka region in Zambia. Therefore, the researchers stated that intervention tactics is urgently required to decrease the levels of AF contamination in peanuts [93].

In another Asian country, in Punjab major city of Pakistan, the focused on the assessment of the frequency of aflatoxin contamination in peanut and peanut products (peanut butter, roasted peanut, peanut bran and groundnut nimko) on the market. The researchers reported that the survey is of high importance to create the awareness among consumers, policy makers and law enforcement agencies to establish permissible limits for these toxins. As a result of their study, they told that the level of Aflatoxins in peanut and peanut products is high and poses a significant threat for the health of people [95].

One of the studies about aflatoxins in peanuts comes from Nigeria, which was planned to show the presence of aflatoxigenic *Aspergillus* populations and AFB1 profile in sold peanut cake in Nigeria. In this study in order to measure the awareness of consumers for the incidence of aflatoxin in the snack and public health threats of its steady consumption, was used questionnaire method. As a result *Aspergillus* section *Flavi* populations were recovered from 83% of the peanut cake samples. It was found that all analysed cake samples contained AFB_1 in concentrations exceeding the NAFDAC recommended level for AFB_1 in food and reaching up to 2824 mg/kg [92].

As seen before, most studies have showed us that aflatoxin contamination of peanuts can occur in the field (pre-harvest) when severe late-season drought stress occurs and poor agricultural practice and during storage (post-harvest) when improper conditions of moisture and temperature exist. Moreover, several techniques for aflatoxin controls have been proposed in the scientific literature, but just some are currently used by the peanut producers. So, aflatoxin control strategies are necessary to prevent health risks and economic losses for result from aflatoxin contamination. Besides, the studies and regulations related to Aflatoxins especially in peanuts and the other foods should be improved and carry on.

Author details

Isilay Lavkor[1*] and Isil Var[2]

*Address all correspondence to: lavkor@gmail.com

1 Biological Control Institute, Yüregir, Adana, Turkey

2 Department of Food Engineering, Faculty of Agriculture, Cukurova University, Adana, Turkey

References

[1] Diao E, Hou H, Chen B, Shan C, Dong H. Ozonolysis efficiency and safety evaluation of aflatoxin B_1 in peanuts. Food and Chemical Toxicology. 2013;**55**:519-525. DOI: 10.1016/j.fct.2013.01.038

[2] Shifa H, Tasneem S, Gopalakrishnan C, Velazhahan R. Biological control of pre-harvest aflatoxin contamination in groundnut (*Arachis hypogaea* L.) with *Bacillus subtilis* G1. Archives of Phytopathology and Plant Protection. 2016;**49**:137-148

[3] Woodroof, J.G. Peanuts: Production, processing, products. 3rd edition. Westport Conn. AVI Pub Co. 1983. 414p.

[4] Chiewchan N, Mujumdar AS, Devahastin S. Application of drying technology to control aflatoxins in foods and feeds: A review. Drying Technology. 2015;**33**:1700-1707. DOI: 10.1080/07373937.2015.1068795

[5] Lewis L, Onsongo M, Njapau H, Schurz-Rogers H, Luber G, Kieszak S. Aflatoxin contamination of commercial maize products during an outbreak of acute aflatoxicosis in eastern and central Kenya. Environmental Health Perspectives. 2005 Vol.113:1763-1767

[6] Monyoa ES, Njorogea SMC, Coe R, Osiru M, Madinda F, Waliyar F, Thakur RP, Chilunjika T, Anitha S. Occurrence and distribution of aflatoxin contamination in groundnuts (*Arachis hypogaea* L.) and population density of Aflatoxigenic *Aspergilli* in Malawi. Crop Protection. 2012;**42**:149-155. http://dx.doi.org/10.1016/j.cropro.2012.07.004

[7] Sweeney MJ, Dobson ADW. Mycotoxin production by *Aspergillus, Fusarium* and *Penicillium* species. International Journal of Food Microbiology. 1998;**43**:141-158. DOI: 10.1016/S0168-1605(98)00112-3

[8] Hussein HS, Brasel JM. Toxicity, metabolism, and impact of mycotoxins on humans and animals. Toxicology. 2001;**167**:101-134. http://dx.doi.org/10.1016/S0300-483X(01)00471-1

[9] Lavkor I, Arıoglu HH, Var I, Oztemiz S. A new biopesticide for control of aflatoxin on peanut in Turkey: *Aspergillus flavus* NRRL 21882 (Afla-guard). Turkey 6th Plant Protection Congress with International Participation, September 5-8, 2016. p. 76

[10] Williams JH. Aflatoxin as a public health factor in developing countries and its influence on HIV and other diseases. Peanut Collaborative Research Support Program, University of Georgia, World Bank Report. 2011:1-95

[11] Williams JH, Phillips TD, Jolly PE, Stiles JK, Jolly CM, Aggarwal D. Human aflatoxicosis in developing countries: A review of toxicology, exposure, potential health consequences, and interventions. American Journal of Clinical Nutrition. 2004;**80**:1106-1122

[12] Rijsberman F. PACA (Partnership for Aflatoxin Control in Africa) presents at the global forum for innovations: Middle East and Africa focus (Meridian Institute, Washington, DC), Aflatoxin Partnership Newsletter. 2014

[13] Oladele D. The effects of aflatoxins on animals. Partnership for Aflatoxin Control in Africa (Meridian Institute, Washington, DC), Aflatoxin Partnership Newsletter. 2014;4

[14] Kooprasertying P, Maneeboon T, Hongprayoon R, Mahakarnchanakul W. Exposure assessment of aflatoxins in Thai peanut consumption. Cogent Food & Agriculture. 2016;2:1204683. http://dx.doi.org/10.1080/23311932.2016.1204683

[15] Jonnala RS, Dunford NT, Chenault K. Tocopherol, phytosterol and phospholipid compositions of genetically modified peanut varieties. Journal of the Science of Food and Agriculture. 2006;**86**:473-476. DOI: 10.1002/jsfa.2372

[16] Arnarson A. Peanuts 101: Nutrition facts and health benefits. https://authoritynutrition.com/foods/peanuts/2017

[17] CAES. Reducing Aflatoxins in Corn During Harvest. Atlanta, GA: CAES Publication, University of Georgia. 2014;B 1231

[18] Horn BW, Greene RL, Dorner JW. Inhibition of aflatoxin B_1 production by *Aspergillus parasiticus* using nonaflatoxigenic strains role of vegetative compatibility. Biological Control. 2000;**17**:147-154

[19] Uckun O, Var I. Monitoring of aflatoxins in peanuts. Turkish Journal of Agricultural and Natural Sciences Special Issue of Balkan Agriculture Congress. Cilt:1 Özel Sayı:1. 2014;1310-1314

[20] Blankenship PD, Cole RJ, Sanders TH. Comparative susceptibility of four experimental peanut lines and the cultivar Florunner to preharvest aflatoxin contamination. Peanut Science. 1985;**12**:70-72. http://dx.doi.org/10.3146/pnut.12.2.0006

[21] Azaizeh HA, Pettit RE, Smith OD, Taber RA. Reaction of peanut genotypes under drought stress to *Aspergillus flavus* and *A. parasiticus*. Peanut Science. 1989;**16**:109-113. http://dx.doi.org/10.3146/i0095-3679-16-2-12/

[22] Ozias-Akins P, Yang H, Perry E, Akasaka Y, Niu C, Holbrook C, Lynch RE. Transgenic peanut for preharvest aflatoxin reduction. Mycopathologia. 2002;**155**:98

[23] Guo BZ, Chen X P, Dang P, Scully BT, Liang X Q, Holbroo CC, Yu JJ. Peanut gene expression profiling in developing seeds at different reproduction stages during *Aspergillus parasiticus* infection. BMC Developmental Biology. 2008;**4**:8-12

[24] Guo B Z, Fedorova ND, Chen XP, Wan CH, Wang W, Nierman WC, Yu JJ. Gene expression profiling and identification of resistance genes to *Aspergillus flavus* infection in peanut through EST and microarray strategies. Toxins. 2011;**3**:737-753. http://dx.doi.org/10.3390/toxins3070737

[25] Mutegi CK, Ngugi HK, Hendriks SL, Jones RB. Factors associated with the incidence of *Aspergillus* section *Flavi* and aflatoxin contamination of peanuts in the Busia and Homa bay districts of western Kenya. Plant Pathology. 2012;**61**:1143-1153. http://dx.doi.org/10.1111/j.1365-3059.2012.02597.x.

[26] Torres AM, Barros GG, Palacios SA, Chulze SN, Battilani P. Review on pre- and post-harvest management of peanuts to minimize aflatoxin contamination. Food Research International. 2014;**62**:11-19. http://dx.doi.org/10.1016/j.foodres.2014.02.023

[27] Desai N, Lee J, Upadhya R, Chu Y, Moir RD, Willis IM. Two steps in Maf1-dependent repression of transcription by RNA polymerase III. Journal of Biological Chemistry. 2005;**8**:6455-62. DOI: 10.1074/jbc.M412375200

[28] Waliyar F.,Craufura, P., Reddy, KV, Reddy, SV.,Nigam, SN., Lavakumar, P. Effect of soil application of lime, crop residue and biocontrol agents on preharvest *Aspergillus flavus* infection and aflatoxin contamination in groundnut. In: International conference on groundnut Aflatoxin management and Genomics, 5-9 Nov. 2006, Guangdong, China.

[29] Cotty P, Garcia RJ. Influences of climate on aflatoxin producing fungi and aflatoxin contamination. International Journal of Food Microbiology. 2007;**119**:109-115. DOI: 10.1016/j.ijfoodmicro.2007.07.060

[30] Probst C, Bandyopadhyay R, Price LE, Cotty PJ. Identification of atoxigenic *Aspergillus flavus* isolates to reduce aflatoxin contamination of maize in Kenya. Plant Disease. 2011;**95**:212-218. http://dx.doi.org/10.1094/PDIS-06-10-0438

[31] Alaniz Zanon MS, Chiotta ML, Giaj-Merlera G, Barros G, Chulze S. Evaluation of potential biocontrol agent for aflatoxin in Argentinean peanuts. International Journal of Food Microbiology. 2013;**162**:220-225. Doi: 10.1016/j.ijfoodmicro.2013.01.017

[32] Alaniz Zanon MS, Barros GG, Chulze SN. Non-aflatoxigenic *Aspergillus flavus* as potential biocontrol agents to reduceaflatoxin contamination in peanuts harvested in Northern Argentina. International Journal of Food Microbiology. 2016;**231**:63-68. http://dx.doi.org/10.1016/j.ijfoodmicro.2016.05.016

[33] Power IL, Dang PM, Sobolev VS, Orner VA, Powell JL, Lamb MC. Characterization of small RNA populations in non-transgenic and aflatoxin-reducing-transformed peanut. Plant Science. 2017;**257**:106-125 http://dx.doi.org/doi:10.1016/j.plantsci.2016.12.013

[34] Horn B, Pitt J. Yellow mold and aflatoxin. Compendium of Peanut Diseases. 1997;**2**:40-42

[35] Gachomo EW, Mutitu EW, Kotchoni OS. Diversity of fungal species associated with peanuts in storage and the levels of aflatoxins in infected samples. International Journal of Agriculture and Biology. 2004;**6**:955-959

[36] Youssef MS, El-Maghraby O MO, Ibrahim YM. Mycobiodata and mycotoxins of Egyptian peanut (*Arachis hypogaea* L.) seeds. International Journal of Botany. 2008;**4**:349-360. http://dx.doi.org/10.3923/ijb.2008.349.360

[37] Vaamonde G, Patriarca A, Pinto VF, Comerio R, Degrossi C. Variability of aflatoxin and cyclopiazonic acid production by *Aspergillus* section Flavi from different substrates in Argentina. International Journal of Food Microbiology. 2003;**88**:79-84. http://dx.doi.org/10.1016/S0168-1605(03)00101-6

[38] Dorner J W. Simultaneous quantitation of *Aspergillus flavus* /*A. parasiticus* and aflatoxin in peanuts. Journal of AOAC International. 2002;**85**:911-916

[39] Horn BW, Dorner JW. Soil populations of *Aspergillus* species from section *Flavi* along a transect through peanut growing regions of the United States. Mycologia. 1998;**90**:767-776. http://dx.doi.org/10.2307/3761317

[40] Taylor H, Knoche L, Granville W. Color Harmony Manual. Chicago: Container Corporation of America; 1985

[41] Manabe M, Tsuruta O. 1978. Geographical distribution of aflatoxin-producing fungi inhabiting in Southeast Asia. Japan Agricultural Research Quarterly. 1978; **12**:224-227

[42] Klich MA. Biogeography of *Aspergillus* species in soil and litter. Mycologia. 2002;**94**:21-27. http://dx.doi.org/10.2307/3761842

[43] Ioannou-Kakouri E, Aletrari M, Christou E, Ralli A, Koliou A, Christofidou M, et al. An overview on toxigenic fungi and mycotoxins in Europe, Springer Netherlands, USA. Isolation, Identification and Antimicrobial Activity Archives of Microbial. 2004;**126**:223-230. http://dx.doi.org/10.1007/978-1-4020-2646-1

[44] Wu L, Ding X, Li P, Du X, Zhou H, Bai YZ. Aflatoxin contamination of peanuts at harvest in China from 2010 to 2013 and its relationship with climatic conditions. Food Control. 2016;**60**:117-123. http://dx.doi.org/10.1016/j.foodcont.2015.06.029

[45] Sanders T, Cole R, Blankenship P, Dorner J. Aflatoxin contamination of peanuts from plants drought stressed in pod or root zones 1. Peanut Science. 1993;**20**:5-8. http://dx.doi.org/10.3146/i0095-3679-20-1-2

[46] Blankenship PD, Cole RJ, Sanders TH, Hill RA. Effect of geocarposphere temperature on pre-harvest colonization of drought-stressed peanuts by *Aspergillus flavus* and subsequent aflatoxin contamination. Mycopathologia. 1984;**85**:69-74. DOI: 10.1007/BF00436705

[47] Bankole S, Schollenberger M, Drochner W. Mycotoxins in food systems in Sub Saharan Africa: A review. Mycotoxin Research. 2006;**22**:163—9. http://dx.doi.org/10.1007/BF02959270

[48] Rachaputi N, Krosch S, Wright G. Management practices to minimise pre-harvest aflatoxin contamination in Australian peanuts. Animal Production Science. 2002;**42**:595-605. http://dx.doi.org/10.1071/EA01139

[49] Barnett NM, Naylor A. Amino acid and protein metabolism in Bermuda grass during water stress. Plant Physiology. 1966;**41**:1222-30. http://dx.doi.org/10.1104/pp.41.7.1222

[50] Yellamanda Reddy T, Sulochanamma B, Subramanyam A, Balaguravaiah D. Influence of weather, dryspells and management practices on aflatoxin contamination in groundnut. Indian Phytopathology. 2003;**56**:262-265

[51] Ortiz M, Barros G, Reynoso M, Torres A, Chulze S, Ramirez M, editors. Soil populations of Aspergillus section Flavi from the main and new peanut growing areas in Argentina. In: ISM Conference. March 3-4, 2011 Atlanta, GA

[52] Commission CA. Code of practice for the prevention and reduction of aflatoxin contamination in peanuts. CAC/RCP. 2004;55

[53] Anderson W, Holbrook C, Wilson D, Matheron M. Evaluation of Preharvest Aflatoxin Contamination in Several Potentially Resistant Peanut Genotypes 1. Peanut Science. 1995;**22**:29-32. http://dx.doi.org/10.3146/pnut.22.1.0007

[54] Var I, Uçkun O. Osmaniye İlinde Üretilen Yerfıstıklarının Mikrobiyolojik Florasının Belirlenmesi. (unpublished)

[55] Lee S-E, Campbell BC, Molyneux RJ, Hasegawa S, Lee H-S. Inhibitory effects of naturally occurring compounds on aflatoxin B1 biotransformation. Journal of Agricultural and Food Chemistry. 2001;**49**:5171-5177. http://dx.doi.org/10.1021/jf010454v.

[56] Dorner JW, Cole RJ, Connick WJ, Daigle DJ, McGuire MR, Shasha BS. Evaluation of biological control formulations to reduce aflatoxin contamination in peanuts. Biological Control. 2003;**26**:318-324. http://dx.doi.org/10.1016/S1049-9644(02)00139-1

[57] Bowen K, Mack T. Relationship of damage from the lesser cornstalk borer to *Aspergillus flavus* contamination in peanuts 2. Journal of Entomological Science. 1993;28:29-42

[58] Aguamah GE, Langcake P, Leworthy DP, Page JA, Pryce RJ, Strange RN. Two novel stilbene phytoalexins from *Arachis hypogaea*. Phytochemistry. 1981;**20**:1381-1383. http://dx.doi.org/10.1016/0031-9422(81)80044-1

[59] Vidhyasekaran P, Lalithakumari D, Govindaswamy C. Production of a phytoalexin in groundnut due to storage fungi. Indian Phytopathology. 1972;**25**:240-245

[60] Dorner JW, Cole RJ, Sanders TH, Blankenship PD. Interrelationship of kernel water activity, soil temperature, maturity, and phytoalexin production in preharvest aflatoxin contamination of drought-stressed peanuts. Mycopathologia. 1989;**105**:117-128. http://dx.doi.org/10.1007/BF00444034

[61] Diener UL, Davis ND. Limiting temperature and relative humidity for aflatoxin production by *Aspergillus flavus* in stored peanuts. Journal of the American oil Chemists society. 1970;**47**:347-351. http://dx.doi.org/10.1007/BF02639000/

[62] Cole RJ, Dorner J, Holbrook C. Advances in Mycotoxin Elimination and Resistance: Advances in Peanut Science. Stillwater, OK: American Peanut Research and Education Society, Inc; 1995. pp. 456-474

[63] Dorner J. Management and prevention of mycotoxins in peanuts. Food Additives and Contaminants. 2008;**25**:203-208. http://dx.doi.org/10.1080/02652030701658357

[64] Dickens J. Aflatoxin control program for peanuts. Journal of the American Oil Chemists' Society. 1977;**54**:A225-A228. http://dx.doi.org/10.1007/BF02894413

[65] Davidson JI, Hill JRRA, Cloe JR, Mixon AC, Henning RJ: Field performance of two peanut cultivars relative to aflatoxin contamination. Peanut Science. 1983;**10**:43-47.

[66] Grybauskas AP, Thomson PR, Cassel EU. 1992. Aflatoxins.[Online]. Available: http://www.inform.umd.edu

[67] Mosiello L, Lamberti I. Biosensors for aflatoxins detection. In: Irineo Torres-Pacheco, editor. Aflatoxins - Detection, Measurement and Control. INTECH Open Access; 2011. p. 147-160 DOI: 10.5772/22095

[68] Tothill I. Biosensors and nanomaterials and their application for mycotoxin determination. World Mycotoxin Journal. 2011;**4**:361-374

[69] Büyükcan MB, Türkyılmaz İ, Caner C. Hiperspektral Görüntü İşleme Tekniğinin Gıda Alanında Kullanımı. 7. Gıda Mühendisliği Kongresi, Ankara 2011

[70] Kandpal LM, Lee S, Kim MS, Bae H, Cho B-K. Short wave infrared (SWIR) hyperspectral imaging technique for examination of aflatoxin B_1 (AFB_1) on corn kernels. Food Control. 2015;**51**:171-176. http://dx.doi.org/10.1016/j.foodcont.2014.11.020

[71] Jiang J, Qiao X, He R. Use of near-infrared hyperspectral images to identify moldy peanuts. Journal of Food Engineering. 2016;**169**:284-290. http://dx.doi.org/10.1016/j.jfoodeng.2015.09.013

[72] Sabet FS, Hosseini M, Khabbaz H, Dadmehr M, Ganjali MR. FRET-based aptamer biosensor for selective and sensitive detection of aflatoxin B_1 in peanut and rice. Food Chemistry. 2017;**220**:527-532. http://dx.doi.org/10.1016/j.foodchem.2016.10.004

[73] Güzel E, Özlüoymak B. Aflatoksin (Zehirli Küf Mantarı) Ayrıştırma Makinesi İcat Etti. Çukurova Üniversitesi Haber Merkezi, Adana 2013

[74] Lavkor I, Bicici M. Aflatoxin Occurrence in Peanuts Grown in Osmaniye at Harvest, Post-Harvest, Drying and Pre-Storage Periods. Journal of Agricultural Sciences. 2015;**21**:394-405

[75] Guchi E. Aflatoxin contamination in groundnut (*Arachis hypogaea* L.) caused by *Aspergillus* species in Ethiopia. Journal of Applied & Environmental Microbiology. 2015;3:11-19. DOI: 10.12691/jaem-3-1-3

[76] Canavar Ö, Kaynak MA. Determination of yield and yield components and seed quality of peanuts (*Arachis hypogaea* L.) at different harvest times. International Journal of Agronomy and Plant Production. 2013;4:3791-803

[77] Proctor A, Ahmedna M, Kumar J, Goktepe I. Degradation of aflatoxins in peanut kernels/flour by gaseous ozonation and mild heat treatment. Food Additives and Contaminants. 2004;**21**:786-793. DOI: 10.1080/02652030410001713898

[78] Chen YC, Liao CD, Lin HY, Chiueh LC, Shih DYC. Survey of aflatoxin contamination in peanut products in Taiwan from 1997 to 2011. Journal of Food and Drug Analysis. 2013;**21**:247-252. http://dx.doi.org/10.1016/j.jfda.2013.07.001

[79] Luo X, Wang R, Wang L, Wang Y, Chen Z. Structure elucidation and toxicity analyses of the degradation products of aflatoxin B_1 by aqueous ozone. Food Control. 2013;**31**:331-336. http://dx.doi.org/10.1016/j.foodcont.2012.10.030

[80] Okello D, Kaaya A, Bisikwa J, Were M, Oloka H. Management of Aflatoxins in Groundnuts: A Manual for Farmers, Processors, Traders and Consumers in Uganda. National Agricultural Research Organization, Entebbe, Uganda. 2010

[81] Jubeen F, Bhatti IA, Khan MZ, Zahoor-Ul H, Shahid M. Effect of UVC irradiation on aflatoxins in ground nut (Arachis hypogea) and tree nuts (Juglans regia, prunus duclus and pistachio Vera). Journal of the Chemical Society of Pakistan. 2012;**34**:1366-1374

[82] Sharareh M, Maryam T, Azadeh K, Ali MS. Effect of uv irradiation on quantity and structure of aflatoxıns using hplc, gc/ms and ft-ır technıques. Ludus vıtalıs. 2015;**11**:64-69

[83] Dorner JW, Cole RJ. Effect of application of nontoxigenic strains of *Aspergillus flavus* and *A. parasiticus* on subsequent aflatoxin contamination of peanuts in storage. Journal of Stored Products Research. 2002;**38**:329-339. http://dx.doi.org/10.1016/S0022-474X(01)00035-2

[84] Kabak B, Dobson AD, Var I. Strategies to prevent mycotoxin contamination of food and animal feed: A review. Critical Reviews in Food Science and Nutrition. 2006;**46**(8):593-619. DOI: 10.1080/10408390500436185

[85] Azaman NNM, Kamarulzaman NH, Shamsudin MN, Selamat J. Stakeholders' knowledge, attitude, and practices (KAP) towards aflatoxins contamination in peanut-based products. Food Control. 2016;**70**:249-256

[86] Schwartzbord, JR., Brown, DL. Aflatoxin contamination in Haitian peanut products and maize and the safety of oil processed from contaminated peanuts. Food Control. 2015;**56**:114-118. http://dx.doi.org/10.1016/j.foodcont.2016.05.058

[87] Zheng H, Wei S, Xu Y, Fan M. Reduction of aflatoxin B_1 in peanut meal by extrusion cooking. LWT-Food Science and Technology. 2015;**64**:515-519. http://dx.doi.org/10.1016/j.lwt.2015.06.045

[88] Yao H, Hruska Z, Di Mavungu JD. Developments in detection and determination of aflatoxins. World Mycotoxin Journal. 2015;**8**:181-191. http://dx.doi.org/10.3920/WMJ 2014.1797

[89] Thiel P, Stockenström S, Gathercole P. Aflatoxin analysis by reverse phase HPLC using post-column derivatization for enhancement of fluorescence. Journal of Liquid Chromatography. 1986;9:103-112. http://dx.doi.org/10.1080/01483918608076625

[90] Oplatowska-Stachowiak M, Sajic N, Xu Y, Haughey SA, Mooney MH, Gong YY, et al. Fast and sensitive aflatoxin B_1 and total aflatoxins ELISAs for analysis of peanuts, maize and feed ingredients. Food Control. 2016;**63**:239-245. http://dx.doi.org/10.1016/j.foodcont.2015.11.041

[91] Sun C, Li H, Koidis A, Chen Q. Quantifying aflatoxin B_1 in peanut oil using fabricating fluorescence probes based on upconversion nanoparticles. Spectrochimica Acta Part A: Molecular and Biomolecular Spectroscopy. 2016;**165**:120-126. http://dx.doi.org/10.1016/j.saa.2016.04.040

[92] Ezekiel C, Sulyok M, Babalola D, Warth B, Ezekiel V, Krska R. Incidence and consumer awareness of toxigenic Aspergillus section Flavi and aflatoxin B_1 in peanut cake from Nigeria. Food Control. 2013;**30**:596-601. http://dx.doi.org/10.1016/j.foodcont.2012.07.048

[93] Bumbangi N, Muma J, Choongo K, Mukanga M, Velu M, Veldman F, et al. Occurrence and factors associated with aflatoxin contamination of raw peanuts from Lusaka district's markets, Zambia. Food Control. 2016;**68**:291-6. http://dx.doi.org/10.1016/j.foodcont.2016.04.004

[94] Kamika I, Takoy LL. Natural occurrence of Aflatoxin B_1 in peanut collected from Kinshasa, Democratic Republic of Congo. Food Control. 2011;**22**:1760-1764. http://dx.doi.org/10.1016/j.foodcont.2011.04.010

[95] Iqbal SZ, Asi MR, Zuber M, Akram N, Batool N. Aflatoxins contamination in peanut and peanut products commercially available in retail markets of Punjab, Pakistan. Food Control. 2013;**32**:83-86. http://dx.doi.org/10.1016/j.foodcont.2012.11.024

The Effect on Oxidative Stress of Aflatoxin and Protective Effect of Lycopene on Aflatoxin Damage

Seval Yilmaz, Emre Kaya and Mehmet Ali Kisacam

Additional information is available at the end of the chapter

http://dx.doi.org/10.5772/intechopen.69321

Abstract

Aflatoxin (AF) is polysubstituted bifuranocoumarins that are secondary fungal metabolites produced by *parasiticus/flavus* group of the genus *Aspergillus*. AF is hepatotoxic, nephrotoxic, mutagenic, teratogenic, genotoxic, and immunotoxic, so the International Agency for Research on Cancer has classified AF as class I human carcinogen. AF-mediated cell injury may be associated with the release of free radicals, and these radicals initiate lipid peroxidation and a damaging process in biological systems since all cell membranes contain the polyunsaturated fatty acids (PUFAs), which are substrates for such a reaction. One of the causes for AF-induced toxicity is the oxidative stress, which leads to the improved generation of reactive oxygen species (ROS) and the oxidative DNA damage. Lycopene, a naturally occurring carotenoid, has drawn a particular attention in recent years because of its high antioxidant activity and free radical scavenging capacity and has been shown to be effective against oxidative stress due to AF. Lycopene blocks Phase 1 metabolic enzymes of AFB such as 3A4, 2A6, and 1A2.

Keywords: aflatoxin, oxidative stress, reactive oxygen species, antioxidant, lycopene

1. Introduction

Mycotoxins are toxic products generated by fungi that are present spontaneously in foodstuffs. Mycotoxins may be generated in foodstuffs at different stages from production to transfer and preservation processes. Chemical stability and persistence of mycotoxins make them long-lasting, and even after elimination of fungi, mycotoxins may exist in foodstuffs [1].

Aflatoxin (AF) is the most abundant type of mycotoxins found in foodstuffs. Chemists isolated AF from *Aspergillus flavus* and named it as AF by taking "a" from *Aspergillus* and "fla" from *flavus*. There are at least 20 intermediates of AFs generated from *Aspergillus* species [1, 2]. AFs are found in the chemical construction of the furanocoumarins, and they possess two prominent structures: one of them is difurocoumarocyclopentenone (AFB1, AFB2, AFB2A, AFM1, AFM2, AFM2A, and aflatoxicol), and the other one is difurocoumarolactone (AFG1, AFG2, AFG2A, AFGM1, AFGM2, AFGM2A, and AFB3). AFs are named as AFB or AFG referring to the blue "B" or green "G" fluorescent color emitted by them under UV light on thin-layer chromatography, while the subscript numbers 1 and 2, respectively, show major and minor compounds. Moreover, AFB1 and AFB2 metabolites that show up in body fluids are named as AFM1 and AFM2 (**Figure 1**) [3, 4].

AFs commonly contaminate cereals and cereal-based foods such as rice, maize, sorghum, millet, groundnuts, dried cassava, and many others during the storage and poor processing conditions. AFs not only contaminate foodstuff but are also found in edible tissues, milk, and eggs after consumption of contaminated feed by farm animals [1, 6]. Trout, rats, ducklings, cattle, poultry, and swine are some of the many animals that have been shown to be sensitive to AF [6]. According to AFB1 concentration, the organs may be classed as follows: gonads, liver, kidney, spleen, bursa cloacalis, thymus, endocrine glands, muscles, lungs, and brain [7]. Petr et al. [8] revealed that AFB1 was determined in the blood, kidney, liver, and testis to maximum 8–10 h after a single intraperitoneal (i.p.) injection at 0.1 mg/kg AFB1.

AFs are a group of naturally occurring food-borne poisons that have been associated with death and disease in humans and animals. They are of great worldwide concern due to their toxic effects on human and animal health [9]. Among all AFs, AFB1 is the most toxic, mutagenic, and carcinogenic to both humans and livestock and is classified into group I as human carcinogen by the International Agency for Research on Cancer [10]. The extent of the carcinogenicity of AF depends on the presence of human health factors including hepatitis B virus infection, nutritional status, sex, and age as well as the amount of AF exposure [11, 12]. In transgenic mice, it was shown that overexpression of the hepatitis B virus large peptide

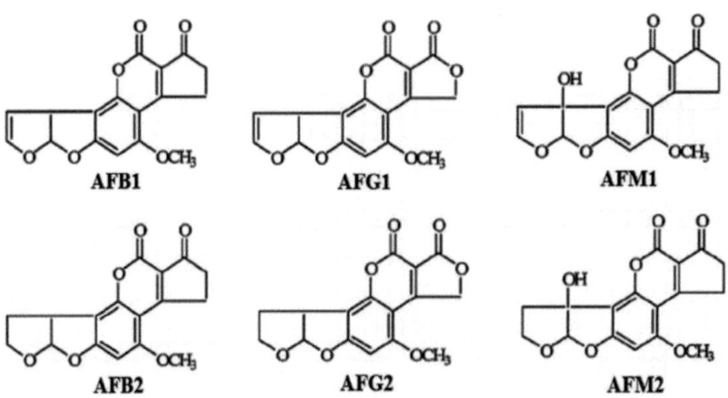

Figure 1. Chemical structure of AFs (Adapted from Marin and Taranu [5]).

envelope acted synergistically with AFB1 to have an effect on neoplastic development and other forms of chronic liver damage [13].

The immunotoxic potential of AF is known in many species, including laboratory and domestic animals [14]. In pigs, AF causes the decreases in blastogenic responses to mitogen, the reductions in complement titers, the decreases in macrophage activation, and the depression of delayed hypersensitivity responses [15]. Poultry is known to be extremely sensitive to the toxic effects of AFB1. Consumption of AFB1-contaminated feed causes a myriad of other effects either directly or indirectly associated with this toxicity: reduced feed utilization and efficiency, reduced growth rate, decreased body and organ weights [16], lowered egg production and reproductivity [17], immunosuppression [18], and increased susceptibility to disease [19].

2. Biotransformation (metabolism) of aflatoxins

AFs undergo biotransformation mainly in the liver. There are two types of biotransformations: Phase 1 and Phase 2. Phase 1 reactions are generally oxidative, reductive, or hydrolytic processes and provide a necessary chemical structure for Phase 2 reactions, which are generally conjugation reactions. Phase 1 reactions may result in activation as well as detoxification of a compound, whereas Phase 2 reactions, depending on conjugated cellular constituents, may lead either to detoxification or formation of biochemical lesions. Phase 1 is mostly mediated by the cytochrome P450 (CYP450) enzyme systems. Phase 2 metabolism involves sulfate, glucuronide, glutathione (GSH), and amino acid conjugation reactions (**Figure 2**) [20].

2.1. Phase 1: Metabolism of aflatoxins

AB1 is oxidized by CYP450 subfamilies and specific isoforms of enzymes to several products. Only one of these, AFB1 epoxide, appears to be mutagenic, and others are detoxification products. The putative AFB1 epoxide is generally accepted as the active electrophilic form of AFB1, which may attack nucleophilic nitrogen, oxygen, and sulfur heteroatoms in cellular constituents [22]. The CYP450-mediated oxidation to the extremely reactive AFB1-8,9-epoxide is considered the primary (Phase 1) bioactivation pathway for AFB1 [23]. This conversion of AFB1, to the epoxide, is the phase of reaction that enables covalent binding to cellular macromolecules (e.g., DNA and/or protein) to occur. This reaction can involve a number of isozymes of CYP450 including 1A2 and 3A4 [24]. The AFB1-8,9-epoxide reacts with the N7 atom of guanine to form a pro-mutagenic DNA adduct (AF-N7-guanine). The DNA adducts are fairly resistant to DNA repair processes, and this causes gene mutation and thus the development of cancers especially the hepatocellular carcinomas (**Figure 2**) [21, 25].

CYP450 3A4, which can both activate and detoxicate AFB1, is found in the liver and small intestine. In the small intestine, the first contact after oral exposure, epoxidation, would not lead to liver cancer. CYP450 3A4 has been shown to play a major role in the activation of AFB1 due to its intrinsic activity toward this substrate, and the high level of this enzyme is present

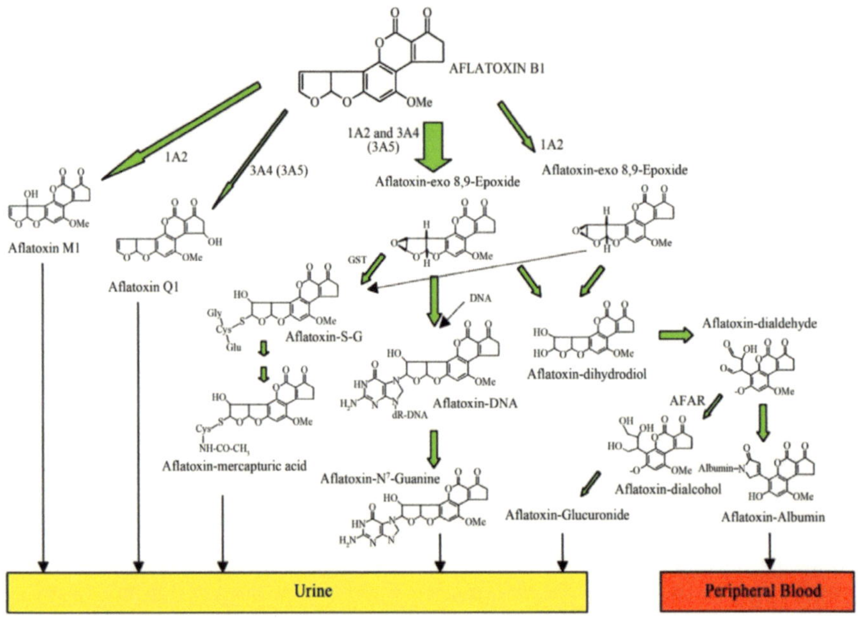

Figure 2. Metabolism of AF in the liver. 1A2, CYP1A2; 3A4, CYP3A4; 3A5, CYP3A5; GST, glutathione-S-transferase; AFAR, AF aldehyde reductase; AF-SG, aflatoxin-glutathione conjugate [21].

in human liver. CYP450 1A2 and some other human CYP450s also contribute, but they play a lesser role, even at relatively low AFB1 concentration [26, 27]. CYP450 3A4 forms mostly the genotoxic AFB-2,3-epoxide, whereas CYP450 1A2 forms both the exo- and nongenotoxic endoisomers [26]. CYP450 1A2 has high affinity for the bioactivation of AFB1 at low substrate concentrations following dietary exposure [21]. Some of the AFB1 intermediates go through far more metabolism in Phase 2 by binding with GSH in order to create the polar and less toxic compound that are simply excreted in urine and bile. However, AFBO and AFB1-dihydroxide intermediates led to carcinogenicity, while AFB2 causes acute toxicity, liver necrosis, and cellular metabolizing enzyme inhibition (**Figure 2**) [28].

2.2. Phase 2: Metabolism of AF role of GSH conjugation in body detoxification of aflatoxins

Phase 2 reactions that lead to the detoxification involve conjugation to glucuronic acid, sulfate, and GSH. The AFB metabolites of Phase 1 metabolism undergo Phase 2 enzymatic metabolism by glutathione-S-transferases (GSTs) that primarily catalyze conjugation reactions. After Phase 1 oxidation, AF can be readily conjugated with SH groups (in Phase 2 reactions) allowing for further detoxification and elimination of the toxin. In a number of mammalian species, the AFB1-8,9-epoxide is efficiently conjugated with reduced GSH in a reaction catalyzed by GST (**Figure 3**) [29, 30].

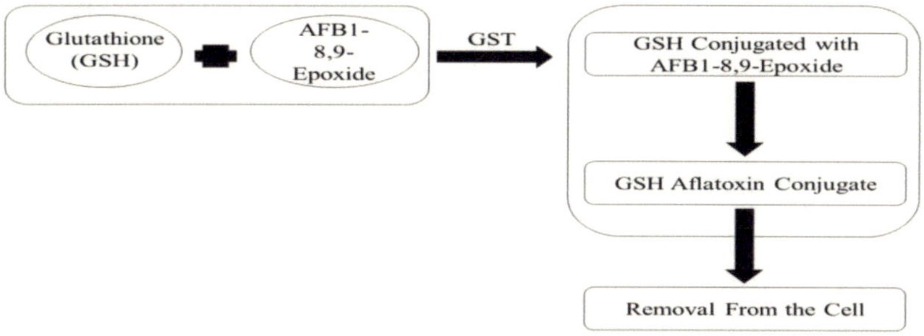

Figure 3. Metabolism of AFB1. Glutathione and glutathione-S-transferase involved in detoxification of activated AFB1.

3. Free radicals and lipid peroxidation

Free radicals are highly reactive species that have an unpaired electron, e.g., hydroxyl (·OH) and superoxide radicals (O^-_2) which have potential to cause tissue damage (**Figure 4**). Although free radicals are highly reactive and potentially damaging, they are also an integral part of some cellular processes. Extracellular secretion of free radicals by leucocytes and microphages evokes immune response against bacteria, viruses, degenerated cells, and other foreign substances. Intracellular secretion of free radicals stimulates different cell signaling pathways and triggers oxidative stress defense response as well as apoptosis [31]. Due to

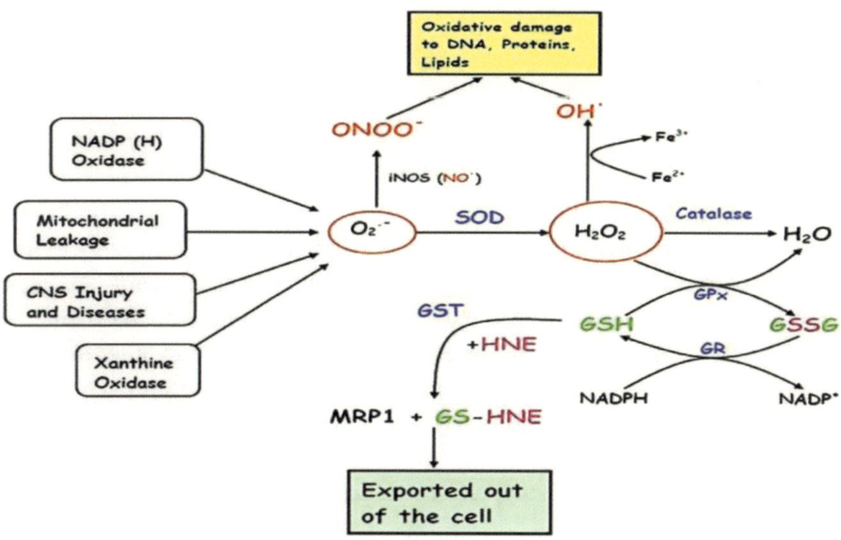

Figure 4. Different endogenous sources for ROS/reactive nitrogen species (RNS), antioxidant defense. Hydroxynonenal (HNE) is one of the end products of lipid peroxidation (adapted from Hardas [35]).

perilous nature of free radicals, cells have a counter mechanism known as antioxidant defense to keep the free radical levels under check. Unfortunately, when certain conditions promote the excess production of free radicals or deplete the antioxidant defense that leads the cell to oxidative damage, oxidative stress is said to exist. Oxygen-derived free radicals are referred as reactive oxygen species (ROS). Oxygen radicals are produced as a consequence of the normal process of reduction of oxygen to water and represent by products of oxidative cellular metabolism. The main sites of ROS produced in living organisms are mitochondrial electron transport system, peroxisomal fatty acid, CYP450, and phagocytic cells [32]. ROS can react with DNA to cause breaks in the DNA chain and mutation, which could initiate carcinogenesis. Free radicals can react with membrane lipids leading to peroxidation of polyunsaturated fatty acid (PUFA) residues (**Figure 4**) [33, 34].

The majority of lipid peroxidation events that occur within the cell are result of free radical chain reaction. Oxidative damage to lipids generally results in formation of cytotoxic aldehyde and ketone derivatives. Typically free radicals have a very short half-life; therefore, the damage caused by them is localized. Unlike free radicals, lipid peroxidation products have a longer half-life, and so they can diffuse into bilayer and can cause oxidative damage away from their site of production. For a given fatty acid, multiple aldehydic or ketonic products can arise as a result of lipid peroxidation, depending upon which allylic carbon gets attacked to initiate the chain reaction [36, 37]. Malondialdehyde (MDA) is a significant final product, which composes via the degeneration of certain primary and secondary lipid peroxidation products [38]. The MDA formation promotes the alteration of membrane fluidity and enhances of membrane fragility. Furthermore, MDA blocks particular enzyme reactions and causes mutagenicity and carcinogenicity by creating DNA adducts [39, 40].

4. Oxidative stress

Although ROS and reactive nitrogen species (RNS) are generated under normal physiological conditions, their levels are efficiently regulated by antioxidant enzymes and molecules to maintain the cellular redox balance. Oxidative stress is defined as a disturbance in the balance between antioxidants and prooxidants, with increased levels of prooxidants leading to potential damage. This imbalance can be due to the decrease of endogenous antioxidants, low intake of dietary antioxidants, and/or increased formation of free radicals and other reactive species. In any case, either of both circumstances occurring together or separately eventually will lead to deleterious modifications of biomolecules and multitude of downstream consequences. Oxidative stress has been implicated in vast array of conditions including cancer, arthritis, cardiovascular diseases, diabetes, aging, and neurodegenerative disorders [33, 41, 42].

4.1. Effects on oxidative stress of aflatoxin

Oxidative stress plays a major role in aflatoxicosis. Oxidative stress may be due to direct effect of AFs themselves or by their metabolites. AFB1, a mutagenic food contaminant, is widely recognized as one of the most potent hepatocarcinogens in humans and experimental animals.

Metabolizing AFB1 increases the production of free radicals and lipid peroxides, resulting in cell damage [43, 44]. AFB1 is activated in the liver by CYP450 to AFB1-8,9-epoxide, which forms adducts with both DNA and protein. The toxic effects of AFs mostly arise from the binding of this particular epoxide derivative to DNA. AFs form after a series of highly organized oxidation-reduction reactions. Several studies provided evidences indicating that CYP450 enzymes generate superoxide hydrogen peroxide (H_2O_2) as intermediate compounds, and these ROS can cause apoptosis and other cell pathologies [45–47]. AFB1 is able to induce ROS generation, which causes oxidative stress. The genetic toxicity of AFB1 is partly due to the accumulation of ROS such as O_2^-, ·OH, and H_2O_2 radical during the metabolic processing of AFB1 by CYP450 in the liver (**Figure 5**). These species may attack soluble cell compounds as well as membranes, eventually leading to the impairment of cell functioning and cytolysis [48].

It has been reported that there is free radical generation during AFB1 metabolism, and oxidative damage is one type of damage caused by AFB1 [49, 50]. Oxidative damage induced by these ROS can, in turn, cause tissue damage by a variety of mechanisms including DNA damage, lipid peroxidation, protein oxidation, and depletion of thiols. The oxidative stress caused by AFB1 may be one of the underlining mechanisms for AFB1-induced cell injury and DNA damage, which eventually lead to tumorigenesis [37]. Studies have revealed that AFB1 alters cell cycle and apoptosis-signaling pathways in liver cell models [43, 47, 51, 52]. AFB1 can cause an increase in ROS formation in animals' target organs including rat liver, duck liver, and mouse lung [37, 44, 53]. It is indicated that AFB1 induced an important liver cell injury, as shown by the significant increase in nitric oxide, but also a strong lipid peroxidation in the liver and kidney, accompanied with a significant decrease in total antioxidant capacity in rats [53], mice [54], and chicken [55]. Also, it was shown that a strong inducible nitric oxide synthase (iNOS) and nitrotyrosine immunoreactivity were observed in the livers of chicks administered with 300 ppb of AF. Moreover, AFB1 carcinogenicity is associated with altered expression of many p53-target genes and induction of mutations, principally the p53 codon 249 hotspot mutation [13, 48].

AFs are claimed as potential risk factor of hepatocarcinoma, and the oxidative stress is considered to be a main factor in the initiation and the progression of liver cirrhosis, which is known

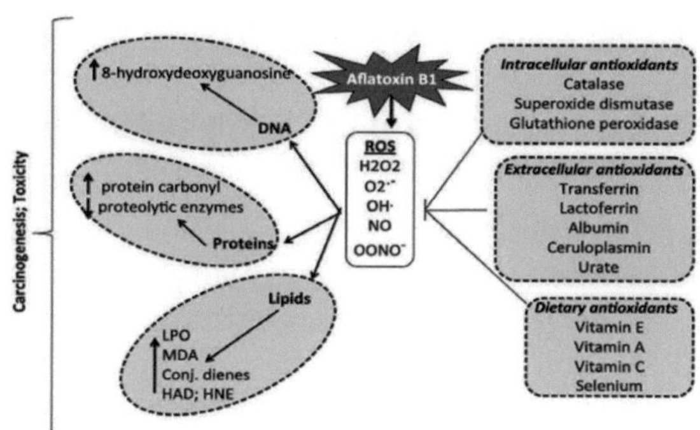

Figure 5. Effect of AFs on the oxidative stress, the alleviating role of antioxidants (adapted from Marin and Taranu [5]).

to be a pioneer of human hepatocellular carcinoma [11]. The oxidative damage caused by AF is considered to be the main mechanism leading to the subsequent hepatotoxicity [56]. AFB1 may disturb the integrity of cell membranes by stimulating phospholipid A2 to initiate lipid peroxidation in cells [57]. Animals fed with AF-contaminated diet suffer from oxidative stress as indicated by the significant increment of lipid peroxidation and the significant reduction of enzymatic antioxidant such as SOD and GSH-Px [54, 58, 59]. According to the pioneering work of Shen et al. [60], AFB1 promotes lipid peroxidation in rat liver, and lipid peroxidation is intimately linked with liver cell injury. A time- and dose-dependent increase in 8-hydroxy-2′-deoxyguanosine (8-OHdG) was observed in DNA after a single intraperitoneal injection of AFB1. It reveals that AFB1 leads to oxidative DNA damage in rat liver, which may participate in ·OH as the initiating species. Therefore, factors having an effect on formation or action of ·OH would affect the generation of 8-OHdG.

It is well known that a possible mechanism of AF cytotoxicity is the induction of oxidative stress. The induction of oxidative stress is commonly related to an imbalance between the oxidants and the antioxidant systems [49]. It is explained by its effect on mitochondria; increased lipid peroxidation; increased adduct formation with DNA, RNA, and protein; or all the three. Damage to mitochondria can lead to mitochondrial diseases and may be responsible for aging mechanisms. The damage can cause mitochondrial DNA (adducts and mutation), mitochondrial membranes, as well as disruption of energy production (production of adenosine triphosphate) [61]. The mycotoxin alters energy-linked functions of adenosine diphosphate (ADP) phosphorylation and flavin adenine dinucleotide (FAD) and nicotinamide adenine dinucleotide (NAD)-linked oxidizing substrates and α-ketoglutarate-succinate cytochrome reductases [62, 63]. It causes ultrastructural changes in mitochondria and also induces mitochondria-directed apoptosis [51]. AFB1 induced the production of free radicals and the reduction of antioxidant defenses in livers of murine, human lymphocytes, and bovine peripheral blood mononuclear cells [51, 64, 65].

4.2. Aflatoxin and carcinogenicity

AFB1 primarily causes hepatocellular carcinoma and cholangiocarcinoma in the liver [11]. Among various types of known AFs, AFB1 is the most potent hepatocarcinogen; however, G1 and B2 also cause cancers but with reduced potency. It causes liver tumors in mice, rats, fish, marmosets, tree shrews, and monkeys following the administration by various routes. The types of cancers described in research animals include hepatocellular carcinoma, cholangiocellular cancer, and adenocarcinoma of the gallbladder [66].

Besides, the liver tumors have also been reported to develop AF feeding in lacrimal glands, squamous cells of the tongue, esophagus, trachea, lung adenomas, osteogenic sarcoma, and carcinoma of the pancreas [66–68]. Carcinoma of the colon has been reported by many researchers [67, 69]. AF exposure contributes to the risk for development of hepatocellular carcinoma in ducklings [70]. AFB1 can cause hepatocarcinogenesis and mutation in rat liver (**Figure 6**) [71]. Ghebranious and Sell [13] proposed that some mutant proteins may act as a promoting agent for AFB1 hepatocarcinogenesis. AF and p53 expressions interact to produce malignant liver tumors transgenic in mice.

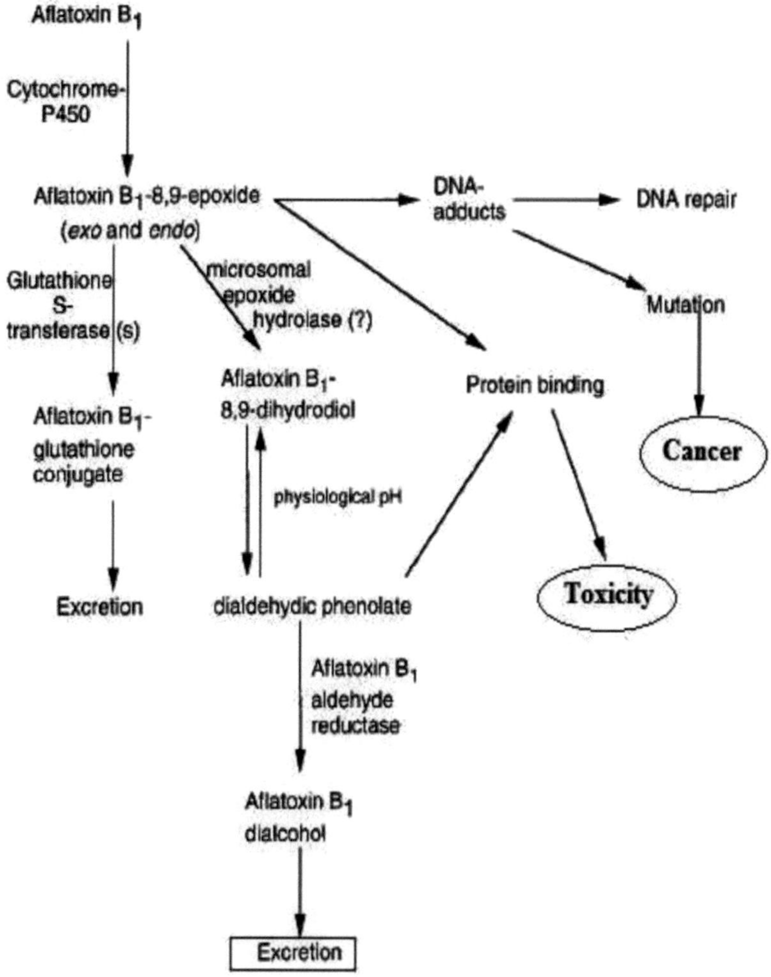

Figure 6. Overview of metabolic pathways leading to toxicity and carcinoma of AFB1 [72].

4.3. Aflatoxins and oxidative lipid damage

Polyunsaturated lipids are essential for cells, being important constituents of cell membranes, endoplasmic reticulum, and mitochondria. Thus, the disruption of their structural properties could have consequences for cellular function. Lipid peroxidation is one of the main factors responsible for structural and functional alterations of the cell membrane following oxidative stress [39] and initiation of carcinogenesis [37, 54].

It remains unknown if the mycotoxins promote the lipid peroxidation directly through the enhancement of the ROS formation or the enhancement of the tissue sensitivity to the peroxidation is the result of the compromised antioxidant defense, but it appears that both processes are taken part. AFB1-mediated cell injury may be due to the release of free radicals that initiate lipid peroxidation. The initiation of lipid peroxidation is caused by the attack of any species that has sufficient reactivity to remove a hydrogen atom from a methylene group upon a PUFA [37, 54]. The peroxidation of PUFAs can be realized not only through nonenzymatic free radical-induced pathways but also through processes that are catalyzed by enzymes as cyclooxygenase and lipoxygenase [39]. It is shown that also 8,9-epoxide increases lipid peroxidation, followed by loss of membrane stability and the blockage of the membrane-bound enzyme activity [73]. Evaluation of the lipid damage is based on measurement of Thiobarbituric acid reactive substances (TBARS) or MDA by the TBA test and conjugated dienes. AFB1 induced an increase in the TBARS concentration in the liver [74] or in human hepatoma cells [75]. The increase of the lipid peroxide synthesis is observed not only in the liver but also in the kidney and brain [7, 35, 41]. This alteration was associated with a significant increase in conjugated diene formation. Concentrations of MDA+ 4-hydroxyalkenals as an index of lipid peroxidation are increased by AFB1 in the liver, lung, brain, and testis, but not the kidney of male Wistar rats [76]. 4-Hydroxynonenal (4-HNE), a major electrophilic by-product of lipid peroxidation caused by oxidative stress, interacts with DNA to form exocyclic guanine products, which have been shown to increase in a rat model of hepatocarcinogenesis. AFB1 induces lipid peroxidation in rat liver, which may be an underlying mechanism of carcinogenesis [44, 77].

4.4. Aflatoxins and oxidative protein damage

ROS can also lead to oxidation of amino acid residue side chains, formation of protein-protein cross-linkages, and oxidation of the protein backbone resulting in protein fragmentation, and the modified forms of proteins will accumulate in organism [78]. By its capacity to generate ROS, AFB1 can promote the ROS-mediated oxidative damages in proteins (**Figure 6**) [79].

AFB1 could inhibit some (serine) proteolytic enzymes responsible for the degradation of damaged proteins, with consequent relevant implications in hepatocarcinogenesis [79, 80]. It has been suggested that numerous action of AFs may be brought about their interactions with the proteasome, the main enzyme family account for the decomposition of most of cytosolic and nuclear proteins in eukaryotic cells. In fact, AFB1 brings about an inhibition of cellular 20S proteasomes, affecting the cellular defense against oxidative stress. Because 20S proteasome is the proteolytic machinery responsible for removing oxidized proteins, its inhibition could contribute to a higher protein carbonyl content observed in cultured hepatoma cell lysates [81].

The reduction of protein synthesis in animals treated with AFs may affect certain metal ions, which play an important role in free radical production and liberation. Inhibition of protein synthesis caused by AFs alters serum protein composition, resulting in suppression of the production of nonspecific humoral substances important to native defense [82]. At higher doses, AFB1 lowers the level of IgG and IgA in chick resulting in decreased acquired immunity. Antibodies to AFB1 have been reported in humans [83, 84].

4.5. Aflatoxins and oxidative DNA damage

Oxidative DNA damage is a general definition for all types of changes (structural or functional) of DNA, due to the interaction of ROS with DNA. The connection of ·OH to the C8 position of DNA guanine forms C8-OH-adduct radical [85], which is eventually altered to

8-OH-guanine (8-OH-Gua) by one-electron oxidation [86]. While impaired lipids and proteins can be removed by metabolic cycle of these compounds, damaged DNA has to be fixed in situ or destroyed by apoptotic processes; conversely, mutations result in the absence of these [87]. In humans, 8-OH-Gua glycosylase is the primary enzyme for the repair of 8-OH-Gua in short-patch base excision repair. The excised form of 8-OH-Gua is a promutagenic adduct, 8-OHdG, which is excreted into urine without further metabolism and is stable for a significant time. 8-OHdG is widely considered as a key biomarker of oxidative DNA damage [60, 88].

The toxic and carcinogenic effects of AFB1 are intimately linked with its biotransformation [12]. There is a tendency for AFs especially AFB1 to convert into the epoxide and produce DNA adducts resulting in the formation of DNA strand breaks and mutations [88, 89]. It is well known that AFB1 is activated by the hepatic CYP450 enzyme system to form a highly reactive product, AFB1-8,9-epoxide, which subsequently connects to nucleophilic sites in DNA and the major adduct 8,9-dihydro-8-(N7-guanyl)-9-hydroxy-AFB1 (AFB1-N7-Gua) is formed. The formation of AFB1-DNA adducts is regarded as a critical step in the initiation of AFB1-induced hepatocarcinogenesis (**Figure 7**).

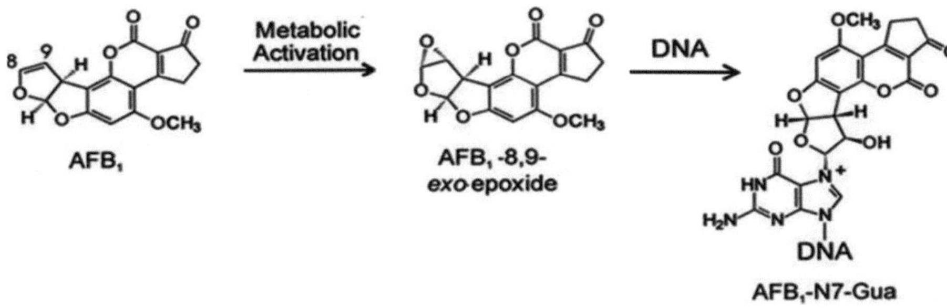

Figure 7. Metabolic activation of AFB1 (adapted from Kobertz et al. [90]).

5. Aflatoxins and the antioxidant defense

The activity of antioxidant enzymes could induce as a result to the oxidative stress or could diminish through direct or indirect action of the mycotoxins. A part of the oxidative metabolism intermediates of AFB1 composes a substrate for the Phase 2 detoxification enzymes. In a vast range of animal species, the fundamental way to detoxify the AFB1 is through the conjugation of AFBO with GSH. This way of detoxification is the principal way of AFB1 excretion in many animal species. The reaction is catalyzed by GST [89]. It is observed that in mice, the reduced sensibility to AFs is correlated with the constitutive increase of GST isoenzyme [29]. GSH and GST are effective antioxidant enzymes that take part in the protection of tissues from harmful effects of AFB1 (**Figure 3**) [90, 91]. GSH is used as a cofactor by GST that conjugates GSH with endogenous substances like estrogens, exogenous electrophiles like AFs and its metabolites, and other various xenobiotics. The increased depletion of GSH leads to abnormally high levels of ROS in cells. AF is one of the main actors in depletion of GSH. The depletion of GSH affects metabolic processes such as catalysis of molecular oxygen (O_2) to H_2O_2 by GSH-Px, and thus the integrity of the cell membranes disrupts. Its reduction further enhances the damage to critical cellular components (DNA, lipids, proteins) by the AFB1-8,9-epoxides that form adducts. GST catalyzes the conjugation of AFB1-8,9-epoxide with GSH to form AFB1-GSH conjugate, thereby decreasing the intracellular GSH content [37]. The AF-GSH product undergoes the sequential metabolism in the liver and kidneys in which it is excreted as a mercapturic acid (AF-N-acetylcysteine) in urine [91, 92]. It has been reported that AF administration results in excessive lipid peroxidation [53] with concomitant decrease in GSH [58], increased protein oxidation, and DNA damage in rat liver. The activity of GSH-Px, which is a constituent of GSH redox cycle, decreases during AFB1 administration. The reduction in GSH-Px activity by AFB1 may be due to a decrease in the availability of GSH and also alterations in their protein structure by ROS. The studies revealed that there were obvious increases in MDA and/or nitric oxide (NO) levels and decreases in both nonenzymatic antioxidant GSH level and enzymatic antioxidant GSH-Px, catalase (CAT), glutathione reductase (GR), and GST activities after administration with AFB1 in vivo or in vitro [41, 51, 64, 65].

The study showed that administration of AFB1 produced a marked oxidative impact as evidenced by a significant increase in MDA in the liver, kidneys, and heart of AF-treated rats. These alterations might have been triggered either by the direct effects of AFB1 or by the metabolites formed by AF and the free radicals, which were generated during the formation of these metabolites. Initiation of LPO by AFB1 is noted as one of the principal appearances of ROS-induced oxidative damage. The mechanism of free radical damage also includes ROS-induced peroxidation of polyunsaturated fatty acids in the cell membrane lipid bilayer which causes a chain reaction of LPO, thus damaging the cellular membrane, causing further oxidation of membrane lipids and proteins, and leading to DNA damage. The study also showed that a significant increase in the oxidative stress was accompanied by a concomitant decrease in the enzyme activities involved in the disposal of O_2^- and peroxides, namely, CAT and SOD, as well as GSH levels and its related enzymes (GST, GSH-Px). A significant increase observed in tissue MDA levels in AFB1-treated animals indicated that AF led to the generation of the high level of free radicals, which could not be tolerated by the cellular antioxidant defense system. A significant decrease in these enzyme activities could be explained by their consumption during the conversion of free radicals into less harmful or harmless metabolites [49].

6. Lycopene

Lycopene is an acyclic hydrocarbon carotenoid responsible for the intense red color of tomatoes (**Figure 8**). Lycopene does not exhibit provitamin A activity since it lacks the β-ionone ring structure which is characteristic in carotenoids that are precursors for vitamin A [93, 94]. Lycopene is a natural pigment and imparts a red color in the foods containing it. In foods,

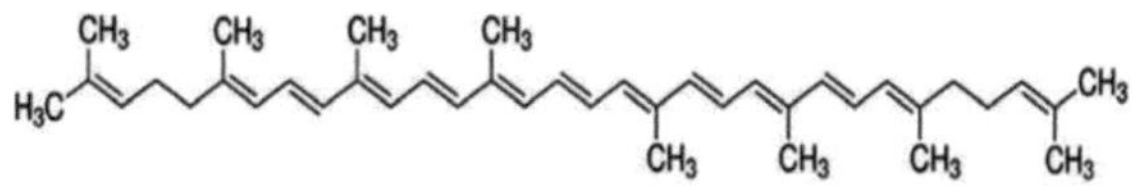

Figure 8. Structure of all-trans lycopene ($C_{40}H_{56}$).

lycopene is found predominantly in its trans-form (approximately 95.4% of total lycopene content), whereas serum and tissues contain more cis-isomers of lycopene [95–97]. Lycopene is nontoxic and Generally Recognized as Safe by the US FDA (21 CFR 73.585) and the European Union (Directive 94/36/EC) for the use as a food additive and colorant [98].

6.1. The role of lycopene as antioxidant and implications

Lycopene acts as an antioxidant by virtue of its conjugated p-electron system, which can react with oxygen radical species such as peroxy and hydroxy radicals as well as non-radical species such as ozone and H_2O_2 [99].

Lycopene has a robust antioxidant defense system, attributed to its acyclic structure, numerous conjugated double bonds, and high hydrophobicity, and thus prevents the onset of carcinogenesis and atherogenesis processes by protecting/stabilizing biomolecules such as DNA, proteins, lipids, and lipoproteins. Lycopene, as the main carotenoid in tomato products, possesses the greatest ability to quench singlet oxygen compared to the other carotenoids. It also scavenges the free radicals *via* three different mechanisms: adduct formation, electron transfer, and hydrogen atom transfer [100, 101]. Galano et al. [102] reported that lycopene and torulene are more reactive scavengers of peroxide radicals than β-carotene.

Lycopene is capable of acting as an antioxidant by virtue of its many conjugated double bonds. It is the most efficient neutralizer of singlet oxygen among all carotenoids and has also been found to be a potent scavenger of free radicals [94, 95]. The lycopene molecule reacts with free radicals to form a short-lived intermediate species, which later end up as lycopene decomposition products including apocarotenals, apocarotenones, and epoxides. Being a highly hydrophobic molecule, the greatest scavenging ability of lycopene is seen in lipophilic environments [94, 103]. After supplementing subjects with lycopene from different dietary sources, serum TBARS (a biomarker for lipid peroxidation) is significantly reduced, whereas nonsignificant reductions are observed in biomarkers for protein and DNA oxidation. Hence, lycopene may be a biologically important antioxidant by protecting membrane lipids from being oxidized which in turn preserves the integrity of cellular membranes [104].

Much of the evidence for the antioxidant function of lycopene comes from studies conducted with in vitro systems, and virtually all of them indicate lycopene to function as a superior dietary antioxidant. Being is a strong antioxidant, lycopene has been shown to reduce the amount of oxidative DNA damage and also decrease lipid peroxidation in cell culture and in rats in vivo [105–107]. Di Mascio et al. [108] compared the singlet oxygen quenching ability of various carotenoids, α-tocopherol, bile acids, and retinoic acid. They found lycopene to be the most efficient quencher among all, with a greater than twofold quenching potency. Lycopene is the most efficient carotenoid in reducing TBARS formation by 75% compared to control in multilamellar liposomes. In a study examining the relative ability of several antioxidants in reducing carotenoid cations, it was found that lycopene was the most superior carotenoid antioxidant and the lycopene cation radical was the most stable carotenoid cation radical [109].

6.2. Protective effect of lycopene on aflatoxin damage

There are many reports indicating that lycopene is effective on inhibition of tumor formation and growth induced by chemical carcinogens in animals [107]. To sum up, in **Figure 9**, AFB1 has two important metabolic pathways: Phase 1 includes metabolism and metabolic activation, and Phase 2 is detoxification [30]. AFM1, AFQ1, AFP1, and AFB1-8,9-epoxide are important Phase 1 metabolites, and also AFB-N7-Gua and AFB-albumin complexes are specific markers formed, respectively, in the tissues and "serum or urine" during the AFB1 metabolic activation. The main Phase 2 detoxification outcome of AFB1-8,9-epoxide is AFB-N-acetyl cysteine (AFB-NAC) complex. AFB1 Phase 1 metabolism and the metabolic activation of AFB1 are inhibited by lycopene. Moreover, lycopene highly activates the enzymes responsible for Phase 2 detoxification and causes to enhance production of AFB-NAC excreted in urine. As shown in decreased urinary levels of AFP1, AFQ1, and AFM1 in lycopene-pretreated or lycopene-intervened animals, lycopene pretreatment or intervention significantly blocks Phase 1 metabolism of AFB1. This indicates that lycopene may selectively inhibit Phase 1 metabolic enzymes such as 3A4, 2A6, and 1A2. Depending on the relative potency in decreasing levels of these specific AFB1 metabolites in urine, lycopene appears to be a moderate competitive inhibitor of 3A4 and 2A6 enzymes and a weak or reversible inhibitor of 1A2

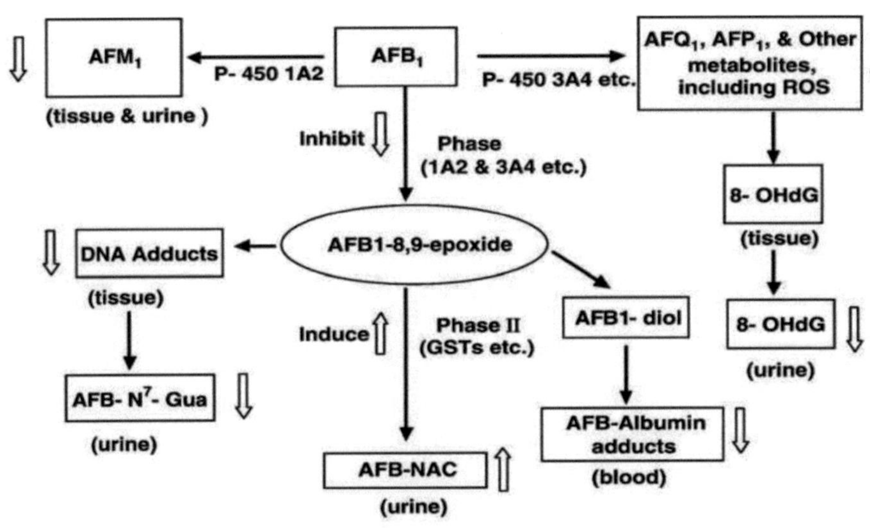

Figure 9. AFB1 metabolic activation, biomarkers, and possible mechanisms of lycopene modulation [110].

enzyme (more potent inhibition of AFP1 and AFQ1 than AFM1). Reducing levels of AFB-albumin adducts in serum, reducing levels of AFB-N7-Gua excreted in urine, and reducing levels of AFB-N7-Gua adduct in the liver, DNA confirmed the inhibitory effect of lycopene on Phase 1 metabolism. These data clearly demonstrate that lycopene pretreatment or intervention effectively blocks AFB1 metabolism and also metabolic activation. AFB-NAC is the major detoxifying metabolic product of AFB1-8,9-epoxide. Lycopene pretreatment and intervention elevated significantly AFB-NAC levels in urine excretion, which suggests that activity of GSTs was greatly induced [30].

AFB1 also induces formation of ROS [44], lipid peroxidation, and formation of 8-OHdG in vivo and in vitro [60]. Lycopene could increase the activity of GSH-Px, GST, and GR in several animal models including rats [111]. The antioxidant capacity of lycopene is at extremely high levels and lessens not only the oxidative damage of DNA in particular rates but also lessens lipid peroxidation both in vitro and in vivo [105–107]. It has also been documented that lycopene intervention reduces the 8-OHdG levels of urine even in recurring exposures to AFB1 (**Figure 10**).

Administration of lycopene alleviates the negative effects of AF. Lycopene removes free radicals produced by AF while improving the body's antioxidant enzymes such as GSH, GSH-Px, and CAT to prevent the oxidative damage caused by AF, enhancing the body antioxidant capacity, reducing the levels of lipid peroxidation, and maintaining cell membrane permeability. For this reason, natural antioxidant lycopene can be regarded as a good therapeutic agent against aflatoxicosis [112].

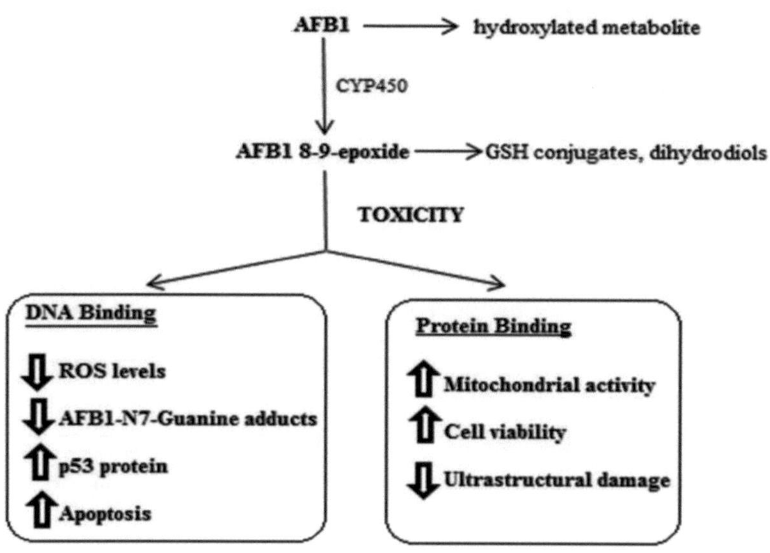

Figure 10. Inhibition of toxicity and cancer by lycopene in AFB1-exposed cells. Chemoprotective effects of lycopene effects are shown by arrows: ↑, increase; ↓, decrease (adapted from Reddy et al. [52]).

7. Conclusion

Current concepts derived from intensive research on biotransformation, mechanisms of toxicity, the effect on oxidative stress of AF, and protective effect of lycopene on AF damage were summarily presented in this chapter. AFB1 exerts its effects after conversion to the reactive compound AFB1 epoxide by means of CYP450-dependent enzymes. This epoxide can form derivatives with cellular macromolecules, including proteins, RNA and DNA. Biomonitoring of AFB1 metabolites such as AFB1-N7-guanine has demonstrated that AFs constitute an important risk factor for hepatocellular carcinoma in highly exposed populations. Oxidative stress formed due to AF is associated with biochemical disturbances in oxidant/antioxidant balance system, which may cause AF toxicity. When administered together with AF, lycopene was determined that it exhibited strong positive effect on AF-induced oxidative stress parameters. It could be concluded that the lycopene being a nontoxic, highly promising natural "eco-friendly" antioxidant compound has a protective effect against AF toxicity. When administered together with AF, lycopene was determined that it exhibited a strong positive effect on AF-induced oxidative stress parameters.

Author details

Seval Yilmaz*, Emre Kaya and Mehmet Ali Kisacam

*Address all correspondence to: sevyilars@yahoo.com

Department of Biochemistry, Faculty of Veterinary Medicine, Firat University, Elazig, Turkey

References

[1] Coulombe RA. Aflatoxins. In: Sharma RP, Salunkhe, editors. Mycotoxins and Phytoalexins. Boca Raton: CRC Press; 1991. pp. 103-143. ISBN 0849388333

[2] Mushtaq M, Sultana B, Anwar F, Khan MZ, Ashrafuzzaman M. Occurrence of aflatoxins in selected processed foods from Pakistan. International Journal of Molecular Sciences. 2012;**13**:8324-8337

[3] Bbosa GS, David Kitya D, Lubega A, et al. Review of the biological and health effects of aflatoxins on body organs and body systems: Aflatoxins-recent advances and future prospects. Intechopen Publisher. 2013;**12**:239-265

[4] Reddy SV, Waliyar F. Properties of aflatoxin and its producing fungi. Aflatoxins. http://www.icrisat.org/aflatoxin/aflatoxin.asp, (2012).

[5] Marin DE, Taranu I. Overview on aflatoxins and oxidative stress. Toxin Reviews. 2012;**31**(3-4): 32-43

[6] Fink-Gremmels J. Mycotoxins: Their implications for human and animal health. Veterinary Quarterly. 1999;**21**:115-120

[7] Marvan F, Vernerova E, Samek M, et. al. Aflatoxin B1 residues in the organs of young poultry. Biologicke Chem Vet (Praha). 1983;**24**:85-92

[8] Petr T, Barta I, Turek B. In vitro effect of mutagenic activity of aflatoxin B1. Journal Hygiene Epidemiology Microbiology Immunology. 1995;**34**:123-128

[9] Hartley RD, Nesbitt BF, O'kelly J. Toxic metabolites of *Aspergillus flavus*. Nature, London. 1963;**198**(4885)

[10] IARC. International Agency for Research on Cancer. IARC Monographs on the Evaluation of Carcinogenic Risks to Humans: Traditional Herbal Medicines, Some Mycotoxins, Naphthalene and Styrene. Lyon: IARC Scientific Publication, No. 82; 2002

[11] Liu ZM, Li LQ, Peng MH, et al. Hepatitis B virus infection contributes to oxidative stress in a population exposed to aflatoxin B1 and high-risk for hepatocellular carcinoma. Cancer Letters. 2008;**263**:212-222

[12] Wu HC, Wang Q, Yang HI, et al. Aflatoxin B1 exposure, hepatitis B virus infection, and hepatocellular carcinoma in Taiwan. Cancer Epidemiology Biomarkers & Prevention. 2009;**18**:846-853

[13] Ghebranious N, Sell S. Hepatitis B injury, male gender, aflatoxin, and p53 expression each contribute to hepatocarcinogenesis in transgenic mice. Journal of Hepatology. 1998;**27**(2):383-391

[14] Bondy G, Pestka JJ. Immunomodulation by fungal toxins. Journal of Toxicology and Environmental Health. 2000;**3**:109-143

[15] Mocchegiani E, Corradi A, Santarelli L, et al. Zinc, thymic endocrine activity and mitogen responsiveness (PHA) in piglets exposed to maternal aflatoxicosis B1 and G1. Veterinary Immunology and Immunopathology. 1998;**62**:245-260

[16] Han XY, Hung QC, Li WF. Changes in growth performance, digestive enzyme activities and nutrient digestibility of cherry valley ducks in response to aflatoxin B1 levels. Livestock Science. 2008;**119**:216-220

[17] Trucksess MW, Stoloff L, Young K. Aflatoxical and aflatoxin B1 and M1 in eggs and tissue of laying hens consuming aflatoxin-contaminated feed. Poultry Science. 1983;**62**: 2176-2182

[18] Jiang Y, Jolly PE, Ellis WO. Aflatoxin B1 albumin adduct levels and cellular immune status in Ghanaians. International Immunology. 2005;**17**:807-814

[19] Sumit R, Ji EK, Roger CJ. Aflatoxin B1 in poultry: Toxicology, metabolism and prevention. Research in Veterinary Science. 2010;**89**:325-331

[20] Monosson, E. Biotransformation. National Library of Medicine (NLM): The Encyclopeadia of earth. http://www.eoearth.org/view/article/150674, (2012).

[21] Wild CP, Turner PC. The toxicology of aflatoxins as a basis for public health decisions. Mutagenesis. 2002;**17**(6):471-481

[22] Swenson DH, Miller EC, Miller JA. Aflatoxin B1-2, 3-oxide: Evidence for its formation in rat liver in vivo and by human liver microsomes in vitro. Biochemical and Biophysical Research Communications. 1974;**60**(3):1036-1043

[23] Swenson DH, Lin JK, Miller EC, Miller JA. Aflatoxin B1-2, 3-oxide as a probable intermediate in the covalent binding of aflatoxins B1 and B2 to rat liver DNA and ribosomal RNA in vivo. Cancer Research. 1977;**37**(1):172-181

[24] Aoyama T, Yamano S, Guzelian PS, Gelboin HV, Gonzalez FJ. Five of 12 forms of vaccinia virus-expressed human hepatic cytochrome P450 metabolically activate aflatoxin B1. Proceedings of the National Academy of Sciences. 1990;**87**(12):4790-4793

[25] Wild CP, Yin F, Turner PC, et al. Environmental and genetic determinants of aflatoxin-albumin adducts in the Gambia. International Journal of Cancer. 2000;**86**:1-7

[26] Ueng YF, Shimada T, Yamazaki H, Guengerich FP. Oxidation of aflatoxin B1 by bacterial recombinant human cytochrome P450 enzymes. Chemical Research in Toxicology. 1995;**8**(2):218-225

[27] Crespi CL, Penman BW, Leakey JA, et al. Human cytochrome P450IIA3: cDNA sequence role of the enzyme in the metabolic of promutagens comparison to nitrosamine activation by human cytochrome P450IIE1. Carcinogenesis. 1990;**11**(8):1293-1300

[28] Dhanasekaran D., Panneerselvam A., Thajuddin N., Shanmugapriya, S. Aflatoxins and aflatoxicosis in human and animals. In: Guevara-Gonzalez RG, Editor. Aflatoxins-Biochemistry and Molecular Biology. Rijeka (Croatia): Intechopen Publisher, 2011:221-254

[29] Neal GE, Green JA. The requirement for glutathione S-transferase in the conjugation of activated aflatoxin B1 during aflatoxin hepatocarcinogenesis in the rat. Chemico-Biological Interactions. 1983;**45**(2):259-275

[30] Wang JS, Shen X, He X, et al. Protective alterations in phase 1 and 2 metabolism of aflatoxin B1 by oltipraz in residents of Qidong, People's Republic of China. Journal of the National Cancer Institute. 1999;**91**(4):347-354

[31] Simkó M, Gazsó A, Fiedeler U, Nentwich M. Nanoparticles, free radicals and oxidative stress. Nano Trust Dossiers. 2011;**12**:1-3

[32] Ray PD, Huang BW, Tsuji Y. Reactive oxygen species (ROS) homeostasis and redox regulation in cellular signaling. Cell Signalling. 2012;**24**:981-990

[33] Sies H. Oxidative stress: From basic research to clinical application. The American Journal of Medicine. 1991;**91**(3):31-38

[34] Halliwell B, Cross CE. Oxygen-derived species: Their relation to human disease and environmental stress. Environmental Health Perspectives. 1994;**102**(10):5

[35] Hardas Sarita S. Investigations of Oxidative Stress Effects and Their Mechanisms in Rat Brain After Systemic Administration of Ceria Engineered Nanomaterials. University of Kentucky, Theses and Dissertations-Chemistry, Paper 7. http://uknowledge.uky.edu/chemistry-e ds/7, 2012.

[36] Muller FL, Lustgarten MS, Jang Y, Richardson A, Remmen HV. Trends in oxidative aging theories. Free Radical Biology and Medicine. 2007;**43**:477-503

[37] Shen HM, Shi CY, Lee HP, Ong CN. Aflatoxin B1-induced lipid peroxidation in rat liver. Toxicology and Applied Pharmacology. 1994;**127**:145-150

[38] Janero DR. Malondialdehyde and thiobarbituric acid-reactivity as diagnostic indices of lipid peroxidation and peroxidative tissue injury. Free Radical Biology and Medicine. 1990;**9**:515-540

[39] Halliwell B, Chirico S. Lipid peroxidation: Its mechanism, measurement, and significance. American Journal of Clinical Nutrition. 1993;**57**:715-725.

[40] Marnett LJ. Lipid peroxidation: DNA damage by malondialdehyde. Mutation Research. 1999;**424**:83-95

[41] Yilmaz S, Kaya E, Comakli S. Vitamin E, α tocopherol, attenuates toxicity and oxidative stress induced by aflatoxin in rats. Advances in Clinical and Experimental Medicine. 2017. In press.

[42] Halliwell B, Gutteridge JMC. Cellular responses to oxidative stress: Adaptation, damage, repair, senescence and death. Free Radicals in Biology and Medicine. 2007;**4**:187-267

[43] Josse R, Dumont J, Fautre A, Robin M, Guillouzo A. Identification of early target genes of aflatoxin B1 in human hepatocytes, inter-individual variability and comparison with other genotoxic compounds. Toxicology and Applied Pharmacology. 2012;**258**:176-187

[44] Shen HM, Shi CY, Shen Y, Ong CN. Detection of elevated reactive oxygen species level in cultured rat hepatocytes treated with aflatoxin B1. Free Radical Biology and Medicine. 1996;**21**:139-146

[45] Bondy SC, Naderi S. Contribution of hepatic cytochrome P450 systems to the generation of reactive oxygen species. Biochemical Pharmacology. 1994;**48**:155-159

[46] Shimamoto NA. A pathophysiological role of cytochrome p450 involved in production of reactive oxygen species. Yakugaku Zasshi. 2013;**133**:435-450

[47] Sun X, Ai M, Wang Y, et al. Selective induction of tumor cell apoptosis by a novel P450-mediated reactive oxygen species (ROS) inducer methyl 3-(4-nitrophenyl) propiolate. Journal of Biological Chemistry. 2013;**288**:8826-8837

[48] Shen HM, Ong CN. Mutations of the p53 tumor suppressor gene and ras oncogenes in aflatoxin hepatocarcinogenesis. Mutation Research/Reviews in Genetic Toxicology. 1996;**366**(1):23-44

[49] Kodama M, Inoue F, Akao M. Enzymatic and non-enzymatic formation of free radicals from aflatoxin B1. Free Radical Research Communications. 1990;**10**:137-142

[50] Amstad P, Levy A, Emerit I, Cerutti P. Evidence for membrane-mediated chromosomal damage by aflatoxin B1 in human lymphocytes. Carcinogenesis. 1984;**5**:719-723

[51] Meki ARM, Abdel-Ghaffar SK, El-Gibaly I. Aflatoxin B1 induces apoptosis in rat liver: Protective effect of melatonin. Neuroendocrinology Letters. 2001;**22**(6):417-426

[52] Reddy L, Odhav B, Bhoola K. Aflatoxin B1-induced toxicity in HepG2 cells inhibited by carotenoids: Morphology, apoptosis and DNA damage. Biological Chemistry. 2006; **387**(1):87-93

[53] El-Nekeety AA, Mohamed SR, Hathout AS, et al. Antioxidant properties of thymus vulgaris oil against aflatoxin-induce oxidative stress in male rats. Toxicon. 2011;**57**(7):984-991

[54] Kanbur M, Eraslan G, Sarica ZS, Aslan O. The effects of evening primrose oil on lipid peroxidation induced by subacute aflatoxin exposure in mice. Food and Chemical Toxicology. 2011;**49**:1960-1964

[55] Sirajudeen M, Gopi K, Tyagi JS, et al. Protective effects of melatonin in reduction of oxidative damage and immunosuppression induced by aflatoxin B1-contaminated diets in young chicks. Environmental Toxicology. 2011;**26**(2):153-160

[56] Preetha SP, Kanniappan M, Selvakumar E, Nagaraj M, Varalakshmi P. Lupeol ameliorates aflatoxin B1- induced peroxidative hepatic damage in rats. Comparative Biochemistry and Physiology. 2006;**143**:333-339

[57] Rastogi S, Dogra RK, Khanna SK, Das M. Skin tumorigenic potential of aflatoxin B1 in mice. Food and Chemical Toxicology. 2005;**44**:670-677

[58] Abdel-Wahhab MA, Hassan NS, El-Kady AA, et al. Red ginseng extract protects against aflatoxin B1 and fumonisins-induced hepatic pre-cancerous lesions in rats. Food and Chemical Toxicology. 2010;**48**:733-742

[59] Abdel-Aziem SH, Hassan AM, Abdel-Wahhab MA. Dietary supplementation with whey protein and ginseng extract counteracts oxidative stress and DNA damage in rats fed an aflatoxin-contaminated diet. Mutation Research. 2011;**723**:65-71

[60] Shen HM, Ong CN, Lee BL, Shi CY. Aflatoxin B1-induced 8-hydroxydeoxyguanosine formation in rat hepatic DNA. Carcinogenesis. 1995;**16**:419-422

[61] Thrasher JD. Are chlorinated pesticides a causation in maternal DNA (mtDNA) mutations? Archives of Environmental Health. 2000;**55**:292-294

[62] Sajan MP, Satav JG, Bhattacharya RK. Alteration of energy-linked functions in rat hepatic mitochondria following aflatoxin B1 administration. Journal of Biochemical and Molecular Toxicology. 1996;**11**(5):235-241

[63] Obasi SC. Effects of scopoletin and aflatoxin on B1 on bovine hepatic mitochondrial respitatory complexes, 2: A-ketoglutarate cytochrome c and succinate cytochrome c reductases. Zeitschrift für Naturforschung C. 2001;**56**(3-4):278-282

[64] Bernabucci U, Colavecchia L, Danieli PP, et al. Aflatoxin B1 and fumonisin B1 affect the oxidative status of bovine peripheral blood mononuclear cells. Toxicology in Vitro. 2011;**25**:684-691

[65] Kotan E, Alpsoy L, Anar M, Aslan A, Agar G. Protective role of methanol extract of Cetraria islandica (L.) against oxidative stress and genotoxic effects of AFB1 in human lymphocytes in vitro. Toxicology and Industrial Health. 2011;**27**:599-605

[66] Butler WH, Greenblatt M, Lijinsky W. Carcinogenesis in rats by aflatoxins B1, G1, and B2. Cancer Research. 1969;**29**(12):2206-2211

[67] Ward JM, Sontag JM, Weisburger EK, Brown CA. Effect of lifetime exposure to aflatoxin B1 in rats. Journal of the National Cancer Institute. 1975;**55**(1):107-113

[68] Dickens JW, Pattee HE. The effects of time, temperature and moisture on aflatoxin production in peanuts inoculated with a toxic strain of *Aspergillus flavus*. Tropical Science. 1966;**8**:11-22

[69] Newberne PM, Butler WH. Acute and chronic effects of aflatoxin on the liver of domestic and laboratory animals: A review. Cancer Research. 1969;**29**(1):236-250

[70] Sell S, Xu KL, Huff WE, et al. Aflatoxin exposure produces serum α-fetoprotein elevations and marked oval cell proliferation in young male Peking ducklings. Pathology. 1998;**30**:34-39

[71] Lee CC, Liu JY, Lin JK, Chu JS, Shew JY. p53 point mutation enhanced by hepatic regeneration in aflatoxin B 1-induced rat liver tumors and preneoplastic lesions. Cancer Letters. 1998;**125**(1):1-7

[72] Bammler TK, Slone DH, Eaton DL. Effects of dietary oltipraz and ethoxyquin on aflatoxin B1 biotransformation in non-human primates. Toxicological Sciences. 2000;**54**(1):30-41

[73] Toskulkao C, Glinsukon T. Hepatic lipid peroxidation and intracellular calcium accumulation in ethanol potentiated aflatoxin B1 toxicity. Journal of Pharmacobio-Dynamics. 1988;**11**(3):191-197

[74] Naaz F, Javed S, Abdin MZ. Hepatoprotective effect of ethanolic extract of Phyllanthus amarus Schum. et Thonn. on aflatoxin B1-induced liver damage in mice. The Journal of Ethnopharmacology. 2007;**113**:503-509

[75] Lee JK, Choi EH, Lee KG, Chun HS. Alleviation of aflatoxin B1-induced oxidative stress in HepG2 cells by volatile extract from Allii Fistulosi Bulbus. Life Sciences. 2005;**77**:2896-2910

[76] Gesing A, Karbownik-Lewinska M. Protective effects of melatonin and N-acetylserotonin on aflatoxin B1-induced lipid peroxidation in rats. Cell Biochemistry and Function. 2008;**26**(3):314-319

[77] Marquez A, Villa-Treviño S, Guéraud F. The LEC rat: A useful model for studying liver carcinogenesis related to oxidative stress and inflammation. Redox Report. 2007;**12**:35-39

[78] Berlett BS, Stadtman ER. Protein oxidation in aging, disease, and oxidative stress. The Journal of Biological Chemistry. 1997;**272**:20313-20316

[79] Peng T, Li LQ, Peng MH, et al. Is correction for protein concentration appropriate for protein adduct dosimetry? Hypothesis and clues from an aflatoxin B1-exposed population. Cancer Sciences. 2007;**98**:140-146

[80] Peng T, Li LQ, Peng MH, et al. Evaluation of oxidative stress in a group of adolescents exposed to a high level of aflatoxin B1–a multi-center and multi-biomarker study. Carcinogenesis. 2007;**28**:2347-2354

[81] Amici M, Cecarini V, Pettinari A, et al. Binding of aflatoxins to the 20S proteasome: Effects on enzyme functionality and implications for oxidative stress and apoptosis. The Journal of Biological Chemistry. 2007;**388**:107-117

[82] Thaxton JP, Tung HT, Hamilton PB. Immunosuppression in chickens by aflatoxin. Poultry Science. 1974;**53**(2):721-725

[83] Tung HT, Wyatt RD, Thaxton P, Hamilton PB. Concentrations of serum proteins during aflatoxicosis. Toxicology and Applied Pharmacology. 1975;**34**(2):320-326

[84] Wang JS, Abubaker S, He X, et al. Development of aflatoxin B1-lysine adduct monoclonal antibody for human exposure studies. Applied and Environmental Microbiology. 2001;**67**(6):2712-2717

[85] Steenken S. Purine bases, nucleosides, and nucleotides: Aqueous solution redox chemistry and transformation reactions of their radical cations and e- and OH adduct. Chemical Reviews. 1989;**89**:503-520

[86] Kasai H, Yamaizumi Z, Yamamoto F, et al. Photosensitized formation of 8-hydroxyguanine (7,8-dihydro-8-oxoguanine) in DNA by riboflavin. Nucleic Acids Symposium Series. 1992;**27**:181-182

[87] Luo H, Tang L, Tang M, et al. Phase IIa chemoprevention trial of green tea polyphenols in high-risk individuals of liver cancer: Modulation of urinary excretion of green tea polyphenols and 8-hydroxydeoxyguanosine. Carcinogenesis. 2006;**27**(2):262-268

[88] Valavanidis A, Vlachogianni T, Fiotakis C. 8-Hydroxy-2′ -deoxyguanosine (8-OHdG): A critical biomarker of oxidative stress and carcinogenesis. Journal of Environmental Science and Health Part C Environmental Carcinogenesis & Ecotoxicology Reviews. 2009;**27**:120-139

[89] Eaton DL, Gallagher EP. Mechanisms of aflatoxin carcinogenesis. Annual Review of Pharmacology And Toxicology. 1994;**34**:135-172

[90] Kobertz WR, Wang D, Wogan GN, Essigmann JM. An intercalation inhibitor altering the target selectivity of DNA damaging agents: Synthesis of site-specific aflatoxin B1 adducts in a p53 mutational hotspot. Proceedings of the National Academy of Sciences. 1997;**94**(18):9579-9584

[91] Larsson P, Busk L, Tjalve H. Hepatic and extrahepatic bioactivation and GSH conjugation of aflatoxin B1 in sheep. Carcinogenesis. 1994;**15**:947-955

[92] Gross-Steinmeyer K. Eaton DL. Dietary modulation of the biotransformation and genotoxicity of aflatoxin B(1). Toxicology. 2012;**299**:69-79

[93] Rao AV, Rao LG. Carotenoids and human health. Pharmacological Research. 2007;**55**(3): 207-216

[94] Gerster H. The potential role of lycopene for human health. Journal of the American College of Nutrition. 1997;**16**(2):109-126

[95] Rao AV, Agarwal S. Role of lycopene as antioxidant carotenoid in the prevention of chronic diseases: A review. Nutrition Research. 1999;**19**(2):305-323

[96] Nguyen ML, Schwartz SJ. Lycopene: Chemical and biological properties. Food Technology (USA). 1999:**53**;38-45.

[97] Nguyen ML, Schwartz SJ. Lycopene stability during food processing. Proceedings of the Society for Experimental Biology and Medicine. 1998;**218**(2):101-105

[98] Trumbo PR. Are there adverse effects of lycopene exposure? The Journal of Nutrition. 2005;**135**(8):2060-2061

[99] Halliwell B, Gutteridge JMC. Free Radicals in Biology and Medicine. New York: Oxford University Press. Inc; 1999

[100] El-Agamey A, Lowe GM, McGarvey DJ, et al. Carotenoid radical chemistry and antioxidant/pro-oxidant properties. Archives of Biochemistry and Biophysics. 2004;**430**(1):37-48

[101] Krinsky NI, Yeum KJ. Carotenoid–radical interactions. Biochemical and Biophysical Research Communications. 2003;**305**(3):754-760

[102] Galano A, Francisco-Marquez M. Reactions of OOH radical with β-carotene, lycopene, and torulene: Hydrogen atom transfer and adduct formation mechanisms. The Journal of Physical Chemistry B. 2009;**113**(32):11338-11345

[103] Stahl W, Sies H. Lycopene: A biologically important carotenoid for human. Archives of Biochemistry and Biophysics. 1996;**336**(1):6

[104] Rao AV, Agarwal S. Effect of diet and smoking on serum lycopene and lipid peroxidation. Nutrition Research. 1998;**18**(4):713-721

[105] Matos HR, Di Mascio P, Medeiros MH. Protective effect of lycopene on lipid peroxidation and oxidative DNA damage in cell culture. Archives of Biochemistry and Biophysics. 2000;**383**(1):56-59

[106] Yilmaz S, Atessahin A, Engin S, Karahan I, Ozer S. Protective effect of lycopene on adriamycin-induced cardiotoxicity and nephrotoxicity. Toxicology. 2006;**218**(2-3):164-171

[107] Velmurugan B, Bhuvaneswari V, Burra UK, Nagini S. Prevention of N-methyl-N′-nitro-N-nitrosoguanidine and saturated sodium chloride-induced gastric carcinogenesis in Wistar rats by lycopene. European Journal of Cancer Prevention. 2002;**11**(1):19-26

[108] Di Mascio P, Kaiser S, Sies H. Lycopene as the most efficient biological carotenoid singlet oxygen quencher. Archives of Biochemistry and Biophysics. 1989;**274**(2):532-538

[109] Mortensen A, Skibsted LH. Relative stability of carotenoid radical cations and homologue tocopheroxyl radicals. A real time kinetic study of antioxidant hierarchy. FEBS Letters. 1997;**417**(3):261-266

[110] Tang L, Guan H, Ding X, Wang JS. Modulation of aflatoxin toxicity and biomarkers by lycopene in F344 rats. Toxicology and Applied Pharmacology. 2007;**219**(1):10-17

[111] Breinholt V, Lauridsen ST, Daneshvar B, Jakobsen J. Dose-response effects of lycopene on selected drug-metabolizing and antioxidant enzymes in the rat. Cancer Letters. 2000;**154**(2):201-210

[112] Karaca A. The Effect of Lycopene on Hepatotoxicity of Aflatoxin B1 in Rats. Director: Seval Yilmaz. University of Firat, Health Sciences Institute-Biochemitry, Elazig/Turkey, Master's Thesis; 2015

Chapter 5

Aflatoxin: A Risky Menace for African's Food Commodities

Nitin Mahendra Chauhan

Additional information is available at the end of the chapter

http://dx.doi.org/10.5772/intechopen.69302

Abstract

Aflatoxins contamination of African food and food commodities exhibits a serious threat to human and animal health over the past few decades. To protect the safety of food commodities, regular monitoring for afltoxins has began to implicate in developing countries. The food contaminating species *Aspergillus flavus* and *Aspergillus parasiticus* are responsible for production of aflatoxins. Various studies have followed ELISA, TLC, HPLC, immunoassay, etc for quantification of aflatoxins. The data from different reports demonstrate that staple foods in most countries are particularly vulnerable to attack by aflatoxigenic fungi and found contaminated with aflatoxins. In our study from Ethiopia, we have utilized a quick and precise biosensor and thin layer chromatography method to measure contamination of aflatoxins in maize. Our data revealed that all the samples tested were greater than the safety level of aflatoxins as recommended by Food and Drug Administration (FDA) and European Union (EU). Utilization of internationally developed biosensor for presence of fungal toxin in food samples is the first approach that was applied in the developing country like Ethiopia. In the end, we conclude that fungal contaminants and there toxic products are potential threat to the agro and food industry in Africa and require immediate control measures.

Keywords: aflatoxins, food commodities, *Aspergillus*, cancer, Africa

1. Introduction

Africa's crop agriculture is very complex, involving substantial variation in crops cultivated across various countries as well as involving different regions and ecologies among each country. Among these, crops that constitute the staple food of African countries are at risk to fungal infections, which results in aflatoxin contamination due to the poor agronomic practices, storage condition of crops and more importantly processing of food materials under favourable temperature and humidity conditions [1]. The extent of contamination of food commodities by aflatoxin also varies with different geographical locations among the country. According to United Nations Food and Agriculture Organization, 25% of world's agriculture commodities are contaminated with fungal toxins, which leads to severe economic and health loss to the affected country [2].

Mycotoxins, i.e., aflatoxins represent the class of fungal polyketide secondary metabolites that are mainly produced by two fungi viz. *Aspergillus flavus* and *Aspergillus parasiticus* [3]. These fungi are known to produce four major kinds of aflatoxins, i.e., aflatoxin B_1 (AFB_1), aflatoxin B_2 (AFB_2), aflatoxin G_1 (AFG_1) and aflatoxin G_2 (AFG_2). Among these four principle classes of aflatoxins, AFB_1 is found to be predominant in natural environment and reported carcinogenic in animal models if the toxicity exceeds beyond the safety level [3, 4]. The agricultural commodities that are prone to aflatoxins toxicity are corn and corn products, peanuts, cottonseed, milo, animal feed and majority of tree nuts [5, 6]. Aflatoxins toxicity has always been a topic of debatable interest in international market and economic development of country, which are the part of trade market. To overcome this problem, many countries have set standard safety levels of aflatoxins in food and food products and animal feed [7, 8]. Increased risk of hepatocellular carcinoma in the presence of hepatitis B virus infection [9] and esophageal cancer [10] has been associated with aflatoxins contamination of food in most of the developing countries from Africa. Intensive exposures of AFB_1 at a concentration in excess of 2 ppm are reported to cause non-specific liver problems and death within few days, whereas chronic effect of AFB_1 leads to immunosuppression and nutritional deficiency [11].

Various food commodities like maize and maize products, peanuts, cottonseed, milo, animal feed and majority of tree nuts are considered as one of the best substrates for the fungi to grow and produce toxicgenesis. Many surveys across the globe showed that the food commodities that constitute the staple food of African countries could be highly contaminated with aflatoxins. Aflatoxins in feed also possesses negative impacts on the production of healthy livestock, affecting a decrease in milk and egg yield, which results in toxic residues in dairy, meat and poultry products. Aflatoxins are reported to be prevalent among various parts of Africa. Some of the previous studies reported that 90% of East African maize samples showed the evidence of high level of aflatoxins, and some parts of West Africa showed the exposure of aflatoxins is as high as 99% [12]. Aflatoxins not only support severe health risk, but also favour significant economic loss to farmers due to the rejection of their crops by international buyers if it is contaminated with fungal toxins. For example, in Kenya, two World Food Programs of the United Nations purchased maize samples that were confiscated and destroyed because of the lack of acceptable levels of aflatoxins in the purchased crops [13]. This is of particular concern to smallholder farmers as aflatoxins toxicity primarily occurs where there is a high moisture content and high temperature, which is supported by inadequate storage structures. Implementation of national prevention and control strategies like proper pre- and pro-harvest treatment of various infected food commodities and standard storage facilities are required to reduce the risk of aflatoxin contamination by fungi.

2. Chemical and biological basis of aflatoxins

Aflatoxins are the class of mycotoxins that have been well-known for their delirious outbreak of 'Turkey 'X' disease' in England and were first isolated and characterized from *A. flavus* which is reported to be a common contaminant of poorly stored grains [14]. Aflatoxins are secondary metabolites, which are naturally occurring contaminants of food and elaborate the toxins under favourable conditions of temperature, relative humidity and poor storage conditions. They are now known to be mainly produced by *A. flavus, A. parasiticus, Aspergillus nomius* and two different *Emericella* species [15]. Aflatoxins have received more attention due to their effects on agricultural production loss, threats to human health because of their high toxicity and carcinogenic nature as well as potential threats to food safety [16]. Till date, there are roughly 20 known aflatoxins reported based on chromatographic and fluorescence characteristics but only six of these aflatoxins, i.e., AB_1, AB_2, AG_1, AG_2, aflatoxin (AFM_1) and aflatoxin M_2 (AFM_2) (**Figure 1**) are widely studied because of severe toxicity and more prevalence

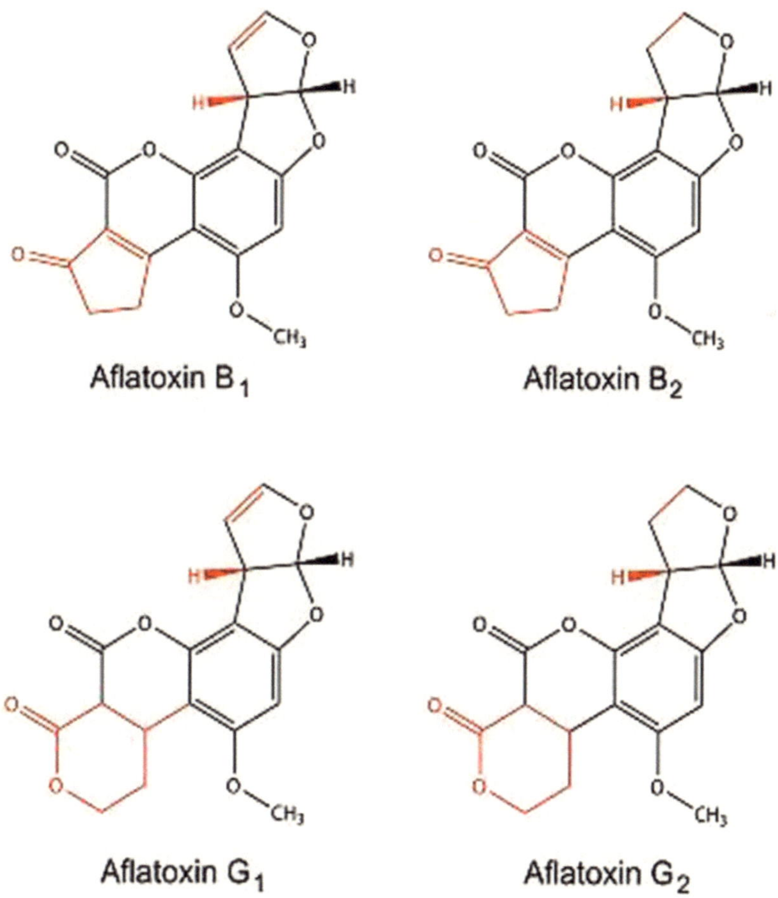

Figure 1. Chemical structure of major class of aflatoxins (Source: www.istockphoto.com).

in food and food products. Other aflatoxins have paid less attention as they exist very rare in nature, since they are metabolic derivatives mostly found in pure cultures [17]. AFB_1 is the most dangerous among these toxins; however, the order of acute and chronic toxicity is $AFB_1 > AFG_1 > AFB_2 > AFG_2$ [18].

2.1. Chemical basis of major aflatoxins

The major aflatoxins have been classified into B and G groups due to their fluorescence properties in the presence of UV to give blue and green colourations, respectively [19]. The B series aflatoxins, AFB_1 and AFB_2 are chemically known as difurocoumarocyclopentenones and the G series aflatoxins, AFG_1, AFG_2 are difurocoumarolactone series (**Figure 1**). Structurally, the dihydrofuran moiety, containing a double bond and the constituents linked to the coumarin moiety play an important role in producing biological effects. For the B series aflatoxins, cyclopentenone was reported to be responsible for the major toxicity [20]. On the other hand, M groups of aflatoxins are chemically called as methoxycyclopenta. It is usually considered that AFM_1 is a detoxification end product of AFB_1, which is due to the result of mutagenic and carcinogenic process, and is found to be the main mono-hydroxylate derivative of AFB_1 in liver by means of cytochrome P_{450}-associated enzymes [21]. The common aflatoxins are AFB_1, AFB_2, AFG_1 and AFG_2. Their molecular weights are 312.3 g/mol for aflatoxin B_1, 314.3 g/mol for aflatoxin B_2, 328.3 g/mol for aflatoxin G_1 and 330.3 g/mol for aflatoxin G_2. Aflatoxin M_1 and M_2, which are metabolites of aflatoxins were first isolated from milk of lactating animals that were fed on aflatoxin preparations [22].

2.2. Biological basis of aflatoxins

2.2.1. *Aflatoxins producing fungi*

Aflatoxins are difuranocoumarin derivatives produced by a polyketide pathway mainly by strains of *A. flavus* and *A. parasiticus*; in particular, *A. flavus* is a common contaminant in agriculture. In spite of these two fungi, *Aspergillus bombycis*, *Aspergillus ochraceoroseus*, *A. nomius* and *Aspergillus pseudotamarii* are also reported as aflatoxin-producing species, but they are found less predominant in nature [23]. All the aflatoxins-producing fungi exhibits a great variation in terms of qualitative and quantitative differences in the toxicology abilities that are markedly attributes by different strains within each fungal species. For instance, only about half of *A. flavus* strains may produce over 106 μg/kg aflatoxins in comparison to other *Aspergillus* strains [14]. *A. flavus* only produces type B toxins [24] while, other species such as *A. nomius* and *A. parasiticus* produce both B and G types [14]. Some strains of *A. flavus*, which are regarded as the S strains based on the size of the sclerotia are known to produce more toxin than toxicogenic *A. flavus* L strains [25].

2.2.2. *Biosynthesis of aflatoxins*

The aflatoxins constitute a number of structurally related metabolites that differ considerably in their biological effects. However, all of them contain a coumarin ring combined to a bis-dihydrofurano moiety and additionally either a cyclopentenone ring (B series) or a six-membered lactone ring (G series). Among all of these toxins, AFB_1 is the one with the greatest biological activity. Carcinogenic in several animal species, AFB_1 reveals itself as the most potent hepatocarcinogen known in the rat and the rainbow trout [26]. It has been reported that it is probable that the enzymes of aflatoxin biosynthesis and of other polyketides are similarly arranged in discrete particles in the post-mitochondrial fraction [10]. The aflatoxin biosynthesis is also characterized by 29 clustered aflatoxin pathway genes and can be described in two major stages: an early stage from acetate to versicolorin A (VERA) (coloured pigment in brick-red, yellow or orange) and a later stage from dimethyl-sterigmatocystin (DMST) to AFB_1 (colourless under normal light and fluorescent-blue under UV light) [26].

2.2.3. *Modus of operandi of toxicity by aflatoxins in human*

Like many other chemical carcinogens, AFB_1 requires bio-activation to a reactive toxic metabolite-activation as an important stage in its toxicity expression [27]. AFB_1 cannot itself be the toxic molecule but it is metabolized in the animal body in a complex network of reactions and it is the result of this metabolism, which determines both acute and chronic toxicity. Many researchers have studied the relationship between the biological activity of AFB_1 and its metabolism, and have found the evidence that AFB_1 needs metabolic activation to exert its carcinogenic and mutagenic effects [28]. After ingestion, AFB_1 presents a short half-life; 65% of the quantity absorbed after 90 min is removed from the blood and plasma and metabolized by the liver to a reactive epoxide intermediate. It has been estimated that in human liver homogenates, the half-life of AFB_1 is 15 min [20, 29]. In the metabolism, however, the first step of it takes place in the hepatocyte with non-reversible detoxification, which leads to the formation of hydroxylated metabolites followed either by reversible detoxification through aflatoxicol formation, or by activation [30].

However, AFB_1 is mainly bio-activated by cytochrome P_{450}-dependent mono-oxygenase, which results in the production of many metabolic products such as aflatoxin Q_1, aflatoxin P_1, aflatoxin M_1 and aflatoxin $B_{1\text{-}8\text{-}9\text{-epoxide}}$. Aflatoxin $B_{1\text{-}8\text{-}9\text{-expoxide}}$ has been found to be the most toxic metabolite [31]. Cytochrome P_{450} mono-oxygenase has been demonstrated as a key factor in the metabolic activation of several chemical carcinogens such as AFB_1, various heterocyclic and aromatic amines and specific nitro-aromatic compounds [31]. Among these metabolic products, aflatoxin $B_{1\text{-}8\text{-}9\text{-epoxide}}$ has been shown as an important metabolite synthesized in the animal liver and can react with guanine residues in DNA and lead to depurination [26]. The net result is gene mutation. The most regularly induced mutation is the GC→TA transversion, potentially leading to carcinogenesis [32]. In addition, the epoxide occurs in endoforms and exoforms. The exo-epoxide is highly electrophilic and reacts with several macromolecules [32]. The activated AFB_1, aflatoxin $B_{1\text{-}8\text{-}9\text{-epoxide}}$ can bind to glutathione, cellular proteins, RNA and DNA. The binding of this toxic compound to DNA has been investigated in rats and was found to take place at the critical nucleophilic sites of DNA and identified to form 2,3-dihydro-2-(N_7-guanyl)-3-hydroxy-aflatoxin B_1 [20], which is also associated with tumour development in animals [33]. However, when bound to glutathione, aflatoxin $B_{1\text{-}8\text{-}9\text{-epoxide}}$ produces another metabolite that is less toxic [10].

Many mineral elements including Zn^{2+}, Cu^{2+} and Fe^{2+} are also essential for this activation by contributing to the cyclization of the polyketide precursors, and also affecting the induction of the enzymes of secondary metabolism [31]. In light of this, AFB_1 may be seen as a multiple menace by its carcinogenic, teratogenic and mutagenic effects, and also by its immunosuppressive effects [31].

3. Method for detection for aflatoxins

Aflatoxins not only possess severe effects on human health but also cause serious economic losses when tons of foods have to be discarded or destroyed as a result of aflatoxin contamination in developing countries, due to which a rapid and sensitive method has been a pre-requisite for quantification of aflatoxins in food samples. To ensure food safety, maximum levels for aflatoxins in food and feed have been set by national and international organizations and various approaches have been developed for the determination of aflatoxin concentrations in food and feed commodities. Following methods are widely used for quantification estimation of aflatoxins in various food commodities.

3.1. Chromatography method

Chromatography is one of the most widely used as well as the oldest method for quantifying aflatoxins. In the beginning of aflatoxin analysis and research, gas chromatography (GC) was frequently used for detection and quantification of aflatoxins. However, modern biology leads to new chromatography-based techniques for the detection of aflatoxins. Examples of these improvements are liquid chromatography (LC), thin layer chromatography (TLC) [34] and high-performance liquid chromatography (HPLC) [35], which, nowadays, is the most commonly used chromatographic technique for detection of a wide diversity of mycotoxins, especially for aflatoxin derivatives [36]. Frisvad and Thrane [37] described an HPLC method for the detection of 182 mycotoxins and other fungal metabolites based on their alkylphenone retention indices and diode array spectra. Nowadays, coupling of HPLC with mass spectroscopy or tandem mass spectroscopy allows for highly accurate determination of toxin concentrations and identification of different types of toxins in a single analysis [38]. Alternatively, fluorescence property is also used for the detection of unmodified aflatoxins in HPLC applications as well as in thin layer chromatography. Furthermore, there are combinations of the above described methods with pre-process techniques, which can detect the concentration of aflatoxin in a solution in a better way. For example, immune-affinity column sample clean-up followed by a normal or reverse phase of HPLC separation along with fluorometric detection is mostly used for quantitative determination of AFM_1 due to the characteristics of specificity, high sensitivity and simplicity of operation [39].

3.2. Immunoassay method

Immunochemical detection for aflatoxins is based on the principle of antibody-antigen reactions (Ab-Ag) [40]. Since different kinds of aflatoxin molecules possess antigenic properties, it is possible to detect them by raising antibodies against them. Most of the immunological methods are based on enzyme-linked immunosorbent assays (ELISA), which have good sensitivity, speed and simplicity. In addition, some lateral flow immunoassays (LFIAs) are also applied for the qualitative and semi-quantitative detection of aflatoxins in food, feed and milk [41]. Even though several studies have been published on the immunochemical determination of aflatoxins in food, only a few validation protocols are available to show that the results comply with certain regulations because of the requirement for expensive instrumentation.

3.3. Biosensor and other methods

Biosensors, an alternative to overcome the disadvantages of the previous described methods, are multidisciplinary tools with an enormous potential in detection and quantification of aflatoxin with minimum cost. There are different kinds of biosensors that base their performance on different physical or biochemical principles, such as optical, optoelectronic, electrochemical, piezoelectric, DNA and combined. Thus, such devices have a huge impact on healthcare, food management, agronomical economy and bio-defence [42]. Different types of biosensors are applied to detect aflatoxins in various food commodities. However, they mainly work on the principle of conjunction with various immunochemical methods. Such junctions are based on simple principle that employs the property of high affinity of antigen-antibody interaction, which automatically increased the sensitivity and thus reducing the detection time of toxic element [43]. For example, Chauhan et al. [44] used 150 different maize samples that were collected from different Gedeo zones of Ethiopia. Commodity samples included dry maize flour, freshly harvested corn fruits and dry maize kernels. For quantification of aflatoxin in maize samples from Ethiopia, we followed biosensor approach. The assay is based on a single-step lateral flow immunochromatographic principle with competitive immunoassay format. Use of such technique is the first approach utilized in the developing country like Ethiopia.

Further methods also exist which are less common than the previously described methods but have a tremendous potential for detection of fungal toxins. The most important are those ones that utilize the principle of electrochemistry, spectroscopy and fluorescence. Compared with traditional methods for aflatoxin determination, electrochemical techniques offer some advantages such as reliability, low cost, *in situ* measurements, fast processes and easier methodology over common chromatography techniques through a similar performance. Especially, for measurement of AFM_1, the disposable immunosensors have been applied directly in milk following a simple centrifugation step without dilution or other pre-treatment steps. Exhibition of a good working range with linearity between 30 and 240 ng/ml makes this method very useful for AFM_1 monitoring in milk (maximum acceptable level of AFM_1 in milk is 50 ppt) [45]. Spectroscopy techniques are also popularized due to the characteristics of fast, low-cost and non-destructive analytical methods suitable to work with solid and liquid samples. Among them, near infrared spectroscopy (NIRS) is an excellent option for a rapid and low cost detection of aflatoxin in cereals [46]. When incorporated with a bundle reflectance fibre-optic probe, NIRS was successfully applied to quantify AFB_1, ochratoxin A and total aflatoxins in paprika [47]. Aflatoxins have a native fluorescence due to their oxygenated pentaheterocyclic structure, which forms the basis of most analytical and microbiological methods for detection and quantification of aflatoxins [48].

4. Occurrence of aflatoxins in various food commodities

Aflatoxins are toxic secondary metabolites produced by various *Aspergillus* species growing in susceptible agricultural commodities. Many African countries had begun to implicate prevention, control and surveillance strategies to reduce the incidence of aflatoxin in foods. The main mycotoxins, i.e., aflatoxins, have been reported to be widespread in major dietary food products in African countries. These mycotoxins occur mostly in maize, spices and groundnuts and many more food commodities. Our data demonstrate that all the maize samples tested were beyond the safety level of aflatoxins as determined by Food and Drug Administration (FDA) and European Union (EU). Many studies are reported on contamination of food and food products in African countries. The food and food commodities that are prone to aflatoxin contaminations is briefly highlighted in **Table 1**, adapted from various literature.

Country	Food and food commodities	Concentration	Reference
Ethiopia	Shiro and red pepper	100–525 ppb	[49]
	Sorghum, barley, teff and wheat	00–26 ppb	[50]
	Maize	5 μg/kg	[51]
	Pre- and Post-harvest maize	18.38–43.4 μg/kg	[52]
	Maize	40–90 ppb	[44]
	Sorghum	1.17–344 μg/kg	[53]
Nigeria	Groundnut	2000 g/kg	[54]
	Pre-harvest maize	3–138 μg/kg	[55]
	Dried yam chips	27.1 μg/kg	[56]
	Maize	770 ppb	[57]
	Melon seed	2.3–47.7 μg/kg	[58]
	Bush mango seed	0.2–4.2 μg/kg	[59]
	Millet	1.370–28 μg/kg	[60]
	Maize	0–1874 μg/kg	[61]
	Roasted groundnut	3–106 μg/kg	[62]
	Smoke dried fish	1.5–8.11 μg/kg	[62]
	Powdered soy milk	4.58–19.76 μg/kg	[63]
	Mouldy sorghum	0–1164 μg/kg	[64]
	Beans	59.29–106 μg/kg	[65]
	Wheat	85.66–198.4 μg/kg	[65]
	Wheat	17.10–20.53 μg/kg	[66]
	Poultry/Live stock feed	0–67.9 μg/kg	[67]
	Food thickeners	4–9 μg/kg	[68]
	Dried beef	0.003–0.004 μg/kg	[69]
	Fresh beef	0.02–0.03 μg/kg	[69]
	Rice	28–372 ppb	[70]
	Weaning food	4.6–530 ppb	[71]
	Rice	37.26–113.2 μg/kg	[72]
	Maddi	0.2–125 μg/kg	[72]
	Dry sesame	14–140 μg/kg	[72]
	Maize and maize products	102–213 ppb	[73]
	Okra fruits	0.08–8.5 μg/kg	[74]
	Fruits	3.8 μg/kg	[75]
	Suya spices	2.65–43 μg/kg	[76]

Country	Food and food commodities	Concentration	Reference
Egypt	Meat products	2–150 ppb	[77]
	Spices	2–35 ppb	[78]
	Cereal grains	36 ppb	[79]
	Chicken and chicken products	1–4 ppb	[80]
	Nuts and seed	24 ppb	[81]
	Medicinal plants	49 ppb	[81]
Tunisia and Morocco	Poultry feed	0.03–5.38	[21]
	Barley	3.5–11.5 μg/kg	[82]
	Wheat	4.0–12.9 μg/kg	[82]
	Sorghum	0.34–52.9 μg/kg	[83]
	Pistachio	0.24–12.24 μg/kg	[83]
	Cereals and cereals products	5.5–66.7 ppb	[84]
Sudan	Animal feeds	4.1–579 μg/kg	[85]
	Sesame oil	0.2–0.8 ppb	[86]
	Groundnut oil	0.6 ppb	[87]
	Peanut butter	21.17 ppb	[87]
Tanzania	Maize	158 ppb	[88]
	Red chilli	<4 ppb	[89]
Uganda	Maize	1–1000 μg/kg	[90]
	Groundnuts, cassava, millet, etc. Maize	0–55 ppb	[91]
	Cassava	7–12 μg/kg	[92]
	Groundnut and groundnut paste	0–5 μg/kg	[92]
		0–940 μg/kg	[92]
Kenya	Wheat	0–7 μg/kg	[93]
	Animal feed and milk	<5 ppb	[94]
	Groundnut	0–7525 μg/kg	[95]
	Maize	<20 ppb	[96]
	Grains	<10 ppb	[97]
	Groundnut	0–2377 ppb	[97]
Ghana	Maize	0.7–335 ppb	[98]
Benin	Chips	2.2–220 ppb	[99]
	Store maize	14–58 g/kg	[54]
	Maize	5 ppb	[100]
	Dried vegetables	3.2–6.0 ppb	[101]
	Cowpea	3.58 μg/kg	[102]
Mali and Togo	Dried vegetables	3.2–6.0 ppb	[101]
Botswana	Raw peanut	12–329 μg/kg	[103]
Senegal	Peanut oil	40 ppb	[104]
South Africa	Traditionally brewed beers	200–400 μg/l	[105]
	Wheat and products	0.5–2.0 μg/kg	[106]
	Animal feeds	0.8–156 μg/kg	[107]
	Cotton seed meal	0.3–75 μg/kg	[108]
	Grains	<20 ppb	[109]

Country	Food and food commodities	Concentration	Reference
Cameroon	Cow pea	0.2–6.2 μg/kg	[110]
	Soy bean	0.2–3.9 μg/kg	[110]
	Egg	0.002–7.68 μg/kg	[111]
Morocco	Maize flour	0.23–11.2 μg/kg	[21]
	Dried figs	0.28 μg/kg	[112]
	Dried raisins	3.2–13.9 μg/kg	[112]
	Pistachio	0.04–14.30 μg/kg	[112]
Congo	Groundnut	1.5–937 μg/kg	[18]
	Grains	<20 ppb	[109]
Malawi	Groundnut	0–3871 μg/kg	[113]
	Maize		[113]
	Sorghum		[114]
	Local beer		[114]
	Groundnut		[115]
		0–1335 μg/kg	
		1.7–33.0 μg/kg	
		8.8–34.5 μg/kg	
		0.2–4.3 ppb	
Algeria	Wheat and products	0.13–37.42 μg/kg	[116]
Zambia	Peanut butter	20–10740 μg/kg	[117]
Zimbabwe	Ground nut	6.6–247 ppb	[118]
	Peanut and peanut butter	75 ppb	[118]
	Groundnut	1–175 μg/kg	[119]
	Cowpea	1.4–103.4 μg/kg	[119]
Gambia	Groundnut	8.22–813.86 μg/kg	[120]
Burkina Faso	Groundnuts	170 ppb	[121]

Table 1. Incidence of aflatoxins contamination in various foods and food commodities from different parts of Africa.

5. Aflatoxins safety level set up by African countries

Only few African countries are known to have regulations for aflatoxins in food and/or feed. These are summarized in **Table 2**, which was adapted from Anonymous [122] and van Egmond [123].

Country	Food commodity	Aflatoxins type	Regulatory level (ng/g)
Ivory Coast	Feedstuffs	B_1, B_2, G_1, G_2	100
	Mixed feeds	B_1, B_2, G_1, G_2	10
	Mixed feeds: pigs/poultry	B_1, B_2, G_1, G_2	38
	Mixed feeds: ruminants	B_1, B_2, G_1, G_2	75
	Mixed feeds: dairy cattle	B_1, B_2, G_1, G_2	50

Country	Food commodity	Aflatoxins type	Regulatory level (ng/g)
Egypt	Peanuts and products; oil seeds and	B_1, B_2, G_1, G_2	10
	products; cereals	B_1	5
	Peanuts and products; oil seeds and	B_1, B_2, G_1, G_2	20
	products; cereals	B_1	10
	Maize (food)	B_1, B_2, G_1, G_2	0
	Maize (food)	B_1	0
	Starch and derivatives (food)	M_1, M_2, G_1, G_2	0
	Starch and derivatives (food)	M_1	0
	Milk, dairy products	B_1, B_2, G_1, G_2	20
	Milk, dairy products	B_1	10
	Animal and poultry feeds		
	Animal and poultry feeds		
Kenya	Peanuts and products, vegetable oils (food)	B_1, B_2, G_1, G_2	20
Malawi	All foods	B_1, B_2, G_1, G_2	35
	Peanuts for export (food)	B_1	5
Nigeria	All foods	B_1	20
	Infant foods	B_1	0
	Milk	M_1	0
	Feedstuffs	B_1	50
Senegal	Peanut product feeds	B_1	50
	Peanut product feed components	B_1	300
South Africa	All foods	B_1, B_2, G_1, G_2	10
	All foods	B_1	5
	Feed components	B_1, B_2, G_1, G_2	50
	Mixed feeds for beef cattle, sheep	B_1, B_2, G_1, G_2	50
	and goats	B_1, B_2, G_1, G_2	20
	Mixed feeds for lactating cows,	B_1, B_2, G_1, G_2	10
	swine, calves, lambs	B_1, B_2, G_1, G_2	0
	Mixed feeds for unweaned piglets,		
	broilers and pullets		
	Mixed feeds for trout		
Zimbabwe	Foods	B_1	5
	Foods	G_1	4
	Groundnuts, maize, sorghum	B_1	5
	Groundnuts, maize, sorghum	G_1	4
	Poultry feed	B_1, B_2	10
	Peanut butter, cereal flour	B_1, B_2, G_1, G_2	20
Mauritius	Peanuts	B_1, B_2, G_1, G_2	15
	Peanuts	B_1	5
	Other products	B_1, B_2, G_1, G_2	10
	Other products	B_1	5
Algeria	Nut, cereals	B_1	20

Table 2. Aflatoxins safety level in several countries of Africa.

6. Strategy to control aflatoxins in Africa

Measure for control of aflatoxins in Africa is not only crucial for implications of health safety, but also required to enhance the economy in the affected countries. According to Cassel et al. [125], the number of different approaches has been implicated to diminish and eradicate mycotoxins from different African countries. For example, control strategies include delaying of mould growth in crops and other feedstuffs, decontamination of mycotoxins affected foods and continuous monitoring of aflatoxins in agricultural crops, animal feedstuffs and human food. Apart from these measures, other prevention measures include separation of infected peanuts in Malawi, reduction of toxicity in peanut meal in Senegal for export, regulation of aflatoxins proportion in animal feed according to the susceptibility of respective animal species in Zimbabwe, selection of groundnut varieties less susceptible to aflatoxin contamination in Burkina Faso and improvement in handling and storage practices during production around 1960s in Nigeria and in 1990s in Gambia [124]. According to Cassel et al. [125], time of harvest is an important factor in influencing the occurrence and levels of aflatoxin. For example, harvesting maize above 20% moisture content followed by rapid drying to at least 14% within 24–48 hours of harvest minimizes aflatoxin level efficiently. Chulze [126] reported that it is possible to control aflatoxins in stored commodities by maintaining good atmosphere and use of preservatives or natural inhibitors in the form of antioxidants and essential oils can be applicable but the cost can be prohibitive on a large scale.

In recent times, there have been initiatives undertaken by international bodies with the aim to control aflatoxins in developing countries, especially from Africa. One of the best initiatives initiated is the Partnership for Aflatoxin Control in Africa (PACA), which is based on a Memorandum of Understanding that was undersigned between the African Union Commission and Mars Incorporated with a vision of sharing food safety resources and expertise to control aflatoxins contamination in food crops, which constitutes a significant threat and a major problem to African agricultural commodities as well as raw materials in global market [127]. Another initiative includes various projects that aim to control aflatoxin contamination in maize and peanuts. These projects aimed in developing and implementing control strategies by scaling up different bio-techniques intervention to improve the health and income of farmers and their families as well as to generate wealth in the crop value chain [128]. The project is funded by Bill and Melinda Gates Foundation and African Agricultural Technology Foundation (AATF) through the International Institute of Tropical Agriculture (IITA) and UK aid from the UK government, respectively.

7. Conclusion

The literature reviewed reveals that African population is highly exposed to food borne aflatoxins, due to the tropical climate that is present in most of the African countries and provides optimal conditions for fungus to grow happily. These fungal toxins have been shown to cause a variety of toxic and severe health effects in humans and thus lead to reduced life expectancy in Africa automatically. However, where quality control is absent, unsafe levels of aflatoxin are present. AFB_1 was identified as the most predominant and toxic among all the aflatoxins types. Their vicinity in African foods and feeds is unavoidable due to which, humans and animals are suffering from aflatoxins contamination on various and regular bases that lead to a wide range of health effects. Particularly, AFB_1 has been directly correlated to hepatocarcinoma and deaths among humans and animals across the world. Although, this may be the case globally, the status in sub-Saharan Africa is very critical, as rising levels of aflatoxins exposure through different dietary products are a common problem as evidence by various literatures highlighted in **Table 1**. Again, the problem is further exacerbated by increased prevalence of AFB_1 in this continent, as such endemic diseases like malaria, hepatitis and HIV/AIDS are identified in peoples who consumed aflatoxins contaminated food. In Africa, we have already experienced the most fatal aflatoxin poisoning outbreaks including two episodes especially one in Kenya and other in Nigeria.

It is obvious that impoverished and less privileged people of developing countries of Africa are indirectly linked to greater risk of further poverty and food scarcity, if control measures are not undertaken for the regulation of aflatoxins contamination in agriculture commodities. Utilization of recommended prevention and control strategies may make food more costly and less usable, since farmers will have to focus in drying and storage equipment to protect food that is directly related to more investment. Even though there are various methods available for detection of aflatoxins, their plight is worsened by the absence of well-equipped state of art laboratories for testing mycotoxins levels, which are economically and financially inaccessible. However, it will be better to confirm that contamination levels of fungal toxins are minimal to safeguard the health of people in developing countries whose lifespan is relatively short. It is unfortunate for the people in developing countries that international bodies like the World Health Organization (WHO) do not consider aflatoxins as a high priority risk; hence, little attention has been paid to the health issues resulting from the consumption of contaminated food.

Developed countries and international agencies should come forward for necessary financial and technical assistance to support developing countries to carry out research and education. This will also directly benefit to developing countries in terms of increased foreign exchange earnings, from the sale of products that meet required standards and better health through the consumption of safer food, which are not beyond the safety level of mycotoxins.

In the end, implementation of national prevention and control strategies like proper pre- and pro-harvest treatment of infected maize and standard storage facilities are required to reduce the risk of aflatoxin contamination by fungi in foods from African countries. Since, very few countries have set the safety level of aflatoxins in food, more studies are required from different parts of Africa to generate data for different governments to work on policy making decision strategy and required to set the safety level for aflatoxins in foods. The quantity of aflatoxins reported in various researches as shown above possesses a potential threat to agro as well as food industry in Africa and require immediate control measures. It is also important to implement control strategies to differentiate the food samples that are safe for human and animal consumptions for saving lives.

Conflict of interest

The author declared no conflict of interest.

Author details

Nitin Mahendra Chauhan

Address all correspondence to: nitinchauhan25@gmail.com

College of Natural and Computational Sciences, Dilla University, Dilla, SNNPR, Ethiopia

References

[1] Chauhan YS, Wright GC, Rachaputi NC. Modelling climatic risks of aflatoxin contamination in maize. Australian Journal of Experimental Agriculture. 2008;**48**:358-366

[2] FAO. Worldwide Regulations for Mycotoxins in Food and Feed in 2003. Rome: Food and Nutrition Papers, Food and Agriculture Organization of the United Nations; 2004

[3] Bennett JW, Klich M. Mycotoxins. Clinical Microbiology Review. 2003;**16**:497-516

[4] Council for Agricultural Science and Technology (CAST). Mycotoxins: Risks in Plant, Animal and Human Systems. Task Force Report No. 139, Ames, IA, USA, 2003

[5] Beatriz LS, Alessandra BR, Paulo AM, Miguel MJ. Aflatoxins, ochratoxin A and zearalenone in maize-based food products. Brazilian Journal of Microbiology. 2005;**36**:289-294

[6] Binder EM, Tan LM, Chin LJ, Hadle J, Richard J. Worldwide occurrence of mycotoxins in commodities feeds and feed ingredients. Animal Feed Science and Technology. 2007;**137**:265-282

[7] Diener U, Cole R, Sanders T, Payne G, Lee L, Klich M. Epidemiology of aflatoxin formation by *Aspergillus flavus*. Annual Review of Phytopathology. 1987;**25**:249-270

[8] European Commission (EC). Commission Regulation (EC) No. 1881/2006 of 19 December 2006 setting maximum levels for certain contaminants in foodstuffs. Official Journal of European Union. 2006;**L364**:5-24

[9] Henry SH, Bosch FX, Troxell TC, Bolger PM. Reducing liver cancer—global control of aflatoxin. Science. 1999;**286**:2453-2454

[10] Wild CP, Turner PC. The toxicology of aflatoxins as a basis for public health decisions. Mutagenesis. 2002;**17**:471-481

[11] Peraica M, Radic B, Lucic A, Pavlovic M. Toxic effects of mycotoxins in human health. Bulletin WHO. 1999;77:754-766

[12] Rodrigues I, Handl J, Binder EM. Mycotoxin occurrence in commodities, feeds and feed ingredients sourced in the Middle East and Africa. Food Additives and Contaminants Part B Surveillance. 2011;**4**:168-179

[13] Hassan RM, Onyango R, Rutto JK. Relevance of maize research in Kenya to maize production problems perceived by farmers. In: Hassan RM, editor. Maize technology development and transfer: A GIS approach to research planning in Kenya. London: CAB International; 1998

[14] Klich MA. Environmental and developmental factors influencing aflatoxin production by *Aspergillus flavus* and *Aspergillus parasiticus*. Mycoscience. 2007;**48**:71-80

[15] Frisvad JC, Thrane U, Samson RA, Pitt JI. Important mycotoxins and the fungi which produce them. In: Hocking AD, Pitt J., Samson RA, Thrane U, editors. Advances in Food Mycology. Vol. 57. Advances in Experimental Medicine and Biology. New York: Springer Science and Business Media; 2006. pp. 3-3

[16] Ellis WO, Smith JP, Simpson BK. Aflatoxins in food-occurrence, biosynthesis, effects on organisms, detection and methods of control. Critical Reviews in Food Science and Nutrition. 1991;**30**:403-439

[17] Franco CM, Fente CA, Vazquez BI, Cepeda A, Mahuzier G, Prognon P. Interaction between cyclodextrins and aflatoxins Q_1, M_1 and P_1: Fluorescence and chromatographic studies. Journal of Chromatography A. 1998;**815**:21-29

[18] Kamika I, Takoy LL. Natural occurrence of aflatoxin B_1 in peanut collected from Kinshasa, Democratic Republic of Congo. Food Control. 2011;**22**:1760-1764

[19] Pavao AC, Neto LAS, Neto JF, Leao, MBC. Structure and activity of aflatoxins B and G. Journal of Molecular Structure (Theochem). 1995;**337**:57-60

[20] Fung F, Clark RF. Health effects of mycotoxins: A toxicological overview. Clinical Toxicology. 2004;**42**:217-234

[21] Zinedine A, Juan C, Soriano JM, Molto JC, Idrissi L, Maries J. Limited survey for the occurrence of aflatoxins in cereals and poultry feeds from Rabat, Morocco. International Journal of Food Microbiology. 2007;**115**:124-127

[22] Sassahara M. Netto PD, Yanaka EK. Aflatoxin occurrence in foodstuff supplied to dairy cattle and aflatoxin M_1 in raw milk in the North of Paraa state. Food Chemistry and Toxicology. 2005;**43**:981-984

[23] Klich MA. Identification of Common *Aspergillus* spp, Ponson and Looijen, Wageningen. The Netherlands. Centraalbureau Voor Schimmelcultures; 2002. pp. 1-107. ISSN/ISBN: 978-9070351465

[24] Klich M, Pitt JI. A laboratory guide to *Aspergillus* species and their teleomorphs. North Ryde, NWS, Australia: CSIRO Division of Food Processing; 1988

[25] Cotty PJ, Jaime-Gracia R. Influence of climate on aflatoxin producing fungi and aflatoxin contamination. International Journal of Food microbiology. 2007;**119**:109-115

[26] Yu J, Cleveland TE. *Aspergillus flavus* genomics for discovering genes involved in aflatoxin biosynthesis. In: Rimando AM, Baerson SR, editors. Polyketide: Biosynthesis, Biological Activity and Genetic Engineering. Washington: ACS; 2007. pp. 246-260

[27] Donnelly PJ, Stewart RK, Ali SL, Conlan AA, Reid KR, Petsikas D, Massey TE. Biotransformation of aflatoxin B_1 in human lung. Carcinogenesis. 1996;**17**:2487-2494

[28] Niu G, Wen Z, Rupasinghe SG, Zeng RS, Berenbaum MR, Schuler MR. Aflatoxin B_1 detoxification by CYP321A1 in *Helicoverpa zea*. Insect Biochemistry and Physiology. 2008;**68**:32-45

[29] Bastaki SA, Osman N, Kochiyil J, Shafiullah M, Padmanabhan R, Abdulrazzaq YM. Toxicokinetics of aflatoxin in pregnant mice. Internal Journal of Toxicology. 2010;**29**: 425-431

[30] Neal GE. Participation of animal biotransformation in mycotoxin toxicity. Revue de Medecine Veterinaire. 1998;**149**:555-560

[31] Do JH, Choi D. Aflatoxins: Detection, toxicity, and biosynthesis. Biotechnology and Bioprocess Engineering. 2007;**12**:583-593

[32] Smela ME, Sophie S, Curier E, Bailey A, John ME. The chemistry and biology of aflatoxin B_1. Carcinogenesis. 2001;**22**:535-545

[33] Dashwood RH, Arbogast DN, Perieira C, Hendricks JD, Bailey GS. Quantitative interrelationships between aflatoxin B_1 carcinogen dose, indole-3-carbinol anti-carcinogen dose, target organ DNA adduction and final tumor response. Carcinogenesis. 1989;**10**: 175-181

[34] Stroka J, Van Otterdijk R, Anklam E. Immunoaffinity column clean-up prior to thin layer chromatography for the determination of aflatoxins in various food matrices. Journal of Chromatography A. 2000;**904**:251-256

[35] Bacaloni A, Cavaliere C, Cucci F, Patrizia F. Determination of aflatoxins in hazelnuts by various sample preparation methods and liquid chromatography-tandem mass spectrometry. Journal of Chromatography A. 2008;**1179**:182-189

[36] De Rijk TC, Arzandeh S, Kleijnen HCH, Zomer P, Van der Weg G, Mol JGJ. Aflatoxins determination using in-line immunoaffinity chromatography in foods. Food Control. 2011;**26**:42-48

[37] Frisvad JC, Thrane U. Standardized high-performance liquid chromatography of 182 mycotoxins and other fungal metabolites based on alkylphenone retention indices and UV–VIS spectra (diode array detection). Journal of Chromatography A. 1987;**404**:195-214

[38] Sobolev VS. Simple, rapid and inexpensive cleanup method for quantitation of aflatoxins in important agricultural products by HPLC. Journal of Agricultural and Food Chemistry. 2007;**55**:2136-2141

[39] Muscarella M, Lo Magro S, Palermo C, Centonze D. Validation according to European Commission Decision '02/657/EC of a confirmatory method for aflatoxin M_1 in milk

based on immunoaffinity columns and high performance liquid chromatography with fluorescence detection. Analytica Chimica Acta. 2007;**594**:257-264

[40] Lee NA, Wang S, Allan RD, Kennedy IR. A rapid aflatoxin B_1 ELISA: Development and validation with reduced matrix effects for peanuts, corn, pistachio and soyabeans. Journal of Agricultural Food Chemistry. 2004;**52**:2746-2755

[41] Anfossi L, Darco G, Calderara M, Baggiani C, Giovannoli C, Giraudi G. Development of a quantitative lateral flow immunoassay for the detection of aflatoxins in maize. Food Additives and Contaminants: Part A: Chemistry, Analysis, Control, Exposure and Risk Assessment. 2011;**28**:226-234

[42] Nayak M, Kotian A, Marathe S, Chakravortty D. Detection of microorganisms using biosensors-A smarter way towards detection techniques. Biosensors and Bioelectronics. 2009;**25**:661-667

[43] Dinckaya E, Kinik O, Sezginturk MK, Altug C, Akkoca A. Development of an impedimetric aflatoxin M_1 biosensor based on a DNA probe and gold nanoparticles. Biosensors and Bioelectronics. 2011;**26**:3806-3811

[44] Chauhan NM, Washe AP, Minota T. Fungal Infection and Aflatoxin Contamination in Maize Collected from Gedeo Zone. Vol. 5. Ethiopia: SpringerPlus; 2016. p. 573

[45] Micheli L, Grecco R, Badea M, Moscome D, Palleschi G. An electrochemical immunosensor for aflatoxin M_1 determination in milk using screen-printed electrodes. Biosensors and Bioelectronics. 2005;**21**:588-596

[46] Fernández-Ibanez V, Soldado A, Martínez-Fernandez A, De la Roza-Delgado B. Application of near infrared spectroscopy for rapid detection of aflatoxin B_1 in maize and barley as analytical quality assessment. Food Chemistry. 2009;**113**:629-634

[47] Hernandez-Hierro JM, Garcia-Villanova RJ, Gonzalez-Martin I. Potential of near infrared spectroscopy for the analysis of mycotoxins applied to naturally Aflatoxins – Recent Advances and Future Prospects 312 contaminated red paprika founding the Spanish market. Analytica chimica acta. 2008;**622**:189-194

[48] Rasch C, Böttcher M, Kumke M. Determination of aflatoxin B_1 in alcoholic beverages: Comparison of one-and two-photon-induced fluorescence. Analytical and Bioanalytical Chemistry. 2010;**397**:87-92

[49] Fufa H, Urga K. Screening of aflatoxins in Shiro and ground red pepper in Addis Ababa. Ethiopian Medical Journal. 1996;**34**:243-249

[50] Ayalew A, Fehrmann H, Lepschy J, Beck R, Abate D. Natural occurrence of mycotoxins in staple cereals from Ethiopia. Mycopathologia. 2006;**162**:57-63

[51] Ayalew A. Mycotoxins and surface and internal fungi of maize from Ethiopia. African Journal of Food Agriculture and Nutrition Development. 2010;**10**:4109-4123

[52] Assaye MA, Gemeda N, Weledesemayat GT. *Aspergillus* species and aflatoxin contamination of Pre and Post-Harvest maize grain in West Gojam, Ethiopia. Journal of Food Science and Nutrtion. 2016;**2**:013

[53] Wondimeneh T, Ayalew A, Alemayehu C, Mashilla D. Aflatoxin B and total fumonisin contamination and their producing fungi in fresh and stored sorghum grain in East Hararghe, Ethiopia. Food Additives and Contaminants Part B. 2016;**9**:237-245

[54] Bankole SA, Adebanjo A. Mycotoxins in food in West Africa: Current situation and possibilities of controlling it. African Journal of Biotechnology. 2003;**2**:254-263

[55] Bankole SA, Mabekoje OO. Occurrence of aflatoxins and fumonisins in preharvest maize from south-western Nigeria. Food Additives and Contaminants. 2004;**21**:251-255

[56] Bankole SA, Mabekoje OO. Mycoflora and occurrence of aflatoxin B_1 in dried yam chips from markets in Ogun and Oyo States, Nigeria. Mycopathologia. 2004;**15**:111-115

[57] Williams JH, Phillips TD, Jolly PE, Stiles JK, Jolly CM, Aggarwal D. Human aflatoxicosis in developing countries a review of toxicology, exposure potential health consequences and intervention American Journal of Clinical Nutrition. 2004;**80**:1106-1122

[58] Bankole SA, Ogunsanwo BM, Osho A, Adewuyi GA. Fungal contamination and aflatoxin B_1 of egusi melon seeds in Nigeria. Food Control. 2006;**17**:814-818

[59] Adebayo-Tayo BC, Onilude AA, Ogunjobi AA, Gbolagade JS, Oladapo MO. Detection of fungi and aflatoxin in shelved bush mango seeds (*Irvingia spp.*) stored for sale in Uyo, Nigeria African. Journal of Biotechnology. 2006;**5**:1729-1732

[60] Makun HA, Gbodi TA, Akanya HO, Sakalo AE, Ogbadu HG. Fungi and some mycotoxins contaminating rice (*Oryza sativa*) in Niger state, Nigeria. African Journal of Biotechnology. 2007;**6**:99-108

[61] Atehnkeng J, Ojiambo PS, Donner M, Ikotun K, Sikora RA, Cotty PJ, Bandypadhyay. Distribution and toxicity of *Aspergillus* species isolated from maize kernels from three agro-ecological zones of Nigeria. International Journal Food Microbiology. 2008;**122**: 74-84

[62] Adebayo-Tayo, BC, Onilude AA, Ukpe GP. Mycofloral of Smoke-Dried fishes sold in Uyo, Eastern Nigeria. World Journal of Agricultural Sciences. 2008;**4**:346-350

[63] Adebayo-Tayo BC, Adegoke AA, Akinjogunla OJ. Microbial and physicochemical quality of powdered soymilk samples in Akwa Ibom, South Southern Nigeria. African Journal of Biotechnology. 2009;**8**:3066-3071

[64] Makun HA, Gbodi TA, Akanya HO, Salako EA, Ogbadu GH. Fungi and some mycotoxins found in mouldy Sorghum in Niger State, Nigeria. World Journal of Agricultural Sciences. 2009;**5**:5-17

[65] Makun HA, Anjorin ST, Moronfoye B, Adejo FO, Afolabi OA, Fagbayibo G, Balogun BO, Surajudeen AA. Fungal and aflatoxin contaminations of some human food commodities in Nigeria. African Journal of Food Sciences. 2010;**4**:127-135

[66] Odoemelam SA, Osu CI. Aflatoxin B_1 contamination of some edible grains marketed in Nigeria. E-Journal of Chemistry. 2009;**6**:308-314

[67] Adebayo-Tayo BC, Ettah AE. Microbiological quality and aflatoxin B_1 level in poultry and livestock feeds. Nigerian Journal of Microbiology. 2010;**24**:2145-2152

[68] Okwu GI, Achar PN, Sharma SK. Quantification of aflatoxin B_1 in ready-to-use food thickeners in South-east geo-political zone in Nigeria. African Journal of Microbiology Research. 2010;**4**:1788-1793

[69] Oyero GO, Oyefolu AB. Natural occurrence of aflatoxin residues in fresh and sun-dried meat in Nigeria. Pan African Medical Journal. 2010;7:14

[70] Makun HA, Dutton MF, Njobeh PB, Mwanza M, Kabiru AY. Natural multi- mycotoxin occurrence in rice from Niger State, Nigeria. Mycotoxin Research. 2011;**27**:97-104

[71] Oluwafemi F, Ibeh IN. Microbial contamination of seven major weaning foods in Nigeria. Journal of Health Population and Nutrition. 2011;**29**:415-419

[72] Makun HA, Apeh DO, Adeyemi HRY, Nagago T, Okeke Jo, Musthapha AS, Oyinloye BY. Determination of aflatoxins in Sesame, Rice, Millet and Acha from Nigeria using HPLC. Chemical Science Transactions. 2014;**3**:1516-1524

[73] Enyisi S, Orukotan I, Ado A, Adewumi AAJ. Total aflatoxin level and fungi contamination of maize and maize products. African Journal of Food Science and Technology. 2015;**6**:229-233

[74] Jonathan SG, Dare M, Abe OA, Olawuyib OJ, Daniel A. Food value, fungi and aflatoxin detection in stored 'Orunla' *Abelmoschus esculentus* L. (Moench) from Ibadan, Nigeria. Researcher. 2016;**8**:7-11

[75] Pepple GA, Chukunda FA, Ukoima HN. Detection of fungi and aflatoxin contamination in shelved fruits of *Dialium guineense* wild, Rivers State, Nigeria. Asian Journal of Plant Science and Research. 2016;**6**:1-5

[76] Segun GJ, Mary AA, Michael DA. Fungal biodeterioration, aflatoxin contamination, and nutrient value of "Suya Spices". Scientifica. 2016. doi: 10.1155/2016/4602036

[77] Aziz NH, Youssef YA. Occurrence of aflatoxins and aflatoxin-producing moulds in fresh and processed meat in Egypt. Food Additives and Contaminants. 1991;**8**:321-331

[78] Selim MI, Popendorf W, Ibrahim MS, el Sharkawy S, el Kashory ES. Aflatoxin B_1 in common Egyptian foods. Journal of AOAC International. 1996;**79**:1124-1129

[79] Motawee MM, Bauer J, Mc Mahon DJ. Survey of Aflatoxin M_1 in Cow, Goat, Buffalo and Camel Milks in Ismailia-Egypt. Bulletin of Environment Contamination and Toxicology. 2009;**83**:766-769

[80] Darwish WS, Ikenaka Y, Nkayama SMM, Ishizuka M. An overview of mycotoxin contamination of foods in Africa. Journal of Veterinary and Medical Science. 2014;**76**:789-797

[81] El-Tras WF, El-Kady NN, Tayel AA: Infants exposure to aflatoxin M as a novel food borne zoonosis. Food Chemistry and Toxicology. 2011;**49**:2816-2819

[82] Ghali R, Khlifa KH, Ghorbel H, Maaroufi K, Hedilli A. Incidence of aflatoxins, ochratoxin A and zearalenone in Tunisian foods. Food Control. 2008;**19**:921-924

[83] Ghali R, Belouaer I, Hdiri S, Ghorbel H, Maaroufi K, Hedilli A. Simultaneous HPLC determination of aflatoxins B_1, B_2, G_1 and G_2 in Tunisian sorghum and pistachios. Journal of Food Composition and Analysis. 2009;**22**:751-755

[84] Serrano AB, Font G, Ruiz MJ, Ferrer E. Co-occurrence and risk assessment of mycotoxins in food and diet from Mediterranean area. Food Chemistry. 2012;**135**:423-429

[85] Elzupir AO, Younis MH, Himmat Fadul M, Elhussein AM. Determination of aflatoxins in animal feed in Khartoum State, Sudan. Journal of Animal and Veterinary Advances. 2009;**8**:1000-1003

[86] Idris YM, Mariod AA, Elnour IA, Mohamed AA. Determination of aflatoxin levels in Sudanese edible oils. Food Chemistry and Toxicology. 2010;**48**:2539-2541

[87] Elshafie SZ, ElMubarak A, El-Nagerabi SA, Elshafie AE. Aflatoxin B_1 contamination of traditionally processed peanuts butter for human consumption in Sudan. Mycopathologia. 2011;**171**:435-439

[88] Kimanya ME, De Meulenaer B, Tiisekwa B, Ndomondo-Sigonda M, et al. Co-occurrence of fumonisins with aflatoxins in home-stored maize for human consumption in rural villages of Tanzania. Food Additives and Contaminants. 2008;**25**:1353-1364

[89] Temu GE. Molecular Identification of *Aspergillus* Strains and quick detection of aflatoxin from selected common spices in Tanzania. Journal of Scientific Research and Reports. 2016;**10**:1-8

[90] Kaaya NA, Warren HL. A review of past and present research on aflatoxin in Uganda African. Journal of Food Agriculture Nutrition and Development. 2005;**5**:18

[91] Kitya D, Bbosa GS, Mulogo E. Aflatoxin levels in common foods of South Western Uganda: A risk factor to hepatocellular carcinoma. European Journal of Cancer Care. 2010;**19**:516-521

[92] Jimmy O, Geofrey M, Trasias M, Abdullah AH Archileo KN, Ssempebwa JC, Jia-Sheng W. Aflatoxin contamination of selected staple foods sold for human consumption in Kampala markets, Uganda. Journal of Biological Science. 2016;**16**:44-48

[93] Muthomi JW, Ndungu JK, Gathumbi JK, Mutitu EW, Wagacha JM. The occurrence of *Fusarium* species and mycotoxins in Kenyan wheat. Crop Protection. 2008;**27**:1215-1219

[94] Kangethe EK, Langa KA. Aflatoxin B_1 and M_1 contamination of animal feeds and milk from urban centers in Kenya. African Health Science. 2009;**9**:218-226

[95] Mutegi CK, Ngugi HK, Hendriks SL, Jones RB. Prevalence and factors associated with aflatoxin contamination of peanuts from Western Kenya. International Journal of Food Microbiology. 2009;**130**:27-34

[96] Daniel JH, Lewis LW, Redwood YA, Kieszak S, Breiman RF, et al. Comprehensive assessment of maize aflatoxin levels in Eastern Kenya, 2005-2007. Environment and Health Perspective. 2011;**119**:1794-1799

[97] Ndungu JW, Makokha AO, Onyango CA, Mutegi CK, Wagacha JM, Christie ME, Wanjoya AK. Prevalence and potential for aflatoxin contamination in groundnuts and peanut butter from farmers and traders in Nairobi and Nyanza provinces of Kenya. Journal of Applied Bioscience. 2013;**65**:4922-4934

[98] Kpodo K, Sorensen AK, Jakobsen M. The occurrence of mycotoxins in fermented maize products. Food Chemistry. 1996;**56**:147-153

[99] Bassa S, Mestres C, Hell K, Vernia P, Cardwell K. First report of aflatoxin in dried yam chips in Benin. Plant Disease. 2001;**85**:1032

[100] Hell K, Fandohan P, Bandyopadhyay R, Kiewnick, S, Sikora R, Cotty PJ. Pre-and post-harvest management of aflatoxin in maie: An African perspective. In: Leslie JF, Bandyopadhyay R, Visconti A, editors. Mycotoxins: Detection Methods, Management, Public Health, and Agricultural Trade. Wallingford, UK: CAB International; 2008. pp. 219-229

[101] Hell K, Gnonlonfin GJ, Kodjogbe G, Lamboni Y, Abdourhamane IK. Mycoflora and occurrence of aflatoxin in dried vegetables in Benin, Mali and Togo, West Africa. International Journal of Food Microbiology. 2009;**135**:99-104

[102] Houssou PA, Ahohuendo BC, Fandohan P, Kpodo K, Hounhouigan DJ, Jakobsen M. Natural infection of cowpea (*Vigna unguiculata* (L.) Walp.) by toxigenic fungi and mycotoxin contamination in Benin, West Africa. Journal of Stored Products Research. 2009;**45**:40-44

[103] Mphande FA, Siame BA, Taylor JE. Fungi. Aflatoxins and cyclopiazonic acid associated with peanut retailing in Botswana. Journal of Food Protection. 2004;**67**:96-102

[104] Diop YM, Ndiaye B, Fall M, Diouf A, Sall A, Ciss M, Ba D. Aflatoxin contamination level in artisanal and industrial peanut butter food in Dakar (Senegal). Dakar Medical. 2000;**45**:134-137

[105] Naicker V, Mohanlall V, Odhav B. Mycotoxins in South African traditionally brewed beers. Food Additives and Contaminants. 2002;**19**:55-61

[106] Mashinini K, Dutton MF. The incidence of fungi and mycotoxins in South Africa wheat and wheat-based products. Journal of Environmental Science and Health Part B. 2006;**41**:285-296

[107] Mngadi PT, Govinden R., Odhav B. Co-occurring mycotoxins in animal feeds. African Journal of Biotechnology. 2008;**7**:2239-2243

[108] Reiter EV, Dutton MF, Mwanza M, Agus A, Prawano D, Haggblom P, et al. Quality control of sampling for aflatoxins in animal feedingstuffs: Application of the Eurachem/CITAC guidelines. Analyst. 2011;**136**:4059-4069

[109] Kamika I, Mngqawa P, Rheeder JP, Teffo SL, Katerere DR. Mycological and aflatoxin contamination of peanuts sold at markets in Kinshasa, democratic republic of Congo and Pretoria, South Africa. Food Additives and Contaminants Part B. 2014;**7**:120-126

[110] Njobeh BP, Dutton FM, Makun HA. Mycotoxins and human health: Significance, prevention and control In: Ajay K, Mishra AT, Shivani BM, editors. Smart Biomolecules in Medicine. India: VBRI Press; 2010. pp. 132-177

[111] Tchana AN, Moundipa PF, Tchouanguep FM. Aflatoxin contamination in food and body fluids in relation to malnutrition and cancer status in Cameroon. International Journal of Environmental Research and Public Health. 2010;**7**:178-188

[112] Juan C, Zinedine A, Molto JC, Idriss L, Manes J. Aflatoxins levels in dried fruits and nuts from Rabat-Sale area, Morocco. Food Control. 2008;**19**:849-853

[113] Manyo ES, Waliyar F, Osiru M, Siambi M, Chinyamunyamu B . Assessing occurrence and distribution of aflatoxins in Malawi. Technical report by ICRISAT/ The McKnight Foundation, USA/NASFAM Malawi; 2009. pp. 1-40

[114] Matumba L, Monjerezi M, Khonga, EB, Lakudzala DD. Aflatoxins in sorghum, sorghum malt and traditional opaque beer in southern Malawi. Food Control. 2011;**22**:266-268

[115] Matumba L, Monjerezi M, Biswick T, Mwatseteza J, Makumba W, Kamangira D, Mtukuso A. A survey of the incidence and level of aflatoxin contamination in a range of locally and imported processed foods on Malawian retail market. Food Control. 2014;**39**:87-91

[116] Riba A, Bouras N, Mokrane S, Mathieu F, Lebrihi A, Sabaou N. *Aspergillus flavus* and aflatoxins in Algerian wheat and derived products. Food and Chemical Toxicology. 2010;**48**:2772-2777

[117] Njoroge SC, Matumba L, Kanenga K, Siambi M, Waliyar F, Maruwo J, Monyo E. A case for regular aflatoxin monitoring in peanut butter in Sub-Saharan Africa: Lessons from a 3-Year survey in Zambia. Journal of Food Protection. 2016;**79**:795-800

[118] Mupunga I, Lebelo SL, Mngqawa P, Rheeder JP, Katerere DR. Natural occurrence of aflatoxins in peanuts and peanut butter from Bulawayo, Zimbabwe. Journal of Food Protection. 2014;77:1814-1818

[119] Maringe DT, Chidewe C, Benhura MA, et al. Natural postharvest aflatoxin occurrence in food legumes in the smallholder farming sector of Zimbabwe. Food Additives and Contaminants Part B. 2017;**10**:21-26

[120] Xu A, Doel A, Watson S, Routledge MN, Elliott CT, Moore SE, Gong YY. Study of an educational Hand Sorting Intervention for Reducing Aflatoxin B_1 in Groundnuts in Rural Gambia. Journal of Food Protection. 2016;**80**:44-49

[121] Yameogo RT, Kassamba B. *Aspergillus flavus* and aflatoxin on tropical seeds used for snacks *Arachis hypogaea, Balanites aegyptiaca* and *Sclerocarya birrea*. Tropical Science. 1999;**39**:46-49

[122] Anonymous. Worldwide Regulations for Mycotoxins 1995. A Compendium. Rome: FAO; 1997

[123] van Egmond HP, Schothorst RC, Jonker MA. Regulations relating to mycotoxins in food: Perspectives in a global and European context. Annals of Bioanalytics and Chemistry. 2007;**389**:147-157

[124] Bhat RV, Vasanthi S. Food Safety in Food Security and Food Trade. Mycotoxin Food Safety Risk in Developing Countries. International Food Policy Research Institute, Indian Council of Medical Research, New Delhi, India; 2003

[125] Cassel EK, Campbell B, Draper M, Epperson B. Aflatoxin F907 Hazards in Grain Aflatoxicosis and Livestock. South Dakota State University Cooperation Extension Service/College of Agriculture and Biological Sciences/USDA, Brookings, South Dalota, USA; 2012

[126] Chulze SN. Strategies to reduce mycotoxin levels in maize during storage: A review. Food Additives and Contaminants Part A. 2010;**27**:651-657

[127] African Union Commission. AUC and Mars, Incorporated Sign Agreement to Tackle Aflatoxins. 2015. Available from: http://www.au.int

[128] African Agricultural and Technology Foundation. Aflatoxin Control in Maize and Peanuts Project. 2015. Available from: http://www.aatf-africa.org

Chapter 6

A Focus on Aflatoxins in Feedstuffs: Levels of Contamination, Prevalence, Control Strategies, and Impacts on Animal Health

Andrea Molina Alvarado,
Rebeca Zamora-Sanabria and
Fabio Granados-Chinchilla

Additional information is available at the end of the chapter

http://dx.doi.org/10.5772/intechopen.69468

Abstract

Aflatoxins are mold-synthetized secondary metabolites that are ubiquitously present in agricultural commodities, such as cereals which in turn are substantial part of feed formulation. These toxins are capable of causing disease, exert severe toxic effects, and even death in humans and other animals. Aflatoxins are the only mycotoxins with the regulatory framework, hence we present the legal threshold uphold till now by international and regional control organizations. Additionally, herein we discuss worldwide prevalence of aflatoxins in feeds to demonstrate a global issue and major risks involved in toxin contamination. Furthermore, we present recent data regarding negative effects usually presented by food-producing and companionship animals when ingested. Also, we discuss briefly practical approaches to mitigate aflatoxin burden during feed processing focusing in Good Manufacturing Practice (GMP) and hazard analysis critical control point (HACCP) and we include novel approaches reported in literature to decontaminate feed-containing aflatoxins. Finally, we cite the literature so far published describing the effects of changing climate on aflatoxin production and contamination.

Keywords: aflatoxins, risk factors, prevalence, animal health effects, mycotoxin sorbents, toxicity, climate change

1. Introduction

Livestock, aquaculture animals, and pets are exposed through dietary contact (i.e., through feedingstuff) to toxic fungal metabolites such as mycotoxins. Mycotoxins are low-molecular-weight natural products (i.e., small molecules) produced as secondary metabolites by filamentous fungi. Diseases produced by these means are collectively called mycotoxicoses. As with other toxicological syndromes, exposure to mycotoxins may be acute or chronic, veterinary health problems associated with mycotoxin exposure are usually the result of prolonged contact. This review chapter focuses specifically on aflatoxins. Aflatoxins are a group of biochemical substances produced especially by *Aspergillus* species [1]. They are usually found in cereals and grains such as rice, corn, sorghum, millet, and groundnuts during the harvesting, storage, and poor processing conditions [1].

Aflatoxin contamination associated with food or feed is a global problem especially in the tropical and subtropical regions of the world, where warm temperatures and humidity favor the growth of the fungi [2]. Considering its economic and health relevance, we will discuss certain aspects of the relationship of the contaminant with feeds and feed ingredients. Emphasis will be on the fact that animal feed, and ingredients thereof, are situated at the start the food chain and contaminated feed will, therefore, have an adverse impact on the rest of the alimentary web. Animal feedingstuffs quality directly affect animal productivity, health and can have drastic effects on food that is later consumed by humans as final products [3, 4]. Within the context of aflatoxins, we will discuss food chain safety, prevalence in animal feed and regulations. We will also mention risk factors and health effects of aflatoxins on animals, and control and management approaches to reduce them.

2. Aflatoxins in animal feed

Aflatoxins can be found worldwide in a variety of food and feed commodities especially cereals; the contamination with aflatoxin-producing fungi and the production of the toxin in the products can occur in the field, during storage, transportation at almost all stages of the production chain. In finished animal feed, the contamination of an ingredient could cause the contamination of an entire feed batch [5]. Furthermore, the introduction of a feedstuff contaminated with aflatoxin-producing fungi could lead to the spoilage of other feed shipments and serves as a fungi source in the feed industry environment difficult to eliminate. This deterioration effect has a significant repercussion in association with the global trade and the international exchange of animal feed and feed ingredients [6]. Co-occurrence of different mycotoxins in finished feed could have profound negative effects on animal and human health, due to the synergistic or additive effect among toxins [6]. The global production of animal feed reached 964 million tons in 2014 [7]. Cereal grains, primarily corn, are widely used as energy source in animal feed for different species. These raw materials represent 50–80% of the animal diet in America and Europe. USA and Brazil are the major corn exporter countries, and Japan and Mexico are the largest importer countries [8]. For example, most of the ingredients used in Malaysia for the production of animal feeds such as cereal grains, soybean meal, corn gluten meal, and soybean meal are imported from Thailand, China, India, Argentina, USA, Australia, and Canada. Mycotoxin contamination of feed caused by poor storage conditions during production and transportation are frequent [9]. In Costa Rica, the animal feed produced is based on corn products and only during 2015 over 764 254 tons of corn products were imported [10].

These are only examples of the importance of global trade for the animal feed industry; in this scenario, the origin of the ingredients and the place and length of storage must be taken into account to make a conclusion about mycotoxin contamination in a region. Furthermore, frequently agricultural commodities (peanuts, corn, and rice) used as feed ingredients originating from tropical and subtropical regions contain high amounts of aflatoxins [6, 11].

2.1. Major risk factors for aflatoxins in feedstuffs

As already mentioned above, the contamination of animal feedstuff could take place at different stages throughout the entire food chain. Mycotoxins in feedstuff and finished feed should be monitored from farm-to-fork to assure a safety product for animal and humans. The contamination of cereal grains and other agricultural commodities used in animal feed could occur in the field during the pre-harvest phase during harvest, or in processing stages (postharvest).

In the pre-harvest period, the presence of aflatoxin-producing fungi (and then the production of the toxin) could be influenced and potentiated by different factors such as the plant genetics, e.g. the use of corn germplasm not adapted to local conditions [12]. After that, during the growing and harvesting stages, toxin evolution is predisposed by agricultural practices, including the use of fungicides and pesticides, the use of open-pollinated varieties [13], the contact with aflatoxin-producing fungi or its spores, weather conditions and climate during planting and growing and, finally, insect damage.

Moisture and temperature play a significant role in fungi growth and the production of aflatoxins. Mycotoxin-producing fungi frequently need higher moisture levels (20.0–25.0 g/100 g) for infection during the pre-harvest phase in the field than fungi that proliferate during storage (13.0–18.0 g/100 g) [14]. Agricultural practices that have bearing over crop susceptibility toward infection and contamination include the variety of crops that are planted, the planting date, crop rotation (e.g., avoiding corn as a pre-crop for wheat), and tillage (plowing reduces inoculum from plant residues) [15].

It is worth clarifying that the presence of aflatoxin-producing fungi such as *Aspergillus parasiticus* or *Aspergillus flavus* in plants or the field environment does not necessarily imply the contamination of the crops with the toxin. For the production of aflatoxins, the molds need some stress factors such as nutritional imbalance, drought, or water surplus [16].

Climate plays a relevant role in fungal development and aflatoxin production in crops in the field and during storage [16]. However, in an epidemiological study conducted in our laboratory, 968 samples of animal feed and feed ingredients produced or stored (imported products) in Costa Rica were analyzed for aflatoxins (AFs), in the period 2010–2016. We did not found a direct correlation between aflatoxin concentration and the mean temperature, relative humidity, average rain precipitation, and the number of rainy days for a specific month during the same period in this country [17]. These findings together with the descriptions made by others authors [18] show how difficult it is to predict aflatoxin contamination starting from weather conditions only. The substrate or the ingredient that comprises an animal feed is the most important factor in the fungi growth and mycotoxin production mainly due to its nutritional composition [19].

The fungi growth in cereals and animal feeds after harvest during transportation or storage are also influenced by the temperature, humidity, water activity (a_w), the integrity of the grain, insect damage, and the quantity and type of the mycobiota [16]. The increase of the humidity in cereals and feeds during transportation and storage could favor an increment of aflatoxin concentration in these products [2]. Furthermore, the geographic origin, the transportation route, and the area where the feedstuff is stored, and the length of storage together with particular climate conditions will have a significant impact on aflatoxins concentration and animal exposure to this toxin. Due to this, conditions such as geographic region, temperature, humidity, and duration should be taken into account when comparing mycotoxins analysis from raw feed ingredients or in the prediction of aflatoxins contamination in finished feed [19].

Not only cereals *per se* are necessary components of the animal diets but also the by-products of these grains are commonly used to feed animals [20, 21]. Mycotoxins are resistant to majority of food processing techniques. Nevertheless, food processing such as milling, production of ethanol fuels, and beer brewing could affect mycotoxins distribution and concentration [22–24]. These mycotoxin concentrated fractions are usually employed in animal diets as is the case in rice milling process where several by-products (e.g. rice hulls, rice bran, chipped rice, rice polishings) are used as animal feed ingredient [21]. Also, we demonstrated that during the production of cheese, the aflatoxins M_1 is concentrated in whey which is frequently used to feed young animals or as a feed ingredient by its own right [25].

2.2. Aflatoxins regulations and surveillance in feedstuffs

Worldwide many countries have regulations concerning the maximum concentration of mycotoxins that could be present in food and feed. However, there are no regulations or guidance levels for all mycotoxins known so far. Aflatoxins, some type A and B trichothecenes, zearalenone, fumonisins, and ochratoxin, compounded the mycotoxins with regulatory or guidance levels, due to their demonstrated toxic effects on animals and humans.

Many aflatoxin regulatory levels are set depending on the particular agricultural commodity or compound feed/food, the type, and age of animal which will consume it and the intended use. Many countries base their regulations on the guidelines established by the European Union (EU) (**Table 1**) or by the United States Food and Drug Administration (FDA) (**Table 1**). Guidelines sometimes differ from each other; in most of the cases, the maximum allowed

US FDA		
Intended use	***Grain, grain by-product, feed or other products***	***AFB_1 maximum level ($\mu g\ kg^{-1}$)***
Immature animals	Corn, peanut products, and other animal feeds and ingredients, excluding cottonseed meal	20
Dairy animals, animals not listed above, or unknown use	Corn, peanut products, cottonseed, and other animal feeds and ingredients	20
Breeding cattle, breeding swine and mature poultry	Corn and peanut products	100

US FDA		
Intended use	***Grain, grain by-product, feed or other products***	***AFB_1 maximum level ($\mu g\ kg^{-1}$)***
Finishing swine 100 pounds or greater in weight	Corn and peanut products	200
Finishing (i.e., feedlot) beef cattle	Corn and peanut products	300
Beef, cattle, swine or poultry, regardless of age or breeding status	Cottonseed meal	300

European Commonwealth	
Matrix	*AFB_1 maximum level ($\mu g\ kg^{-1}$)*
All feed materials	20
Complete feedingstuffs for cattle, sheep and goats (except dairy animals)	20
Complete feedingstuffs for dairy animals	5
Complete feedingstuffs for calves and lambs	10
Complete feedingstuffs for pigs and poultry (except young animals)	20
Other complete feedingstuffs	10
Complementary feedingstuffs for cattle, sheep, and goats (except complementary feedingstuffs dairy animals, calves, and lambs)	20
Complementary feedingstuffs for pigs and poultry (except young animals)	20
Other complementary feedingstuffs	5

Table 1. FDA and EU aflatoxin regulatory guidance for feed and feed ingredients.

content of aflatoxins is lower in the regulations given by the EU than in those granted by the FDA. For example, the limit for aflatoxin in dairy feed is set by de EU in 5 $\mu g\ kg^{-1}$ and by the FDA in 20 $\mu g\ kg^{-1}$.

Finally, other international standards have been implemented by several organizations such as Codex Alimentarius Commission (CAC). There is no CAC standard dealing with aflatoxins in animal feeds but three main policies are included in this matrix including Codex General Standard for Contaminants and Toxins in Food and Feed (CODEX STAN 193-1995) concerned with hazards in feeds, CAC Codes of Practice for Reduction of Aflatoxins for Milk-producing Animals (CAC/RCP 45-1997), and CAC Codes of Practice for Good Animal Feeding (CAC/RCP 54-2004).

On the other hand, regional legal limits for aflatoxins have also been established; for example, the Southern Common Market (MERCOSUR) and Australia/New Zealand have harmonized maximum limits. Other regional bodies such as the Association of Southeast Asian Nations (ASEAN), the Economic Community of West African States (ECOWAS), and the Common Market for Eastern and Southern Africa (COMESA) are in the process of harmonizing legal thresholds.

2.3. Aflatoxins prevalence in animal feed and animal feed ingredients

In the analyses of the aflatoxin prevalence in finished feed, the difference in the raw material available in the diverse world regions, the difference in the nutritional requirements (energy, proteins, mineral, and vitamins) of each animal species, and the global trade of feedstuff should be taken into account. Ingredient diversity in a feed formulation is crucial for the livestock industry. Feed costs account for two-thirds or more of total live costs in pig and poultry production [19, 26, 27].

Country	Commodity	Number of samples	Total AF incidence, % (AFB_1)	Total AF mean concentration, μg kg^{-1} (AFB_1)	Maximum, μg kg^{-1} (AFB_1)	Reference
Global survey (America/Europa/Asia)	Corn, soybean, wheat and finished feed	4627	33	21	6105	[28]
Global survey (Myanmar)	Various feed	11,967	26	57	6323	[148]
Global survey (Vietnam)	Corn	10,172	27	16	6105	[149]
Africa						
Africa (South Africa, Nigeria, Kenya and Ghana)	Grains, feed	177	47	42	556.4	[150]
Ethiopia	Dairy feed	156	(100)	–	(419)	[151]
Jordan	Poultry feed ingredients	105	(19.04)	–	(17.06)	[40]
Jordan	Poultry feed	52	(24)	–	(12.7)	[40]
D.R. Congo	Corn	50	32 (32)	10.33–20.64	103.89 (51.23)	[152]
Kenya	Dairy feed and forages	74	(56)	47.84	147.86	[153]
Rwanda	Animal feed	27	–	100.4–168.6	265	[154]
South-Western Nigeria	Fish feed	94	(92)	–	(826.98)	[155]
South Africa	Compound feeds	92	30	9.0	(71.8)	[156]
Malawi	Corn	90	20.1	8.3	140	[157]
America						
North America	Finished feed	21	24	7	56	[28]
South America	Finished feed	203	26	2	83	[28]
Argentina	Poultry feed	49	86	2.68	37.67	[158]
Argentina	Fish feed	28	50	2.82	8.91	[159]
Brazil	Corn	148	4–23	3.1–16.37	49.9	[160]

Country	Commodity	Number of samples	Total AF incidence, % (AFB_1)	Total AF mean concentration, μg kg^{-1} (AFB_1)	Maximum, μg kg^{-1} (AFB_1)	Reference
Brazil	Corn	74	(16)	(<0.8)	(3)	[161]
Brazil	Poultry feed	36	(0)	(<0.8)	(<0.8)	[161]
Brazil	Fish feed, soybean bran, corn bran, other cereals	54	16.7–60	1.1–7.4	19.1	[162]
Costa Rica	Feed and feed ingredients	968	23.9	–	290.4	[36]
Costa Rica	Dairy feed	112	21	20.6	439.2	[17]
Venezuela	Pig feed	23	65 (26)	–	6.84	[163]
Asia						
North Asia	Finished feed	622	20	5	225	[150]
South-East Asia	Finished feed	465	81	23	431	[150]
South Asia	Finished feed	127	95	91	2454	[150]
China	Feed and feed ingredients	127	(63–100)	3.4–20	18.1	[164]
India	Livestock feed	48	(33.3)	32	60	[165]
India	Feed ingredients	49	(24.5)	62	–	[165]
Korea	Poultry feed	20	100 (100)	0.56 (0.38)	1.86 (1.70)	[81]
Pakistan	Poultry feed ingredients	77	(60)	(37.62)	(56)	[166]
Pakistan	Poultry feed	410	(44.39)	(23.75)	(78)	[166]
Pakistan	Poultry feed					
Europe						
Central Europe	Finished feed	45	2	0	1	[28]
Southern Europe	Finished feed	47	66	3	103	[28]
Turkey	Feedstuff	76	(26.32)	(1.02)	(11.37)	[33]
Turkey	Feed	30	(56.66)	(0.26)	(3.31)	[33]
Turkey	Dairy cow feed	76	26.3 (26.3)	2.74 (2.25)	8.43 (6.90)	[29]
Turkey	Cattle and lamb-calf feed	180	60	10.72	116.83	[30]
Oceania						
Oceania	Finished feed	75	9	0	9	[28]
Oceania	Wheat	109	5	2.0	30	[28]
Oceania	Corn	11	18	3.0	5	[28]

Table 2. Aflatoxin occurrence in feed and feed ingredients worldwide (data published 2012–2017).

There are highly sensitive methods for the analysis of aflatoxins; this could lead to the observation of a high percentage of aflatoxin positive samples in surveys that are not always directly related with a high risk for animals and human health. However, the synergistic/additive effect of some mycotoxins should be taken into account even in the case of low aflatoxin concentrations. **Table 2** shows a summary of aflatoxin surveys data worldwide in feed and feed ingredients published between January 2012 and February 2017.

Between January 2009 and December 2011, Rodrigues and Naehrer carried out a survey on mycotoxins occurrence worldwide in which 4 627 samples of corn, soybean meal, wheat, and finished feed were analyzed [28]. The global prevalence of aflatoxin positive samples and the mean concentration in this survey were 33% and 21 μg/kg, respectively; some of the results of this study are shown in **Table 2**. In this review, the major percentage of positive samples in finished feed found in South Asia and South-East Asia were 95 and 81% with a mean concentration of 91 and 23 μg kg^{-1}, respectively. Furthermore, in finished feed in South Asia, an extremely high level of aflatoxin (2 454 μg kg^{-1}) was found. In addition, in some regions of Asia the presence of aflatoxins in corn has been found to be as high as 82% of positive samples. Soybean meal showed a relatively minor susceptibility to aflatoxin contamination.

Another example of a global mycotoxins survey was carried out by Kovalsky et al., between 2012 and 2015, in which 1 113 samples of finished feed, corn, and corn silage were analyzed [6]. The authors found that the majority of samples showed an aflatoxin concentration below established guidelines for animal feed, and only a few samples from Africa and Europe presented levels exceeding the 20 μg kg^{-1} limit.

There also a few recent national surveys in regards to mycotoxins occurrence in animal feed; some of their most relevant results are summarized here and in **Table 2**. A recent study in Turkey by Sahin et al. found that from $n = 76$ cattle feed samples, 26.3% of them exhibited some level, 26.3% of samples exhibited some level of contamination [29], with only two samples exceeding 5 μg kg^{-1}. They did not detect any aflatoxins in ingredients such as sugar beet pulp, alfalfa silage, vetch silage, wheat bran, straw, and cottonseed samples. Kocasari et al. analyzed several toxins including aflatoxins in dairy cattle, beef cattle, and lamb-calf feed ($n = 180$ each) and found that 61.7% ($n = 37$), 55% ($n = 33$), and 63.3% ($n = 38$), respectively, contained considerable levels of aflatoxins ranging from 3.82 to 116.83 μg kg^{-1} [30]. However, it is important to indicate that the data were gathered using a screening assay. There is evidence, including our own, that seems to indicate that when ELISA is substituted by a confirmatory method such as HPLC, prevalence both in percentage and maximum values attained usually decrease probably due to issues with sensitivity and removal of possible false positive results. For example, Ghali et al. detected aflatoxins in 76.4% ($n = 58$) of the sorghum samples analyzed with an average level of 22.3–20.4 μg kg^{-1} using ELISA [31]. Meanwhile, the same research group found 62% prevalence in sorghum ($n = 58/93$) using HPLC [32].

In another study, AFB_1 was detected in 34.9% ($n = 37/106$) feedstuff and feed samples up to levels of 11.4 μg kg^{-1} [33]. A study conducted by Warth et al. in Burkina Fasso and Mozambique found a prevalence of feed samples assayed of 100% (n = 4/4) and 60 (n = 6/10), respectively [34]. The same research group also analyzed corn and sorghum samples from this region with incidences as high as 50% (n = 13/26). It is relevant to note that, in this type of assays, a small sample number may hinder reaching a conclusion regarding the region tested. However, it should be taken into account that minor subsets are usual during these types of surveys considering the costs of such analysis, especially those based on HPLC assays. Other research groups have also reported prevalence data from different countries in dairy feed including: Portugal (22% [35]), Costa Rica (33% [36]), China (42% [37]), Tanzania (65% [38]), and Iran (82.5% [39]) (see **Figure 1**). These differences might be due to geographical differentiation, climate, and seasonal variations, feeding systems applied, farm management, and feed storage practices. Research indicates that stricter vigilance systems encourage feed industry to have control over the ingredients used and better administration and prevalence to diminish [17, 40].

Elevated levels of contamination can be achieved if wrong management of feed ingredients has happened at any point during harvesting, storing, or processing. For example, when Kana et al. analyzed corn and feeds in central Africa, in this study, corn was found to be a relevant source of aflatoxins and the mean values of moisture (14.1 g/100 g) for this ingredient was significantly higher when compared to other commodities tested [2]. In the case of Costa Rica, for example, n = 15 samples, recollected along the country during the first trimester of 2016, were found to average (13.29 ± 0.28) g/100 g of the nutrient. In this regard, current climate change is expected to affect the behavior of aflatoxigenic fungi and contamination of crops, an excellent review regarding how climate changes mycotoxin behavior was written by Paterson and Lima [41].

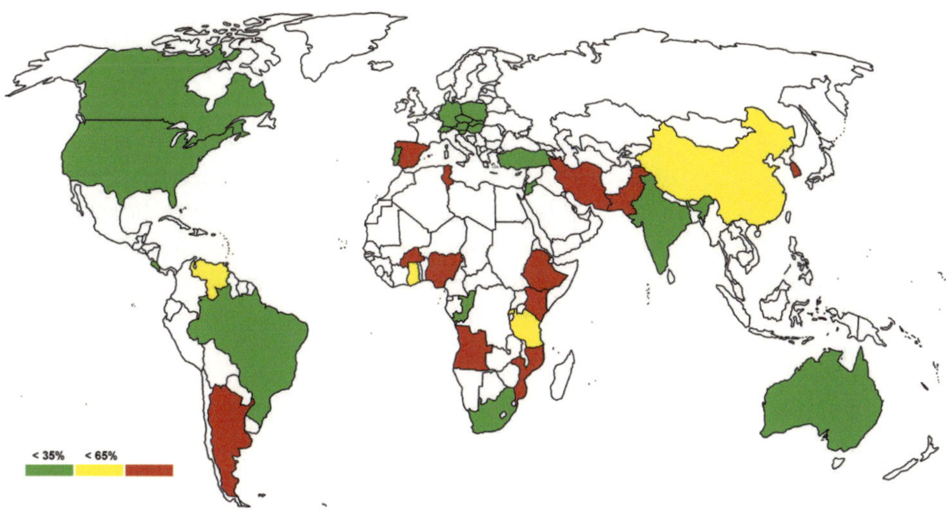

Figure 1. Worldwide prevalence for aflatoxin, expressed as percentages. Based on scientific reports from each country.

3. Effects of aflatoxins on food-producing animals

Dietary aflatoxins have shown detrimental effects on animal health and production. The most common exposure route occurs by ingestion of contaminated food. For example, fungal growth under right conditions may carry the genetic battery for toxin production and can contaminate cereals (e.g., corn kernels) which are used as a feed ingredient and, in turn, reach animal farms (**Figure 2A**). Other exposition routes include dermal contact and inhalation. Aflatoxins can affect animals either individually or additively (in the presence of more than one mycotoxin) and may affect various organs and systems [42].

Mycotoxins have a substantial economic impact because all participants of the production chain as farmers, cereals and grains producers, handlers and distributors, crop processors, and consumers suffer losses. Direct effects include increased veterinary care costs, reduced livestock production, and the continuous detriment of food and feed safety features. Also, public health should be another consideration because of the presence of dangerous and undesirable contaminants in animal products.

The disease called aflatoxicosis causes acute and chronic presentation in animals. Acute aflatoxicosis causes death and chronic aflatoxicosis results in cancer, toxicity, and immune suppression. The liver is the primary target organ. AFB_1 is a potent carcinogen [43] by bioactivation of cytochrome P_{450} in the liver and AFB_1-8,9-epoxide (AFBO) production. AFBO is needed for carcinogenic and toxic activity [44].

Aflatoxins susceptibility depends on species, age, gender, and nutrition; there are individual variations in the rate of activation of aflatoxins in various species. Metabolism of AFB_1 involves oxidative reactions by members of the CYP450 family of isoenzymes. There is a variety of metabolizing enzymes in animal species. In poultry species, CYP2A6, CYP3A37, CYP1A5, and CYP1A1 play a significant role in the biotransformation of AFB_1 [45, 46]. In humans, CYP3A4 in the liver and CYP2A13 in the lung have significant activity in metabolizing AFB_1 to AFBO (**Figure 2B**). The rate of AFBO formation and its conjugation with glutathione to reduce the toxicity by glutathione-S-transferase (**Figure 2B**), seem to be an important parameter in interspecies and individual differences [47, 48]. Hence, AFB_1 can cause hepatocellular carcinomas (**Figure 2B**). Cytochrome P_{450} involvement, 1A2 (responsible for AFM_1 biosynthesis) and 3A4 result in epoxide formation that leads to non-enzymatic oxidations which turn DNA into a mutagenic prone DNA adduct (encompassing mutations of p53 [activation of ras-protooncogenes], leading to mutagenicity) (**Figure 2B**). Ultimately, the DNA adduct is unstable and suffers renal elimination, for example, through conversion to aflatoxin *N*-acetylcysteine.

Rabbits are among the most sensitive animals to the toxic effects of this contaminant, followed by ducks, turkeys, and chickens which are still very sensitive, fish and swine are somewhat susceptible, and cattle and sheep are the most resistant. There are differences between genders, Lozano and Díaz reported male birds to produce more AFBO than females; turkey and duck yield more than chickens and quails [49]. Younger animals are more sensitive to AFB_1 than older individuals [46].

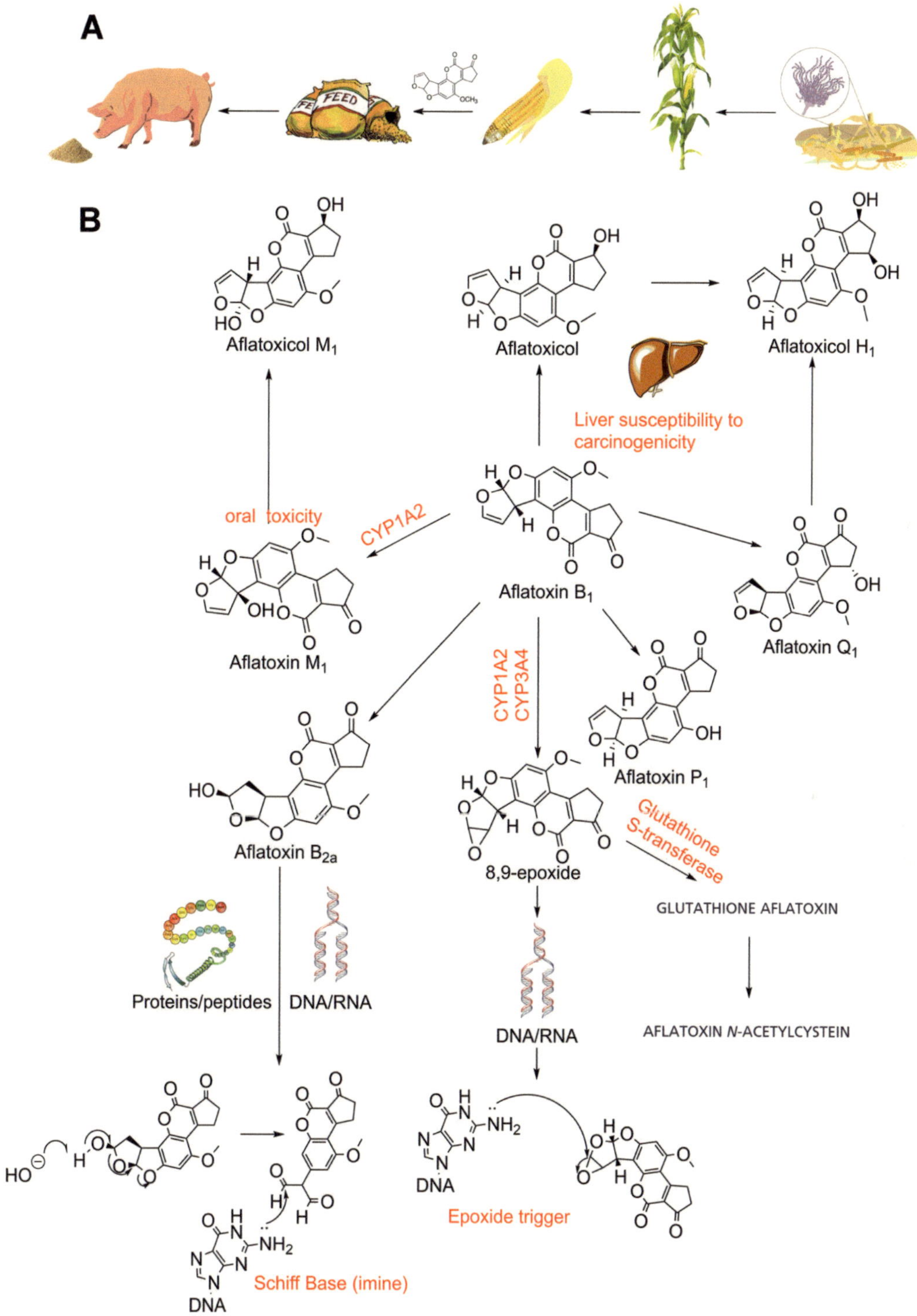

Figure 2. (A) Representation of the usual aflatoxin contamination route for grains and (B) several steps of aflatoxin metabolism.

Diet may have both positive and adverse effects on aflatoxin toxicity. Unfavorable results vary and depend on the frequency and source of the contaminated feed ingredients used, the inclusion percentage in the feed, the exposition period, animal species, gender, and age. Some diet components can act positively by exclusion, sorbent mechanisms, and reduction of AFB_1 bioavailability in the gastrointestinal tract [50]. Burkina et al. reported some phytochemicals in nutrition may act inhibiting the enzymes catalyzing AFBO synthesis [51].

The diagnosis of aflatoxins as etiological agents is trying even when mycotoxins are detected. Isolation and confirmation of mycotoxigenic fungal species in food and feeds do not, necessarily, indicate the presence of mycotoxins. Techniques for qualitative and quantitative analysis of mycotoxins vary in sensitivity and accuracy. Sampling could be complicated because there are myriad of factors affecting the production, distribution, or presence of mycotoxins; several products can be contaminated and sometimes it is not easy to identify which one is involved specifically. Also lesions and symptoms in acute and chronic aflatoxicosis are unspecific (immunosuppression, decreased weight gain, hepatic and kidney lesions, and death) and could be caused by other types of agents.

Appropriate diagnostic criteria, reliable sampling, and laboratory testing are still needed to select a correct approach. Prevention of mycotoxins contamination in animal feed is required to avoid losses in animal production and effects in public health.

3.1. Effects on pigs

Aflatoxins cause detrimental effects in health and production in swine. Reduction in weight gain and feed intake are among the first symptoms reported. Many researchers have also described diarrhea, bloody feces, and an increase in liver, kidney, spleen, and pancreas size [52–55].

Immune response to aflatoxins has been variable; intake between 120 and 180 μg of AFB_1 kg^{-1} of feed in combination with deoxynivalenol may not result in altered immune health [54, 56]. However, altered serum globulin patterns were reported by Mok et al. [55]. Low level of AFB_1 dysregulates the antigen-presenting capacity of porcine dendritic cells; it could explain the immunotoxicity of this mycotoxin [57].

Increased activities of liver-specific enzymes, abnormal histology, increased serum alkaline phosphatase, and γ-glutamyltransferase has been observed in exposed pigs [54, 55].

Pregnant sows treated with 1–3 mg kg^{-1} of AFB_1 showed anorexia, jaundice, loss of body weight atrophied spleen, and depletion of lymphocytes in germinal epithelium area. Liver revealed hypertrophy of the bile duct epithelium, fibrosis, and adenoma, kidney showed intertubular hemorrhages and atrophy of the glomeruli [58]. A great review exploring the effects of aflatoxins on swine reproduction was written by Kanora and Maes [59].

Stojanac et al. reported acute intoxication in a commercial farm [60]. From Piglets of 21–23 days old, died in 7 days, researchers found 960 μg kg^{-1} of AFB_1 in the compound feed and 870 μg kg^{-1} in sow's milk. After removal of the contaminated feed, the number of deaths began to reduce; the clinical symptoms were apathy, depression, cachexia, move reluctance, and death.

Finally, Azevedo demonstrated that pigs fed 1.0 mg AFB_1 kg^{-1} feed for 21 days had reduced growth performance associated with altered hepatic gene expression (specifically, cytochrome P450-2A19/CYP2A19 and glutathione S-transferase theta 1/GSTT1 [61]. Furthermore, the authors concluded that supplementation of 100 mg curcumin kg^{-1} to diets containing AFB_1 had a protective effect on changes in gene expression in liver of pigs.

3.2. Effects on ruminants

Ruminants are more resistant to the mycotoxins than non-ruminants animals because the rumen microbiota is capable of degrading toxins. However, aflatoxins are only partly degraded by ruminal flora resulting in a secondary toxic and carcinogenic metabolite called aflatoxicol.

In the case of cattle, sheep, goats, and deer, aflatoxins consumption cause reproductive problems, immune suppression, decrease on milk, beef or wool yield, and reduced feed utilization.

Aflatoxins have been shown reduced feed efficiency in cattle; growth can be altered when ruminants consume contaminated feed for extended periods of time. AFB_1 (600 μg kg^{-1}) was shown to depress feed efficiency and rate of gain in steers [56]. It has been attributed to compromise ruminal function by reducing cellulose digestion, volatile fatty acids production, and rumen motility. Acute exposure to aflatoxins causes inappetence and lethargy [62].

Aflatoxin levels between 100 and 1 000 μg kg^{-1} within the diet, cause a decrease in rumen motility, feed efficiency, growth inhibition, and an increase in liver and kidney weight. In lactating dairy cows, researchers report milk production decrease and reduced reproduction efficiency [5]. Embryotoxicity has been reported in animals consuming low dietary concentrations of mycotoxins [56].

In cattle, aflatoxins affect the immune system function by many mechanisms such as inhibition of lymphocyte blastogenesis; AFB_1 suppress mitogen-induced stimulation of peripheral lymphocytes. Chronic exposure can interfere with vaccine-induced immunity [62].

Aflatoxins affect the milk quality. Cows metabolize AFB_1 to form the monohydroxy derivative, aflatoxin M_1 (AFM_1), which is secreted into the cow's milk. AFM_1 is a potential human carcinogen very resistant to thermal treatments such as pasteurization and freezing. The European Commission Regulation 1881/2006 sets a maximum limit of 0.05 μg kg^{-1} for AFM_1 in raw milk, heat-treated milk, and milk for the manufacture of milk-based products (EC 2006). Nevertheless, higher levels have been found [63], for example, Škrbić et al. detected the maximum AFM_1 level of 1.44 μg kg^{-1} with a mean value of 0.30 μg kg^{-1} in commercial milk samples in Serbia [64].

In sheep, high levels of aflatoxins resulted in heptotoxicosis, nephritic lesions, and mineral metabolism alterations. In lambs, 2.5 mg $kg^{-1}$$AFB_1$/diet have been reported low feed intake, weight gain, and altered blood parameters [5].

3.3. Effects on poultry

Aflatoxin B_1 has a high range of effects in poultry including acute hepatic toxicity, teratogenicity, carcinogenicity, mutagenicity, hematological problems [65], and immunosuppression.

Poultry is sensitive to low levels of AFB_1, in order of sensitivity: ducks > turkeys > Japanese quail (*Coturnix japonica*) > chickens [45].

Exposure to aflatoxins has been demonstrated to suppress the immune response in poultry. Both, Rawal et al. and Xi Peng et al. have reported impaired T cell production, decreased phagocytosis and apoptosis in thymus, and bursa of fabricius and spleen [66, 67]. Kumar and Balachandran reported spleen lymphoid and erythroid depletion, enlargement, pallor or yellowish livers, crop and proventricular changes, enlarge, pale and congested kidneys in broiler fed with 1 mg kg^{-1} AFB_1 [68].

Aflatoxins exposition could be a serious risk to animal health, increasing susceptibility to infections, or reducing vaccination efficacy. Epidemiological data indicate a high correlation between outbreaks of Newcastle disease and AF contamination of broiler rations [69].

Changed serum biochemical parameters, impaired hepatic antioxidant functions, and severe lesions in hepatic tissues were found by Yang et al. in broilers fed with 36.9–95.2 μg kg^{-1} ABF_1 [70]. They also observed focal necrosis of hepatocytes, biliary hyperplasia, Kupffer cell hypertrophy, microvesicular fatty degeneration, and apoptosis.

Gross findings in broilers, include paralysis and lying down could be observed, the growth of affected birds is retarded. Additional findings include the yellowish to a yellow-earth color of the liver, the multiple hemorrhages, and a characteristic reticular appearance of the capsular surface. In severe intoxications, the kidneys are enlarged and filled with urates.

Our data also demonstrate abnormal fatty tissue accumulation and hepatic lesions including a suggestive increase in liver size, with the loss of usual color (dark brown), pallor, with visible areas of hemorrhage primarily on the left lobule without gallbladder distension (**Figure 3A**), when chickens were subjected to feeds contaminated with aflatoxin. On the other hand, chicks that were fed with an aflatoxin/T-2 toxin diet exhibited a reduced liver size, greater hepatic paleness, and nodular appearance, without bleeding, cholestatic pattern, or gallbladder distension (**Figure 3B**).

Clinical symptoms seen in poultry are diverse. Hussain et al. reported experimental birds intoxicated with 400–800 μg kg^{-1} AFB_1 showed depression, ruffled feathers, watery feces, decrease in water and feed consumption, and nervous signs as torticollis and mortality [71].

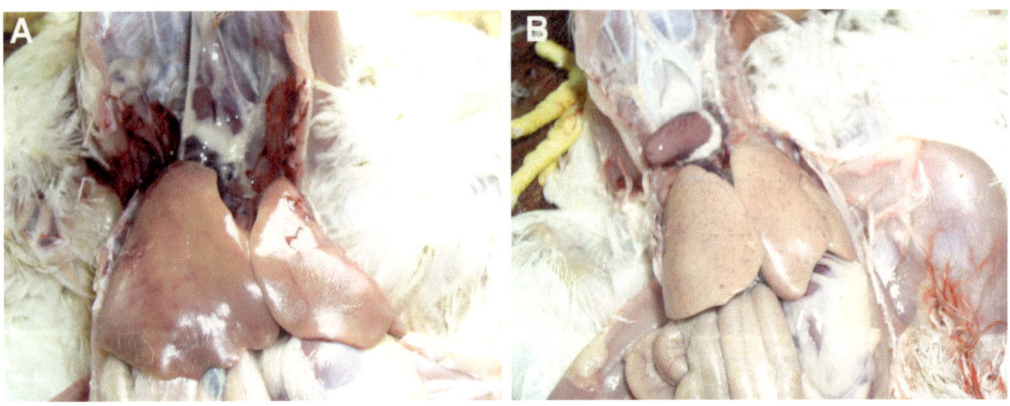

Figure 3. Chicken liver lesions when subjected to (A) 50 μg kg^{-1} aflatoxin diet and (B) 50 μg kg^{-1} aflatoxin plus T-2 toxin diet.

Trebak et al. reported listlessness, anorexia [72]; other symptoms include poor feed utilization, stunted growth, decrease weight gain [73, 74]; reduced egg weight and production. High levels of aflatoxins in broilers and turkeys cause hepatomegaly, fatty degeneration, fatty liver, bile conduct proliferation, periportal fibrosis, renal petechiations, tubular nephrosis, interstitial nephritis, and splenic atrophy [67, 75].

Aflatoxins may cause blood coagulations disorders in broilers characterized by extensive hemorrhagic lesions in the stomach, heart, intestines, lungs, kidneys, and muscles resulting in death. Lesions are causal for condemnations in a slaughterhouse. Prothrombin time (PT) is an indicator of aflatoxin toxicity in chickens, the elongation of which is directly proportional to aflatoxin dose and exposure time. PT is an indicator of the activity of blood coagulations factor V, VII, IX, X, prothrombin, and fibrinogen can serve to diagnose liver lesions in poultry [76].

AFB_1 also affect laying hens; losses are pronounced regarding reduced egg production and egg quality as a result of contamination with aflatoxin residues in eggs and muscles. Feed to egg AFB_1 transmission ratio is approximately 5 000:1 [74]. A substantial percentage of the egg samples (28%) showed AFB_1 levels (0.79 ± 0.45 μg kg^{-1}) in commercial eggs [77]. Several authors, reported excretion of aflatoxin B_1 residues in hen's eggs might occur at relatively low concentrations under long-term exposure of laying hens to AFB_1 at different levels up to 50 μg kg^{-1} in a naturally contaminated feed [78–80]. Interestingly, even though Lee et al. found the prevalence for mycotoxins to range from 85–100% in Korean poultry feed samples (n = 20), but they failed to find contaminated egg samples (n = 275) aflatoxins, ochratoxins, or zearalenone [81]. Thermal processing was not useful for detoxification of AFB_1 in eggs [79, 82]. Some researchers have found a significant decrease in feed consumption, egg production, egg weight, shell weight, shell thickness, and feed conversion ratio value in laying hens fed with 15 μg kg^{-1} of AFB_1 [78, 79]. Aflatoxins disrupt the hypothalamic regulation of neuropeptides involved in feeding behavior and contribute to the lower body weight and decreased weight gain [72]. Aflatoxins in the feed of laying hens may cause a relevant lesion in liver, kidneys, heart, and ovaries. The ovaries show follicular atresia, which has a detrimental effect on egg production [79].

Effects of AFB_1 on the absorption of nutrients have had variable results. Mycotoxins can compromise different functions of the gastrointestinal tract such as decreased surface area available for nutrient absorption, modulation of nutrient transporters, loss of barrier function, and facilitating persistence of intestinal pathogens inflammation [83]. However, it is still unclear how the intestinal lesions affect growth and feed efficiency in poultry.

Kalpana et al. found enrofloxacin, and ciprofloxacin residues in liver, kidney, skin, and fat persisted for 10 days in mycotoxin-exposed broiler chickens, whereas it was detectable only in the liver of unexposed broiler chickens, indicating that subchronic AFB_1 exposure markedly influences the residue levels of enrofloxacin in tissues of broiler chickens [84].

Finally, in an interesting report, Iheanacho tested the cytotoxic effect of cattle and poultry aflatoxin-contaminated compound feed extracts on human lymphocytes [85]. The authors observed that cell viability significantly decreased upon contact with feed extracts, especially those from poultry feed, after just 24 hours of exposure, demonstrating that a direct link may be found between human toxicity and feed.

3.4. Effects on other species

Marine animals could be exposed to AFB_1 contamination through feed chain [86]. The carcinogenic effect of AFB_1 has been studied in fishes such as salmonids, rainbow trout, channel catfish, tilapia, guppy, and Nile tilapia. Consequences of mycotoxin toxicity in fish do not differ from other animal species. Effects are directly related to losses in production, reduced weight gain, feed conversion, and immune impairment. Kidney, liver, and muscles lesions and residues are found in different species of fish [87].

Cagauan et al. found varying levels of aflatoxin contamination did not significantly affect the final average length, weight, and gain in weight of Nile tilapia; aflatoxin negatively influenced percent survival of fingerlings [88]. External manifestations in fish were eye opacity leading to cataract and blindness, lesions on the body surface, fin and tail rot, yellowing of the body surface, abnormal swimming, feeble and stationary on one place, and reduced appetite. In common carp fingerling (*Cyprinus carpio*) levels of 50 and 100 $\mu g\ kg^{-1}$ of aflatoxins in the feed affected growth and accumulate in fish tissues [89]. Interestingly, at least two studies have suggested that mycotoxins, such as AFs, can be present in seafood if fish were exposed to mycotoxin-contaminated feed [87, 90].

In horses, AFB_1 in the contaminated feed (58.4 $\mu g\ kg^{-1}$) cause jaundice, depression, lameness, anorexia, and death. Ponies have shown damage to the skeletal muscles and heart. Post-mortem lesions show enlarged livers, kidney damage, and bile duct hyperplasia [56]. An excellent review regarding equine health implications of the presence of aflatoxin in feed has been essayed by Caloni and Cortinovis [91].

Mycotoxins on companion animals could be severe and can lead to death. AFB_1 in dogs cause hepatitis and severe depression, anorexia, and weakness. Aflatoxins and other mycotoxins have been found in the ingredients and final products of pet food. Gazzotti et al. found aflatoxins contamination in 88% of the dog food samples, showing concentrations of 5 $g\ kg^{-1}$ [92]. Dog food contaminated with aflatoxins is of particular concern due to the bond companionship animals, or pets usually share with their owners. Frehse et al. not only found a high prevalence of aflatoxins in the commercial feed but also found that of AFB_1, AFB_2, AFG_1, and AFG_2 associated positively with mammary tumor growth in female dogs and that neutering was a protective factor for mammary cancer [93].

4. Control and management approaches

Mycotoxins are toxic metabolites that can contaminate various crops before or after harvesting. Aflatoxins are a problem also during storage, transport, processing, and handling steps such as manufacturing.

Prevention measurements are focused on the minimization of crop contamination before harvesting (plant breeding and good agronomic practices) and during storage or postharvest (detoxification). Several methods of prevention and control are available to reduce the contamination with aflatoxins. However, mycotoxin contamination of food and feed is unavoidable [94] mainly because they are ubiquitous nature and current standards are based on regulating the product, not the process. Available approaches are focused on minimizing and mitigating not to eliminate the contamination of both, fungus species and mycotoxins. None of the following methods reduces contamination in high-polluted feed ingredients and foods.

4.1. Pre- and postharvest feed and feed ingredients aflatoxin management: GMP and HACCP practices

Pre-harvest management of aflatoxins in animal feeds requires an approach based on good agricultural practices by the producer, appropriate legislations and regulation enforcement, constant monitoring of aflatoxins in feeds and foods, and adequate management of contaminated feeds.

Agronomic practices have been shown to have a substantial effect on toxin contamination of crops. The primary strategy should be to prevent mycotoxin production by reducing mold proliferation during cultivation and storage. Practices such as selection of seeds and planting of more resistant varieties of cereals; healthy and vigorous plants capable of withstanding pest attack are required. Molecular techniques are now available as a possible strategy to select varieties on their ability to resist mold attack [95]. Ostrý et al. described that Bt corn showed significantly lower concentrations of aflatoxins than non-Bt corn hybrids [96].

Crop residues are often the primary inocula of mycotoxigenic fungi; removal of agricultural waste is effective in preventing the contamination of follow-on crops [97]. Furthermore, selection of harvest seasons could be a critical approach, showing date partly determine the flowering time, if it coincides with spore release, more frequent and more sever attacks are likely. Early harvesting of groundnuts resulted in lower aflatoxin levels and the higher gross return of 27% than in delayed harvesting [98]. Crop planting should be timed to avoid elevated temperatures and drought stress during the period of seed development and maturation [99].

Other practices such as weed control, crop rotation, plowing, avoiding high plant densities and correct fertilization limits mold contamination and mycotoxin production. Appropriate use of pesticides during the manufacturing process could help in minimizing the fungal infection or insect infestations of crops [56]. Insects can act as fungal spore vectors and attack the grain of external teguments of kernel facilitating colonization of mycotoxin-producing fungi [97]. Dorner and Cole reported soil treatment with non-toxigenic strains of *Aspergillus* and use of competitive exclusion using bacteria and fungal strains of *Trichoderma* [100] had a beneficial carry-over effect of reducing aflatoxin contamination in crops.

Containers (e.g., wagons and trucks) to be used for collecting and transporting the harvested grain from the field to drying facilities, and, thereafter, to storage facilities should be clean, dry, and free of insects, birds, rodents, and visible fungal growth before use and reuse [99].

Reduction of grain damage before and during storage is important to avoid fungal invasion. Cereals should be dried in such a manner that damage to the grain is minimized and moisture levels are lower than those required to support mold growth during storage [99, 101]. Mixing grains and a long-time storage should be avoided. Grain damaged by mold should be burnt or buried [101].

Quality check of grain and installation integrity before storage and adequate storage conditions (temperature, humidity, moisture, and insect control) are required and must be monitored. Grains should be stored in less than 15 g/100 g of moisture content, at low temperatures and a low oxygen concentration (< 1 mL/100 mL). In tropical and subtropical conditions, grains are more prone to contamination than temperate regions due to favorable humidity and temperature levels for mold growth (10–40°C, pH range of 4–8 and above, 70% relative humidity) [101]. For example, in Turrialba, Cartago, Costa Rica (9°54′00″N 83°41′00″W), reported a mean temperature and relative humidity of (22.0 ± 0.7)°C and (87.7 ± 2.2)%, respectively.

In storage, many insect species can attack the grain and moisture that can accumulate from their activities providing ideal conditions for fungal activity and management of insect infestations which is required. Prevention of insect pest is desirable but the intensive use of chemical compounds has resulted in the evolution of resistant populations. Phosphine gas is a common and toxic fumigant used for disinfection of storage grains. Essential oils, application of ozone, and use of diatomaceous earth are alternatives to phosphine gas to control insect pest in storage grains.

The addition of antifungal agents, preservatives, antioxidants, essential oils, and controlled atmospheres, may help to reduce fungal growth during storage. Antioxidants such as selenium, vitamins A, C, and E, ethoxyquin, and butylated hydroxytoluene [102] have been recognized as anti-aflatoxigenic agents. Food components (fructose, phenolic compounds, coumarins, and chlorophyll) and food additives (piperine, aspartame, cyproheptadine, and allyl sulfides) have shown toxicity reduction of several mycotoxins [103]. Weak acids are used in animal food and feed to prevent fungal spoilage; the most common are propionic, benzoic, and sorbic acid.

Some essential oils have fungicidal actions such as carvacol, α-p-cymene, terpinolene, anethole, and eugenol. Esper et al. described a considerable AFB_1 reduction in corn, and their efficacy depended mainly on the essential oil concentrations and substrate water activity conditions, concentration, and incubation periods [104]. Hence, essential oils can find a practical and safe application in toxin control [105].

Modified atmospheres (low O_2 and high CO_2 concentrations) are used for fungal growth monitoring and mycotoxin production in stored grains. Silo-bags are also used. They are waterproof and have some level of gas-tightness (O_2 and CO_2). The use of ozone as a strategy to control toxigenic fungi and mycotoxins production needs further evaluations [102].

Hazard analysis critical control point (HACCP) system has been increasingly and successfully applied by the grain and feed industry to prevent and control risks associated with potential contamination with toxins [106]. Mycotoxins can be classified as a biological or a chemical hazard [102]; they fit in an HACCP program at appropriate critical points, and their critical limits must be identified. For example, a critical control point could be at the end of the drying process, and one critical limit would be the water content/water activity [99]. Also, FAO recommends the application of an HACCP program for the systematic control of mycotoxins through the entire food chain from field to consumption including all pre-harvest, harvest, and postharvest stages in the production of animal feed and animal feed ingredients. Additionally, FAO has published a manual to make easier the application of this mycotoxin control program (http://www.fao.org/docrep/005/Y1390S/Y1390S00.HTM).

The efficient and prompt drying of corn for medium- and long-term storage in hygienic silos free of insect pest and fungal populations and accurate and regular moisture content, water activity (a_w), fungal growth, insect presence, bacterial level, the percentage of grain damage, storage time, storage temperature, and humidity measurements must be considered in an HACCP program [101, 102]. Pre- and postharvest measures are paramount to avoid the risk of contamination in both feeds and foods; new trends in the decontamination of aflatoxins [107] should be considered as complete absence of such toxins which is extremely difficult. Lastly, as a case study, we highlight the work of Kamala et al. [108]. The authors examined three agro-ecological zones of Tanzania and determined that local postharvest management practices such as drying corn on a raised platform, sorting (damaged, discolored, and molded grains) and application of synthetic insecticides during storage, associated with less contamination of corn with aflatoxins and fumonisins.

4.2. Decontamination of mycotoxin-contaminated feed

There are different approaches to decontaminate or detoxify a feed or food commodity containing mycotoxins, among them the use of mycotoxin binders in the feed, enzymatic, or microbial detoxification. Some chemical substances have been assayed to reduce aflatoxins, especially ammonia. However, chemical detoxification is expensive and though permitted in some countries, is not so in Europe. Hence, the most common postharvest approach in the feed industry is the inclusion of sorbent materials in the feed to obtain selective removal of toxins by sorption during passage through the gastrointestinal tract [97]. The mycotoxin binders are also called adsorbents, mycotoxin binders, sequestrants, interceptor molecules, trapping agents, or enterosorbents. There are inorganic sorbents principally clay minerals and organic sorbents of microbial origin [42]. In some cases, they have the ability to bind mycotoxins and reduce their absorption across the gastrointestinal tract [109].

Decontamination process should include inactive mycotoxins, generate no toxic products, and guarantee no modification of nutritional properties of the feed or food. The properties of adsorbents are important in the evaluation of their efficacy: physical structure, effectiveness at different gastrointestinal pH levels (acidic and neutral), total charge, distribution, pore size, and surface accessibility should be considered. However, the diversity of mycotoxins chemical structures makes difficult that a single method can decontaminate an animal feed [42].

Mycotoxin characteristics such as polarity, solubility, molecular size, shape, charge distribution, and dissociation constants must be evaluated. Sorbents have been tested using *in vitro* and *in vivo* systems, *in vitro* studies are very common and *in vivo* tests [97] are used to find performance responses or biological markers such as tissue residues or changes in biochemical parameters to determine the effectiveness of binders. A suitable adsorbent or binder should have an unyielding bonding, so no washing or interactions in the digestive tract desorb the bound mycotoxins. Binder use and efficacy should be verified.

Silicate binders are divided into subclasses according to their structure; one group is the phyllosilicate family characterized by the sheet-type framework [97]. Hydrated sodium calcium aluminosilicates (HSCASs) are the most reported; they adsorb aflatoxin selectively during the digestive process, and it involves the formation of a complex by the β-keto-lactone or lactone system.

Other silicates studied are bentonites, zeolites, and clinoptilolites. Other mineral adsorbents include synthetic polymers such as cholestyramine and polyvinylpyrrolidone, indigestible dietary fibers also have absorbance effect. Mineral binders are efficacious against aflatoxins, but they are not very specific and can absorb other molecules such vitamins and others nutrients [110].

Organic substances such as humic acids have the ability to adhere mycotoxins, yeast, and yeast extracts are also able to reduce the aflatoxin effect. Parietal structures of some lactic acid bacteria have the potential to bind mycotoxins; the adsorption is reversible and could be performed with living or dead bacteria. Other biological materials such as fungal conidia have binder effect against AF, zearalenone, and ochratoxin A.

4.2.1. Efficiency of aflatoxin sorbents

The inclusion of different types of adsorbents especially clay minerals has been widely used in the feed and farm industry to counteract the mycotoxins toxic effects in animals [42]. The easy management and low inclusion requirement in feed make the use of adsorbents a standard practice. There are some studies about the protective effect of these sorbents in different animal species especially food-producing animals such as pig, poultry, and cattle using different mycotoxins and different concentrations and testing the various health and productivity parameters. These trials have shown variable results with more or less successful depending on the adsorbent, the mycotoxin, the species, and the parameters tested.

Mitchell et al. have reported that calcium dioctahedral smectite clay has the capability to adsorb mycotoxins in the gastrointestinal tract decreasing toxin bioavailability reducing biomarkers of exposure for AFB_1 as well as FB_1 [111]. Furthermore, other studies have reported the ability of "dioctahedral smectite" clay surfaces to strongly adsorb aflatoxins [112]. This ability is not associated with other clay groups such as kaolinites, attapulgites, zeolites, mica, alumina, and sand [42].

Among the sorbents used by the farm and feed industry are smectite clays, zeolites, kaolinite, mica, silica, and charcoal. Smectite or zeolite minerals with natural or synthetic surfactants giving hydrophobic organoclays or organozeolites are also used [113–115]. There are also sorbents of biological nature such as chlorophyllins, yeast products, lactic acid bacteria, plant extracts, and algae [42].

The aflatoxins adsorbents should be carefully tested trough *in vitro* and *in vivo* studies, and they should fulfill some safety and economic aspects such as stable and high adsorption capability with different mycotoxins, insignificant interactions with vitamins, iron, and zinc, low levels of metals dioxins/furans and other hazardous substances. The European Food Safety Authority (EFSA) has published guidelines pointing out the characteristics that the adsorbents should fulfill [116].

Dos Anjos et al. investigated the efficacy of three different aflatoxins adsorbents: bentonite clay, diatomaceous earth, and turmeric powder in broiler chicks feeding aflatoxins contaminated diets [117]. They found that birds fed with turmeric (without aflatoxins) presented lower body weight gain than control animals. The birds fed with AFB_1 and adsorbent bentonite clay did not experiment the decrease of feed intake and feed gain occurred in the birds fed with AFB_1. Birds fed with diet containing AFB_1, diatomaceous, and tumeric had poorer growth performance than those fed on AFB_1 alone. The toxicity effects and lesions in liver were not counteracted by any of the adsorbent treatments [117].

Commercial products based on this rationale are available, for example, Alltech® Mycosorb A+®. Sun demonstrated that diets with Mycosorb A+® (2 g kg^{-1}) could improve growth performance in swine by increasing average daily gain and average daily feed intake, whereas low-level aflatoxin (20 μg kg^{-1}) had minor effects on hematology without affecting growth performance [118]. On another hand, aluminosilicates, zeolites, and other chemosorptive agents have been assayed against aflatoxins with relative success. In a recent publication, Wongtangtintan et al. demonstrated that thai bentonite exhibited an excellent binding capacity toward AFB_1 surpassing commercial bentonite and activated charcoal *in vitro* [119]. Furthermore, the authors suggest that the adsorption behavior of AFB_1 on these toxin binders represented multilayer/multiple site adsorption on the binders' surfaces. An excellent review of experimental trials demonstrated different detoxification approaches in poultry feed had been written by Oguz et al. [65]. In broilers, a study performed by Denli et al. demonstrated that supplementation of AflaDetox® significantly ameliorated the toxic effects of AFB_1. The authors suggest that the addition of AflaDetox (1, 2, and 5 g kg^{-1} of feed) to diets containing AFB_1 significantly improved performance, counteracted the serum biochemical and histopathological changes, reduced the relative weight of liver, and also appeared to be effective in reducing the relative spleen weight [120]. Some data supporting the effectiveness of adsorbents must be considered with caution as in some cases, chemosorbent developers have participated, to some degree, in the research hence creating an apparent conflict of interest (see, e.g., Ref. [120]).

A study carry out by Neeff evaluated the efficacy of a HSCAS reducing aflatoxin residue in tissues of broiler chicks. The author found that with adding this adsorbent in the diet the concentration of aflatoxins residues in liver was lower than in birds consuming a diet contaminated with AFB_1 without HSCAS [121]. Despite this, as in the study carried out by Dos Anjos et al. [117], this adsorbent could not avoid the lesions in the liver associated with aflatoxicosis in broilers [121]. On the other hand, Fowler et al. did observe an improvement in broilers incorporating 0.2 g/100 g calcium bentonite clay additive (TX4) [122]. The additive effectively reduced the accumulation of AFB_1 in the liver, improving livability in birds fed aflatoxin.

In a previously study carry out by our research group, we evaluate three different mycotoxin adsorbents (HSCAS) in broiler chicken feed aflatoxins contaminated diet. We found little ameliorative effect of some parameters such as creatinine and alanine aminotransferase (ALT) in broilers fed with contaminated diet and the adsorbents compared with broilers fed only aflatoxins diets. However, we found a significant higher liver weight in broilers getting AFB_1 and two of the tested adsorbents in comparison with broilers getting only AFB_1 [123]. From the feed technology standpoint, Maki et al. demonstrated that 6 g calcium montmorillonite clay (Novasil Plus, NSP)/kg feed, can significantly decrease AFM_1 concentrations (up to 55% reduction) in milk without affecting dry matter intake (DMI), milk yield, milk composition, vitamin A, or riboflavin concentrations [124]. Similarly, Mugerva et al. demonstrated that 1 g/100 g of calcium bentonite and charcoal reduced AFM_1 carry-over in goats fed with contaminated feed while DMI and daily milk yield were not altered with treatment [125].

4.2.2. Novel approaches for tackling aflatoxin contamination

Since sorbents have demonstrated a limited capability in toxin management and preventive measurements are difficult to apply, new tactics to control aflatoxins are continually being developed. For example, Wee et al. suggested that use of zinc chelators (e.g., *N,N,N′,N′*-tetrakis(2-pyridylmethyl) ethane-1,2-diamine) has the potential of diminishing the capacity of *A. parasiticus* to produce toxins [126]. In fact, they observed significant inhibition of aflatoxin production but no detectable changes gene expression (i.e., *ver*1 and *afl*R). Furthermore, the authors demonstrated the efficacy of this approach in peanut and sunflower seeds. Weaver et al. used clay and yeast cultures conjointly to improve amelioration in aflatoxin and deoxynivalenol-contaminated swine feed [127]. Interestingly, Das et al. demonstrated that *Pleurotus ostreatus*, a fungus that can grow on different agronomic wastes, can synthesize several ligninolytic enzymes which are capable of degrading compounds including AFB_1 [128]. Additionally, the authors demonstrated that AFB_1 degradation occurs during co-cultivation of *A. flavus* and *P. ostreatus* in rice straw, a common feed for cattle. Similarly, Lee et al. also demonstrated *Aspergillus oryzae* (a microorganism used as a fermentation starter in Meju) capability for detoxification of AFB_1 [129]. Villers [130] detailed field experience governing the exponential growth of aflatoxins during prolonged postharvest storage of grains in tropical countries. In this case, the authors focuses on modern, safe storage methods to ameliorate mold development and subsequent aflatoxin production using UltraHermetic™ structures that generate an atmosphere incompatible with insect and microorganisms' survival, without further use of other additives. Bovo et al. evaluated the capacity of a beer fermentation residue (BFR) containing *Saccharomyces cerevisiae* cells to bind AFB_1 and counteract its toxic effects on performance, serum biochemistry, and histology of broilers. Feed intake, body weight gain and concentrations of albumin, total protein, and globulin increased in broilers fed aflatoxins contaminated diet with BFR in comparison with the broilers that only receive AFB_1. The BFR reduced the severity of histological changes in the liver and kidney caused by AFB_1 but not the effect on kidneys and liver weight [131]. Pizzolito et al. demonstrated a protective capacity of *S. cerevisiae* specifically against aflatoxins in poultry when added to feed and water [132].

Recently, our research group found that the milk proteins casein and the milk whey protein are capable to sequestrate aflatoxins M_1 *in vitro*; this bind capability should be further investigated and could be used in further AFM_1 detoxification intervention in the dairy industry [133].

Yin et al. demonstrated, using poultry feed as a substrate, that carvacol and *trans*-cinnamaldehyde inhibit *A. flavus* and *A. parasiticus* growth and downregulates aflatoxin synthesis genes (*afl*C, *nor*1, *nor*A, and *ver*1). Similarly, Nerilo and et al. demonstrated that *Zingiber officinale* fully inhibited aflatoxin production by *A. flavus* at a concentration of 15 μg mL^{-1} [134].

Furthermore, there are other detoxification approaches based on the transformation of the mycotoxin compounds using microorganism or enzymes. Nowadays, approaches in ameliorating toxin burden have relied heavily on biological methods. An excellent review on the subject was made recently by Ji et al. [135]. An additional point regarding detoxification relies on the fact that they must demonstrate their binding capacity both *in vitro* and *in vivo* through a report on this subject with the most recent advance written by Wielogórska et al. [136]. Jiang et al. demonstrated the efficacy of Bamboo charcoal as an agent capable of ameliorating AFB_1 on an *in vitro* rumen fermentation of a hay-rich feed mixture, the authors assayed 1.0 μg mL^{-1} and compared the effectiveness of this alternative to that of smectite [137]. A novel approach was introduced by Zhao et al. who detoxified peanut meal using solid state fermentation and *Zygosaccharomyces rouxii* from fermented soy paste [138]. The authors demonstrated nonviable cell binding and biotransformation of AFB_1 in which reduction was monitored by LC/MS. A recent pertinent study by Shar suggested that banana peel (*Musa* sp.) may be used as bioadsorbent for AFs and ochratoxin A *in vitro* [139]. Using thermodynamic properties of adsorption, the authors demonstrated that sorption was not affected by low pH, simulating conditions of the gastrointestinal tract, and, even, suggested to incorporate this by-product in animal feeds as economic sorbent.

Finally, evidence suggests that the oxidative stress is a key factor in aflatoxin-related pathology, specifically the role of glutathione [140]. In fact, Jardon-Xicotencatl et al. using neutral electrolyzed oxidizing water demonstrated that lipid peroxidation and oxidative damage (based in glutathione modulation) are reduced when aflatoxin-contaminated corn is treated [141]. Hence, animal antioxidative balance is paramount to counter, detoxify, and ameliorate aflatoxin burden. Then, from the nutritional standpoint, there is room to improve diets and feed formulations using effective antioxidants, which are usually overlooked.

4.3. Aflatoxins and climate change

We already established that aflatoxin production is dependent on multiple environmental factors including temperature and humidity. Hence, climate change intrinsically forces a new dynamic in those naturally produced contaminants. Countries in the tropical fringe, such as Costa Rica, are experiencing an increase in sparing rains during dry seasons increasing relative humidity and rise in overall temperatures. Countries with more proximity to the poles are projecting unusual weather as well, dependant of the region. For example, in an interesting study carried in Southern Norway by Uhlig et al., the authors found *Aspergillus* metabolites (e.g. sterigmatocystin) in concentrations up to 20 μg kg^{-1} [142]. Samples analyzed included barley (n = 20), oats (n = 28), and wheat (n = 28) collected during the wet summer seasons were analyzed using an LC-MS/MS $ESI^{\pm}$. In this regard, some authors already have stated that aflatoxins are among the foodborne risks most susceptible to climate change [143, 144]. Hence, meteorological data should be collected alongside aflatoxin incidence and levels. Several studies have focused on this particular subject [41, 145]. More recently, Mitchel et al. presented an interesting study case which described corn contamination dynamics influenced by weather patterns [146]. As explained before, corn is rather important feed ingredient. Nesic et al. mentioned that plant physiology is also altered as plants are subjected to different photoperiod and temperature regimes, this applies stress to productive species such as corn [143]. Battilani et al. described climate change as a motor force for emerging feed safety issues and elegantly predicted through climate mathematical model aflatoxin contamination in corn and wheat crops [146]. The authors predicted within the next 100 years a +2°C and +5°C climate change scenario, which converts aflatoxin in corn in a food safety issue. Medina et al. described the interaction among a_w, temperature and CO_2 and their effect on the relative expression of AF biosynthetic genes, *A. flavus* growth and aflatoxin production under elevated temperature and drought conditions [147]. The authors concluded that such environmental conditions had limited effect on growth, but significant impact on gene expression (both, structural *afl*D and regulatory *afl*R genes) and significantly arouse the production of AFB_1. The authors demonstrate these effects *in vitro* and on corn grains.

5. Conclusions and perspectives

Conclusive diagnostics regarding aflatoxicosis is difficult, confounding symptoms can cause an animal with aflatoxicosis to be misdiagnosed. In-farm productivity issues caused by toxins can be easily overlooked. On the other hand, farmers may equivocally attribute productivity loss to toxin presence where none is found. Herein we presented several approaches to control toxin in feed production and evidence suggest that GMP, and HACCP should be mandatory as a preventive measure to control aflatoxin contamination. Independently of which countermeasures are selected and applied, they should be pragmatic and implemented in conjunction with those designed for prevention. Changing patterns in weather add hindrance in the prediction of aflatoxigenic fungi colonization and toxin production; hence, countries should increase vigilance and take further preventive and control measures to respond swiftly to an eventual increase in toxin incidence due to regional climate change. Finally, considering the relevance of feed in the food chain safety, countries should implement and improve monitoring programs for aflatoxin in foodstuffs; these programs should contemplate risk management to mitigate the economical and health burden aflatoxin contamination generate.

Acknowledgements

The authors would like to thank Kevin Leiva Gabriel for the consultation regarding several findings described in **Figure 3**. Astrid Leiva Gabriel and Geovanna Méndez Hernández for their help during drafting and preparation of **Table 1**. Vicerrectoría de Investigación is acknowledged for funding this work.

Author details

Andrea Molina Alvarado[1,2], Rebeca Zamora-Sanabria[1,2] and Fabio Granados-Chinchilla[2*]

*Address all correspondence to: fabio.granados@ucr.ac.cr

1 Universidad de Costa Rica, Centro de Investigación en Nutrición Animal and Escuela de Zootecnia, San José, Costa Rica

2 Universidad de Costa Rica, Centro de Investigación en Nutrición Animal, San José, Costa Rica

References

[1] Guchi E. Implication of aflatoxin contamination in agricultural products. American Journal of Food and Nutrition. 2015;**3**(1):12-20. DOI: 10.12691/ajfn-3-1-3

[2] Kana JR, Gnonlonfin BGJ, Harvey J, Wainaina J, Wanjuki I, Skilton RA, Teguia A. Assessment of aflatoxin contamination of maize, peanut meal and poultry feed mixtures from different agroecological zones in Cameroon. Toxins. 2013;**5**:884-894. DOI: 10.3390/toxins5050884

[3] [FAO] Food and Agriculture Organization of the United Nations. Impact of Animal Nutrition on Animal Welfare – Expert Consultation. Animal Production and Health Report. No. 1 [Internet]. September 2012. Available from: http://www.fao.org/3/a-i3148e.pdf [Accessed: 03-05-2017]

[4] Salobir J, Frankič T, Rezar V. Animal nutrition for the health of animals, human and environment. Acta Agriculturae Slovenica. 2012;**Supplement 3**:41-49

[5] Gallo A, Giuberti G, Frisvad JC, Bertuzzi T, Nielsen KF. Review on mycotoxin issues in ruminants: Occurrence in forages, effects of mycotoxin ingestion on health status and animal performance and practical strategies to counteract their negative effects. Toxins. 2015;**7**(8):3057-3111. DOI: 10.3390/toxins7083057

[6] Kovalsky P, Kos G, Nährer K, Schwab C, Jenkins T. Co-occurrence of regulated, masked and emerging mycotoxins and secondary metabolites in finished. Toxins. 2016;**8**(363): 1-29. DOI: 10.3390/toxins8120363

[7] [FEFAC] European Feed Manufacturers' Federation. Annual Report 2014-2015 [Internet]. 2015. Available from: http://www.fefac.eu/files/60917.pdf [Accessed: 08-03-2017]

[8] International Trade Centre. I.T. Trade Map—Trade Statistics for International Business Development [Internet]. 2017. Available from: http://www.trademap.org/ [Accessed: 08-03-2017]

[9] Afsah-Hejri L, Jinap S, Hajeb P, Radu S, Shakibazadeh S. A review on mycotoxins in food and feed: Malaysia case study. Comprehensive Reviews in Food Science and Food Safety. 2013;**12**:629-651. DOI: 10.1111/1541-4337.12029

[10] [PROCOMER] Promotora de Comercio Exterior de Costa Rica. Portal estadísticos del comercio exterior [Internet]. 2017. Available from: http://servicios.procomer.go.cr/estadisticas/inicio.aspx [Accessed: 26-03-2017]

[11] Atalla MM, Hassanein NM, El-Beih AA, Youssef YA. Mycotoxin production in wheat grains by different Aspergilli in relation to different relative humidities and storage periods. Molecular Nutrition Food Research. 2003;**47**(1):6-10. DOI: 10.1002/food.200390017

[12] Fountain JC, Scully BT, Ni X, Kemerait R, Lee RD, Chen ZY, Guo B. Environmental influences on maize-*Aspergillus flavus* interactions and aflatoxin production. Frontiers in Microbiology. 2014;**5**(40):1-7. DOI: 10.3389/fmicb.2014.00040

[13] Warburton ML, William WP. Aflatoxin resistance in maize: What have we learned lately?. Advances in Botany. 2014;**2014**:352831. DOI: 10.1155/2014/352831

[14] Bryden WL. Mycotoxin contamination of the feed supply chain: Implications for animal productivity and feed security. Animal Feed Science and Technology. 2012; **173**(1-2):134-158. DOI: 10.1016/j.anifeedsci.2011.12.014

[15] Magan N, Hope R, Cairns V, Aldred D. Post-harvest fungal ecology: Impact of fungal growth and mycotoxin accumulation in stored grain. European Journal of Plant Pathology. 2003;**109**(7):723-730. DOI: 10.1023/A:1026082425177

[16] Tola M, Kebede B. Occurrence, importance and control of mycotoxins: A review. Food and Agriculture. 2016;**2**(1). http://www.tandfonline.com/doi/citedby/10.1080/23311932.2016.1191103?scroll=top&needAccess=true&nav=BI0001-bottomendid. DOI: 10.1080/23311932.2016.1191103

[17] Granados-Chinchilla F, Molina A, Chavarría G, Alfaro-Cascante M, Bogantes-Ledezma D, Murillo-Williams A. Aflatoxins occurrence trough the food chain in Costa Rica: Applying the One Health Approach to mycotoxins surveillance. Forthcoming.

[18] Medina A, Schmidt-Heydt M, Rodríguez A, Parra R, Geisen R, Magan N. Impacts of environmental stress on growth, secondary metabolite biosynthetic gene clusters and metabolite production of xerotolerant/xerophilic fungi. Current Genetics. 2015;**61**(3):325-334. DOI: 10.1007/s00294-014-0455-9

[19] Guerre P. Worldwide mycotoxins exposure in pig and poultry feed formulations. Toxins. 2016;**8**(12):350. DOI: 10.3390/toxins8120350

[20] Fafiolu AO, Oduguwa OO, Jegede AV, Tukura CC, Olarotimi ID, Teniola AA, Alabi JO. Assessment of enzyme supplementation on growth performance and apparent nutrient digestibility in diets containing undecorticated sunflower seed meal in layer chicks. Poultry Science. 2015;**94**(8):1917-1922. DOI: 10.3382/ps/pev136

[21] Pinotti L, Ottoboni M, Giromini C, Dell'Orto V, Cheli F. Mycotoxin contamination in the EU feed supply chain: A focus on cereal byproducts. Toxins. 2016;**8**(2):45. DOI: 10.3390/toxins8020045

[22] Nørgaard JV, Fernández JA, Jørgensen H. Ileal digestibility of sunflower meal, pea, rapeseed cake, and lupine in pigs. Journal of Animal Science. 2012;**90**(4):203-205. DOI: 10.2527/jas.53919

[23] Almeida FN, Htoo JK, Thomson J, Stein HH. Digestibility by growing pigs of amino acids in heat-damaged sunflower meal and cottonseed meal. Journal of Animal Science. 2012;**92**(2):585-593. DOI: 10.2527/jas.2013-6769

[24] Adeola O, Kong C. Energy value of distillers dried grains with solubles and oilseed meals for pig. Journal of Animal Science. 2014;**92**(1):164-170. DOI: 10.2527/jas.2013-6662

[25] Chavarría G, Molina A, Leiva A, Méndez G, Wong-González E, Cortés-Muñoz M, Rodríguez C, Granados-Chinchilla F. Distribution, stability, and protein interactions of Aflatoxin M_1 in fresh cheese. Food Control. 2017;**73**(Part B):581-586. DOI: 10.1016/j.foodcont.2016.09.005

[26] Lammers PJ, Stender DR, Honeyman MS. Managing Feed Costs, Feed Budgets [Internet]. 2007.Available from: https://www.ipic.iastate.edu/publications/840.FeedBudgets.pdf [Accessed: 08-03-2017]

[27] Ravindran V/FAO. Poultry Feed Availability and Nutrition in Developing Countries [Internet]. 2013. Available from: http://www.fao.org/3/a-al705e.pdf [Accessed: 08-03-2017]

[28] Rodrigues I, Naehrer K. A three-year survey on the worldwide occurrence of mycotoxins in feedstuffs and feed. Toxins. 2012;**4**(9):663-675. DOI: 10.3390/toxins4090663

[29] Sahin HZ, Celik M, Kotay S, Kabak B. Aflatoxins in dairy cow feed, raw milk and milk products from Turkey. Food Additives and Contaminants Part B Surveillance. 2016;**9**(2):152-158. DOI: 10.1080/19393210.2016.1152599

[30] Kocasari FS, Mor F, Oguz MN, Oguz FK. Occurrence of mycotoxins in feed samples in Burdur Province, Turkey. Environmental Monitoring and Assessment. 2013;**185**(6):4943-4949. DOI: 10.1007/s10661-012-2915-3

[31] Ghali R, Hmaissia-khlifa K, Ghorbel H, Maaroufi K, Hedili A. Incidence of aflatoxins, ochratoxin A and zearalenone in tunisian foods. Food Control. 2008;**19**(9):921-924. DOI: 10.1016/j.foodcont.2007.09.003

[32] Ghali R, Belouaer I, Hdiri S, Ghorbel H, Maaroufi K, Hedilli A. Simultaneous HPLC determination of aflatoxins B_1, B_2, G_1 and G_2 in Tunisian sorghum and pistachios. Journal of Food Composition and Analysis. 2009;**22**(7-8):751-755. DOI: 10.1016/j.jfca.2009.04.009

[33] Sahin T, Aksakal DH, Sünnetci S, Keser O, Eseceli H. Detection of aflatoxin, zearalenone and deoxynivalenol in some feed and feedstuffs in Turkey. Pakistan Veterinary Journal. 2013;**34**(4):459-463

[34] Warth B, Parich A, Atehnkeng J, Bandyopadhyay R, Schuhmacher R, Sulyok M, Krska R. Quantitation of mycotoxins in food and feed from Burkina Faso and Mozambique using a modern LC-MS/MS multitoxin method. Journal of Agricultural and Food Chemistry. 2012;**60**(36):9352-9363. DOI: 10.1021/jf302003n

[35] Abrunhosa L, Morales H, Soares C, Calado T, Vila-Chã AS, Pereira M, Venâncio A. A review of mycotoxins in food and feed products in Portugal and estimation of probable daily intakes. Critical Reviews in Food Science and Nutrition. 2016;**52**(2):249-265

[36] Chavarría G, Granados-Chinchilla F, Alfaro-Cascante M, Molina A. Detection of aflatoxin M_1 in milk, cheese and sour cream samples from Costa Rica using enzyme-assisted extraction and HPLC. Food Additives and Contaminants Part B Surveillance. 2015;**8**(2):128-135. DOI: 10.1080/19393210.2015.1015176

[37] Zheng N, Wang JQ, Han RW, Zhen YP, Xu XM, Sun P. Survey of aflatoxin M_1 in raw milk in the five provinces of China. Food Additives and Contaminants Part B Surveillance. 2013;**6**(2):110-115. DOI: 10.1080/19393210.2012.763191

[38] Mohammed S, Munissi JJ, Nyandoro SS. Aflatoxin M_1 in raw milk and aflatoxin B_1 in feed from household cows in Singida, Tanzania. Food Additives and Contaminants Part B Surveillance. 2016;**9**(2):85-90. DOI: 10.1080/19393210.2015.1137361

[39] Bahrami R, Shahbazi Y, Nikousefat Z. Occurrence and seasonal variation of aflatoxin in dairy cow feed with estimation of aflatoxin M_1 in milk from Iran. Food and Agricultural Immunology. 2016;**27**(3). DOI: 10.1080/09540105.2015.1109613

[40] Alshawabkeh K, Alkhalaileh NI, Abdelqader A, Al-Fataftah A-RA, Herzallah SM. Occurrence of aflatoxin B_1 in poultry feed and feed ingredients in Jordan using ELISA and HPLC. American-Eurasian Journal of Toxicological Sciences. 2015;**7**(4):316-320. DOI: 10.5829/idosi.aejts.2015.7.4.10172

[41] Paterson RM, Lima N. How will climate change affect mycotoxins in food? Food Research International. 2010;**43**:1902-1914. DOI: 10.1016/j.foodres.2009.07.010

[42] Murugesan GR, Ledoux DR, Naehrer K, Berthiller F, Applegate TJ, Grenier B, Phillips TD, Schatzmayr G. Prevalence and effects of mycotoxins on poultry health and performance, and recent development in mycotoxin counteracting strategies. Poultry Science. 2015;**94**(6):1298-1315. DOI: 10.3382/ps/pev075

[43] [IARC] International Agency for Research on Cancer. IARC Monographs on the Evaluation of Carcinogenic Risks to Humans. Some Organophosphate Insecticides and Herbicides, Volume 112 [Internet]. 2015. Available from: http://monographs.iarc.fr/ENG/Monographs/vol112/mono112.pdf [Accessed: 03-05-2017]

[44] Wu H-C, Wang Q, Yang H-I, Ahsan H, Tsai W-Y, Wang L-Y, Chen S-Y, Chen C-J, Santella RM. Aflatoxin B_1 exposure, hepatitis B virus infection and hepatocellular carcinoma in Taiwan. Cancer Epidemiology, Biomarkers & Prevention. 2009;**18**(3):846-853. DOI: 10.1158/1055-9965.EPI-08-0697

[45] Monson MS, Coulombe RA, Reed KM. Aflatoxicosis: Lessons from toxicity and responses to aflatoxin B_1 in poultry. Agriculture. 2015;**5**:742-777. DOI: 10.3390/agriculture5030742

[46] Yarru LP, Settivari RS, Antoniou E, Ledoux DR, Rottinghaus GE. Toxicological and gene expression analysis of the impact of aflatoxin B_1 on hepatic function of male broiler chicks. Poultry Science. 2009;**88**(2):360-371. DOI: 10.3382/ps.2008-00258

[47] Bbosa GS, Kitya D, Odda J, Ogwal-Okeng J. Aflatoxins metabolism, effects on epigenetic mechanisms and their role in carcinogenesis. Health. 2013;**5**(10A):14-34. DOI: 10.4236/health.2013.510A1003

[48] Dohnal V, Wu Q, Kuča K. Metabolism of aflatoxins: Key enzymes and interindividual as well as interspecies differences. Archives on Toxicology. 2014;**88**(9):1635-1644. DOI: 10.1007/s00204-014-1312-9

[49] Lozano MC, Diaz GJ. Microsomal and cytosolic biotransformation of aflatoxin B_1 in four poultry species. Brazilian Poultry Science. 2006;**47**(6):734-741. DOI: 10.1080/00071660601084390

[50] Yarru LP, Settivari RS, Gowda NKS, Antoniou E, Ledoux DR, Rottinghaus GE. Effects of turmeric (*Curcuma longa*) on the expression of hepatic genes associated with biotransformation, antioxidant, and immune systems in broiler chicks fed aflatoxin. Poultry Science. 2009;**88**(12):2620-2627. DOI: 10.3382/ps.2009-00204

[51] Burkina V, Zlabek V, Halsne R, Ropstad E, Zamaratskaia G. *In vitro* effects of the citrus flavonoids diosmin, naringenin and naringin on the hepatic drug-metabolizing CYP3A enzyme in human, pig, mouse and fish. Biochemical Pharmacology. 2016;**110-111**:109-116. DOI: 10.1016/j.bcp.2016.04.011

[52] Shi YH, Xu ZR, Feng JL, Xia MS, Hu CH. Effects of modified montmorillonite nanocomposite on growing/finishing pigs during aflatoxicosis. Asian-Australasian Journal of Animal Sciences. 2005;**18**(9):1305-1309. DOI: 10.5713/ajas.2005.1305

[53] Rustemeyer SM, Lamberson WR, Ledoux DR, Rottinghaus GE, Shaw DP, Cockrum RR, Kessler K, Austin KJ, Cammack KM. Effects of dietary aflatoxin on the health and performance of growing barrows. Journal of Animal Science. 2010;**88**(11):3624-3630. DOI: 10.2527/jas.2009-2663

[54] Chaytor AC, See MT, Hansen JA, de Souza AL, Middleton TF, Kim SW. Effects of chronic exposure of diets with reduced concentrations of aflatoxin and deoxynivalenol on growth and immune status of pigs. Journal of Animal Science. 2011;**89**(1):124-135. DOI: 10.2527/jas.2010-3005

[55] Mok CH, Shin SY, Kim BG. Aflatoxin, deoxynivalenol, and zearalenone in swine diets: Predictions on growth performance. Revista Colombiana de Ciencias Pecuarias. 2013;**26**(4):243-254

[56] Zain ME. Impact of mycotoxins on humans and animals. Journal of Saudi Chemical Society. 2011;**15**(2):129-144. DOI: 10.1016/j.jscs.2010.06.006

[57] Mehrzad J, Devriendt B, Baert K, Cox E. Aflatoxin B_1 interferes with the antigen-presenting capacity of porcine dendritic cells. Toxicology *In Vitro*. 2014;**28**(4):531-537. DOI: 10.1016/j.tiv.2013.11.015

[58] Shivasharanappa GY, Mundas S, Rao DGK, Tikare V, Shridhar NB. Histopathological changes in pigs exposed to Aflatoxin B_1 during pregnancy. Indian Journal of Animal Research. 2013;**47**(5):386-391

[59] Kanora A, Maes D. The role of mycotoxins in pig reproduction: A review. Veterinarni Medicina. 2009;**54**(12):565-576

[60] Stojanac N, Stevančević O, Savić B, Davidov I, Cincović MR, Potkonjak A, Spasojević J. Forensic findings on acute mortality of piglets after ingestion of aflatoxin. Kafkas Üniversitesi Veteriner Fakültesi Dergisi. 2015;**21**(3):437-440

[61] Azevedo Muraroll R. Effects of aflatoxin B_1 (AFB_1) on hepatic gene expression in pigs and turkeys [dissertation]. Columbia: University of Missouri; 2013. p. 153. Available from: https://mospace.umsystem.edu/xmlui/bitstream/handle/10355/44664/research.pdf?sequence=1

[62] Sulzberger SA, Melnichenko S, Cardoso FC. Levels of aflatoxin M_1 in different types of milk collected in Serbia: Assessment of human and animal exposure. Journal of Dairy Science. 2017;**100**(3):1856-1869. DOI: 10.3168/jds.2016-11612

[63] Tsakiris IN, Tzatzarakis MN, Alegakis AK, Vlachou MI, Renieri EA, Tsatsakis AM. Risk assessment scenarios of children's exposure to aflatoxin M_1 residues in different milk types from the Greek market. Food Chemistry and Toxicology. 2013;**56**:261-265. DOI: 10.1016/j.fct.2013.02.024

[64] Škrbić B, Živančev J, Antić I, Godula M. Levels of aflatoxin M_1 in different types of milk collected in Serbia: Assessment of human and animal exposure. Food Control. 2013;**40**:113-119. DOI: 10.1016/j.foodcont.2013.11.039

[65] Oğuz H, Keçeci T, Birdane YO, Onder F, Kurtoğlu V. Effect of clinoptilolite on serum biochemical and haematological characters of broiler chickens during aflatoxicosis. Research in Veterinary Science. 2000;**69**(1):89-93. DOI: 10.1053/rvsc.2000.0395

[66] Rawal S, Kim JE, Coulombe Jr R. Aflatoxin B_1 in poultry: Toxicology, metabolism and prevention. Research in Veterinary Science. 2010;**89**(3):325-331. DOI: 10.1016/j.rvsc.2010.04.011

[67] Peng X, Bai S, Ding X, Zeng Q, Zhang K, Fang J. Pathological changes in the immune organs of broiler chickens fed on corn naturally contaminated with aflatoxins B_1 and B_2. Avian Pathology. 2015;**44**(3):192-199. DOI: 10.1080/03079457.2015.1023179

[68] Kumar R, Balachandran C. Histopathological changes in broiler chickens fed aflatoxin and cyclopiazonic acid. Veterinary Archives. 2009;**79**(1):31-40

[69] Yanus AW, Nasir MK, Aziz T. Prevalence of poultry diseases in district Chakwal and their interaction with mycotoxicosis: 2. Effects of season and feed. The Journal of Animal & Plant Sciences. 2009;**19**(1):1-5

[70] Yang J, Bai F, Zhang K, Bai S, Peng X, Ding X, Li Y, Zhang J, Zhao L. Effects of feeding corn naturally contaminated with aflatoxin B_1 and B_2 on hepatic functions of broilers. Poultry Science. 2012;**91**(11):2792-2801. DOI: 10.3382/ps.2012-02544

[71] Hussain Z, Khan MZ, Saleemi MK, Khan A, Rafique S. Clinicopathological effects of prolonged intoxication of aflatoxin B_1 in broiler chicken. Pakistan Veterinary Journal. 2016;**36**(4):477-481

[72] Trebak F, Alaoui A, Alexandre D, El Ouezzani S, Anouar Y, Chartrel N, Magoul R. Impact of aflatoxin B_1 on hypothalamic neuropeptides regulating feeding behavior. Neurotoxicology. 2015;**49**:165-173. DOI: 10.1016/j.neuro.2015.06.008

[73] Liu YL, Meng GQ, Wang HR, Zhu HL, Hou YQ, Wang WJ, Ding BY. Effect of three mycotoxin adsorbents on growth performance, nutrient retention and meat quality in broilers fed on mould-contaminated feed. Brazilian Poultry Science. 2011;**52**(2):255-263. DOI: 10.1080/00071668.2011.559453

[74] Alhousein A, Gurbuz Y. Aflatoxins in poultry nutrition. KSU Journal of Natural Science. 2015;**18**:1-15

[75] Valchev I, Marutsova V, Zarkov I, Ganchev A, Nikolov Y. Effects of aflatoxin B_1 alone or co-administered with Mycotox NG on performance and humoral immunity of turkey broilers. Bulgarian Journal of Veterinary Medicine. 2017;**20**(1):38-50. DOI: 10.15547/bjvm.1019

[76] Buzala M, Słomka A, Janicki B, Ponczek MB, Żekanowska E. Review: The mechanism of blood coagulation, its disorders and measurement in poultry. Livestock Science. 2017;**195**:1-8. DOI: 10.1016/j.livsci.2016.11.009

[77] Iqbal SZ, Nisar S, Asi MR, Jinap S. Natural incidence of aflatoxins, ochratoxin A and zearalenone in chicken meat and eggs. Food Control. 2014;**43**:98-103. DOI: 10.1016/j.foodcont.2014.02.046

[78] Herzallah SM. Aflatoxin B_1 residues in eggs and flesh of laying hens fed aflatoxin B_1 contaminated diet. American Journal of Agricultural and Biological Sciences. 2013;**8**(2): 156-161. DOI: 10.3844/ajabssp.2013.156.161

[79] Aly Salwa A, Anwer W. Effect of naturally contaminated feed with aflatoxins on performance of laying hens and the carryover of aflatoxin B_1 residues in table eggs. Pakistan Journal of Nutrition. 2009;**8**(2):181-186. DOI: 10.3923/pjn.2009.181.186

[80] Pandey I, Chauhan SS. Studies on production performance and toxin residues in tissues and eggs of layer chickens fed on diets with various concentrations of aflatoxin AFB_1. Brazilian Poultry Science. 2007;**48**(6):713-723. DOI: 10.1080/00071660701713534

[81] Lee M, Joo Seo D, Jeon SB, Ok HE, Jung H, Choi C, Chun HS. Detection of foodborne pathogens and mycotoxins in eggs and chicken feeds from farms to retail markets. Korean Journal Food Science and Animal Resources. 2016;**36**(4):463-468. DOI: 10.5851/kosfa.2016.36.4.463

[82] Gul H, Khan S, Shah Z, Ahmad S, Israr M, Hussain M. Effects of local sodium bentonite as aflatoxin binder and its effects on production performance of laying hens. Kafkas Üniversitesi Veteriner Fakültesi Dergisi. 2017;**23**(1):31-37. DOI: 10.9775/kvfd.2016.15714

[83] Grenier B, Applegate TJ. Modulation of intestinal functions following mycotoxin ingestion: Meta-analysis of published experiments in animals. Toxins. 2013;**5**(2):396-430. DOI: 10.3390/toxins5020396

[84] Kalpana S, Aggarwal M, Srinivasa RG, Malik JK. Effects of aflatoxin B_1 on tissue residues of enrofloxacin and its metabolite ciprofloxacin in broiler chickens. Environmental Toxicology and Pharmacology. 2012;**33**(2):121-126. DOI: 10.1016/j.etap.2011.11.005

[85] Iheanacho H. Cytotoxic effects of aflatoxin B_1 standard in relation to aflatoxin extracts from South African compound feeds on human lymphocytes. International Journal of Biomedical Data Mining. 2015;**3**(2):108. DOI: 10.4172/2090-4924.1000108

[86] Farabi SMV, Yousefian M, Hajimoradloo A. Aflatoxicosis in juvenile Huso huso fed a contaminated diet. Journal of Applied Ichthyology. 2006;**22**(s1):234-237. DOI: 10.1111/j.1439-0426.2007.00958.x

[87] Anater A, Manyes L, Meca G, Ferrer E, Luciano FB, Pimpãoa CT, Fontb G. Mycotoxins and their consequences in aquaculture: A review. Aquaculture. 2016;**451**:1-10. DOI: 10.1016/j.aquaculture.2015.08.022

[88] Cagauan A, Tayaban R, Somga J, Bartolome R. Effect of aflatoxin-contaminated feeds in Nile tilapia (*Oreochromis niloticus* L.). In: Remedios GC, Mair K, editors. Proceedings of the Sixth International Symposium on Tilapia in Aquaculture; 2004. pp. 172-178

[89] Akter A, Rahman M, Hasan M. Effects of aflatoxin B_1 on growth and bioaccumulation in common carp fingerling in Bangladesh. Asia-Pacific Journal of Rural Development. 2010;**20**(3)

[90] Huang Y, Han D, Zhu X, Yang Y, Jin J, Chen Y, Xie S. Response and recovery of gibel carp from subchronic oral administration of aflatoxin B_1. Aquaculture. 2011;**319**(1-2):89-97. DOI: 10.1016/j.aquaculture.2011.06.024

[91] Caloni F, Cortinovis C. Toxicological effects of aflatoxins in horses. Veterinary Journal. 2011;**188**(3):270-273. DOI: 10.1016/j.tvjl.2010.06.002

[92] Gazzotti T, Biagi G, Pagliuca G, Pinna C, Scardilli M, Grandi M, Zaghini G. Occurrence of mycotoxins in extruded commercial dog food. Animal Feed Science and Technology. 2015;**202**:81-89. DOI: 10.1016/j.anifeedsci.2015.02.004

[93] Frehse MS, Martins MIM, Ono EYS, Bracarense APFRL, Bissoqui LY, Teixeira EMK, Santos NJR, Freire RL. Aflatoxins ingestion and canine mammary tumors: There is an association?. Food and Chemical Toxicology. 2015;**84**:74-78. DOI: 10.1016/j.fct.2015.08.004

[94] Kendra DF, Dyer RB. Opportunities for biotechnology and policy regarding mycotoxin issues in international trade. International Journal of Food Microbiology. 2007;**119**(1-2):147-151

[95] Bhatnagar-Mathur P, Sunkara S, Bhatnagar-Panwar M, Waliyar F, Kumar Sharma K. Biotechnological advances for combating *Aspergillus flavus* and aflatoxin contamination in crops. Plant Science. 2015;**234**:119-132. DOI: 10.1016/j.plantsci.2015.02.009

[96] Ostry V, Malir F, Leszkowicz A. Comparative data concerning aflatoxin contents in Bt maize and non-Bt isogenic maize in relation to human and animal health—A review. Acta Veterinaria Brno. 2015;**84**(1):47-53. DOI: 10.2754/avb201585010047

[97] Jard G, Liboz T, Mathieu F, Guyonvarc'h A, Lebrihi A. Review of mycotoxin reduction in food and feed: From prevention in the field to detoxification by adsorption or transformation. Food Additives & Contaminants: Part A: Chemistry, Analysis, Control, Exposure & Risk Assessment. 2011;**28**(11):1590-1609. DOI: 10.1080/19440049.2011.595377

[98] Waliyar F, Osiru M, Ntare BR, Kumar KVK, Sudini H, Traore A, Diarra B. Post-harvest management of aflatoxin contamination in groundnut. World Mycotoxin Journal. 2015;**8**(2):245-252

[99] Codex Alimentarius. Code of Practice for the Prevention and Reduction of Mycotoxin Contamination in Cereals, CAC/RCP 51-2003 [Internet]. 2003 [Updated: 2014]. Available from: www.fao.org/input/download/standards/406/CXP_051e_2014.pdf

[100] Dorner JW, Cole RJ. Effect of application of nontoxigenic strains of *Aspergillus flavus* and *A. parasiticus* on subsequent aflatoxin contamination of peanuts in storage. Journal of Stored Products Research. 2002;**28**(4):329-339. DOI: 10.1016/S0022-474X(01)00035-2

[101] Lanyasunya TP, Wamae LW, Musa HH, Olowofeso O, Lokwaleput IK. The risk of mycotoxins contamination of dairy feed and milk on smallholder dairy farms in Kenya. Pakistan Journal of Nutrition. 2005;**4**(3):162-169

[102] Chulze SN. Strategies to reduce mycotoxin levels in maize during storage: A review. Food Additives & Contaminants: Part A: Chemistry, Analysis, Control, Exposure & Risk Assessment. 2010;**27**(5):651-657. DOI: 10.1080/19440040903573032

[103] Devreese M, De Backer P, Croubels S. Different methods to counteract mycotoxin production and its impact on animal health. Vlaams Diergeneeskundig Tijdschrift. 2013;**82**:181-190

[104] Esper RH, Gonçalez E, Marques MO, Felicio RC, Felicio JD. Potential of essential oils for protection of grains contaminated by aflatoxin produced by *Aspergillus flavus*. Frontiers in Microbiology. 2014;**2014**(6):269. DOI: 10.3389/fmicb.2014.00269

[105] Soares C. Inhibitory effect of essential oils on growth and on aflatoxins production by *Aspergillus parasiticus*. World Mycotoxin Journal. 2016;**9**(4):525-534. DOI: 10.3920/WMJ2015.1987

[106] Gil L, Ruiz P, Font Pérez G, Manyes L. An overview of the applications of hazards analysis and critical control point (HACCP) system to mycotoxins. Revista de Toxicología. 2016;**33**(1):50-55

[107] Kim S, Lee H, Lee S, Lee J, Ha J, Choi Y, Yoon Y, Choi K-H. Microbe-mediated aflatoxin decontamination of dairy products and feeds. Journal of Dairy Science. 2017;**100**(2): 871-880. DOI: 10.3168/jds.2016-11264

[108] Kamala A, Kimanya M, Haesaert G, Tiisekwa B, Madege R, Degraeve S, Cyprian C, De Meulenaer B. Local post-harvest practices associated with aflatoxin and fumonisin contamination of maize in three agro ecological zones of Tanzania. Food Additives & Contaminants: Part A: Chemistry, Analysis, Control, Exposure & Risk Assessment. 2016;**33**(3):551-559. DOI: 10.1080/19440049.2016.1138546

[109] Pappas AC, Tsiplakou E, Tsitsigiannis DI, Georgiadou M, Iliadi MK, Sotirakoglou K, Zervas G. The role of bentonite binders in single or concomitant mycotoxin contamination of chicken diets. British Poultry Science. 2016;**57**(4):551-558

[110] Mayura K, Abdel-Wahhab MA, McKenzie KS, Sarr AB, Edwards JF, Naguib K, Phillips TD. Prevention of maternal and developmental toxicity in rats via dietary inclusion of common aflatoxin sorbents: Potential for hidden risks. Toxicology Science. 1998;**41**(2):175-182. DOI: 10.1006/toxs.1997.2405

[111] Mitchell NJ, Xue KS, Lin S, Marroquin-Cardona A, Brown KA, Elmore SE, Tang L, Romoser A, Gelderblom WC, Wang J-S, Phillips T. Calcium montmorillonite clay reduces AFB1 and FB1 biomarkers in rats exposed to single and co-exposures of aflatoxin and fumonisin. Toxicology and Applied Pharmacology. 2014;**34**(7):795-804. DOI: 10.1002/jat.2942

[112] Deng Y, Velázquez ALB, Billes F, Dixon JB. Bonding mechanisms between aflatoxin B_1 and smectite. Applied Clay Science. 2010;**50**(1):92-98. DOI: 10.1016/j.clay.2010.07.008

[113] Fruhauf S, Schwartz H, Ottner F, Krska R, Vekiru E. Yeast cell based feed additives: Studies on aflatoxin B_1 and zearalenone. Food Additives & Contaminants: Part A: Chemistry, Analysis, Control, Exposure & Risk Assessment. 2012;**29**(2):217-231. DOI: 10.1080/19440049.2011.630679

[114] Kensler TW, Groopman JD, Egner PA, Muñoz A, Qian G, Chen J. Chemoprevention of hepatic cancer in aflatoxin endemic areas. In: Gu J, editor. Primary Liver Cancer: Challenges and Perspectives. 1st ed. Hangzhou: Zhejiang University Press and Berlin. Heidelberg: Springer-Verlag; 2013. pp. 339-365. DOI: 10.1007/978-3-642-28702-2_11

[115] Miller JD, Schaafsma AW, Bhatnagar D, Bondy G, Carbone I, Harris LJ, Harrison G, Munkvold GP, Oswald IP, Pestka JJ, Sharpe L, Sumarah MW, Tittlemier SA, Zhou T. Mycotoxins that affect the North American agri-food sector: State of the art and directions for the future. World Mycotoxin Journal. 2013;**7**(1):63-82. DOI: 10.3920/WMJ2013.1624

[116] [EFSA] European Food Safety Authority. Statement on the establishment of guidelines for the assessment of additives from the functional group 'substances for reduction of the contamination of feed by mycotoxins'. EFSA Journal. 2010;**8**(7):1693

[117] Dos Anjos FR, Ledoux DR, Rottinghaus GE, Chimonyo M. Efficacy of adsorbents (bentonite and diatomaceous earth) and turmeric (*Curcuma longa*) in alleviating the toxic effects of aflatoxin in chicks. British Poultry Science. 2015;**56**(4):459-469. DOI: 10.1080/00071668.2015.1053431

[118] Sun Y, Guo J, Park I, Kim SW. Efficacy of yeast cell wall based feed additive in prevention of mycotoxicoses in pigs fed a diet with aflatoxin under the EU regulatory levels. In: Midwest American Society of Animal Science; 17/03/2014; Community Choice Credit Union Convention Center. Des Moines, IA: American Dairy Science Association; 2014. pp. 302-303

[119] Wongtangtintan S, Neeratanaphan L, Ruchuwararak P, Suksangawong S, Tengjaroenkul U, Sukon P, Tengjaroenkul B. Comparative study of aflatoxin B_1 adsorption by Thai bentonite and commercial toxin binders at different temperatures *in vitro*. Livestock Research for Rural Development. 2016;**28**(4):2016

[120] Denli M, Blandon JC, Guynot ME, Salado S, Perez JF. Effects of dietary AflaDetox on performance, serum biochemistry, histopathological changes, and aflatoxin residues in broilers exposed to aflatoxin B(1). Poultry Science. 2009;**88**(7):1444-1451. DOI: 10.3382/ps.2008-00341

[121] Neeff DV, Ledoux DR, Rottinghaus GE, Bermudez AJ, Dakovic A, Murarolli RA, Oliveira CA. *In vitro* and *in vivo* efficacy of a hydrated sodium calcium aluminosilicate to bind and reduce aflatoxin residues in tissues of broiler chicks fed aflatoxin B_1. Poultry Science. 2013;**92**(1):131-137. DOI: 10.3382/ps.2012-02510

[122] Fowler J, Li W, Bailey C. Effects of a calcium bentonite clay in diets containing aflatoxin when measuring liver residues of aflatoxin B_1 in starter broiler chicks. Toxins. 2015;**7**:3455-3464. DOI: 10.3390/toxins7093455

[123] Molina A, Holst I, Ramírez M, Zumbado M. Evaluation of the effects of two mycotoxin adsorbents in broilers feeding diets contaminated with aflatoxin and T-2 toxin. Forthcoming

[124] Maki CR, Thomas AD, Elmore SE, Romoser AA, Harvey RB, Ramirez-Ramirez HA. Effects of calcium montmorillonite clay and aflatoxin exposure on dry matter intake, milk production, and milk composition. Journal of Dairy Science. 2016;**9**:1-8. DOI: 10.3168/jds.2015-10242

[125] Mugerwa S, Kabirizi J, Zziwa E. Effect of supplementing lactating goats fed on aflatoxin contaminated feed with calcium bentonite and activated charcoal on aflatoxin M_1 concentration, excretion and carryover in milk. Uganda Journal of Agricultural Sciences. 2015;**16**(1):83-92. DOI: 10.4314/ujas.v16i1.6

[126] Wee J, Day DM, Linz JE. Effects of zinc chelators on aflatoxin production in *Aspergillus parasiticus*. Toxins. 2016;**8**(6):E171. DOI: 10.3390/toxins8060171

[127] Weaver AC, See MT, Hansen JA, Kim YB, De Souza ALP, Middleton TF, Kim SW. The use of feed additives to reduce the effects of aflatoxin and deoxynivalenol on pig growth, organ health and immune status during chronic exposure. Toxins. 2013;**5**(7):1261-1281. DOI: 10.3390/toxins5071261

[128] Das A, Bhattacharya S, Palaniswamy M, Angayarkanni J. Aflatoxin B_1 degradation during co-cultivation of *Aspergillus flavus* and *Pleurotus ostreatus* strains on rice straw. 3 Biotechnology. 2015;**5**(3):279-284. DOI: 10.1007/s13205-014-0228-7

[129] Lee KR, Yang SM, Cho SM, Kim M, Hong S-Y, Chung SH. Aflatoxin B_1 Detoxification by *Aspergillus oryzae* from Meju, a Traditional Korean Fermented Soybean Starter. Journal of Microbiology Biotechnology. 2017;**27**(1):57-66

[130] Villers P. Aflatoxins and safe storage. Frontiers in Microbiology. 2014;**5**:158. DOI: 10.3389/fmicb.2014.00158

[131] Bovo F, Franco LT, Kobashigawa E, Rottinghaus GE, Ledoux DR, Oliveira CA. Efficacy of beer fermentation residue containing *Saccharomyces cerevisiae* cells for ameliorating aflatoxicosis in broilers. Poultry Science. 2015;**94**(5):934-942. DOI: 10.3382/ps/pev067

[132] Pizzolitto RP, Armando MR, Salvano MA, Dalcero AM, Rosa CA. Evaluation of *Saccharomyces cerevisiae* as an antiaflatoxicogenic agent in broiler feedstuffs. Processing, Products and Food Safety. 2013;**92**:1655-1663. DOI: 10.3382/ps.2012-02846

[133] Yin HB, Chen CH, Kollanoor-Johny A, Darre MJ, Venkitanarayanan K. Controlling *Aspergillus flavus* and *Aspergillus parasiticus* growth and aflatoxin production in poultry feed using carvacrol and *trans*-cinnamaldehyde. Poultry Science. 2015;**94**(9):2183-2190. DOI: 10.3382/ps/pev207

[134] Nerilo SB, Rocha GHO, Tomoike C, Mossini SAG, Grespan R, Mikcha JMG, Machinski Jr M. Antifungal properties and inhibitory effects upon aflatoxin production by *Zingiber officinale* essential oil in *Aspergillus flavus*. International Journal of Food Science and Technology. 2016;**51**:286-292. DOI: 10.1111/ijfs.12950

[135] Ji C, Fan Y, Zhao L. Review on biological degradation of mycotoxins. Animal Nutrition. 2016;**2**:127-133. DOI: 10.1016/j.aninu.2016.07.003

[136] Wielogorska E, MacDonald S, Elliott C. A review of the efficacy of mycotoxin detoxifying agents used in feed in light of changing global environment and legislation. World Mycotoxin Journal. 2016;**9**(3):1-16. DOI: 10.3920/WMJ2015.1919

[137] Jiang Y-H, Wang P, Yang H-J, Chen Y. The efficacy of bamboo charcoal in comparison with smectite to reduce the detrimental effect of aflatoxin B_1 on *in vitro* rumen fermentation of a hay-rich feed mixture. Toxins. 2014;**6**:2008-2023. DOI: 10.3390/toxins6072008

[138] Zhou G, Chen Y, Kong Q, Ma Y, Liu Y. Detoxification of aflatoxin B_1 by *Zygosaccharomyces rouxii* with solid state fermentation in peanut meal. Toxins. 2017;**9**(1):42. DOI: 10.3390/toxins9010042

[139] Shar ZH, Fletcher MT, Sumbal GA, Sherazi ST, Giles C, Bhanger MI, Nizamani SM. Banana peel: An effective biosorbent for aflatoxins. Food Additives & Contaminants: Part A: Chemistry, Analysis, Control, Exposure & Risk Assessment. 2016;**33**(5):849-860. DOI: 10.1080/19440049.2016.1175155

[140] Guilford FT, Hope J. Deficient glutathione in the pathophysiology of mycotoxin-related illness. Toxins. 2014;**6**:608-623. DOI: 10.3390/toxins6020608

[141] Jardon-Xicotencatl S, Díaz-Torres R, Marroquín-Cardona A, Villarreal-Barajas T, Méndez-Albores A. Detoxification of aflatoxin-contaminated maize by neutral electrolyzed oxidizing water. Toxins. 2015;**7**(10):4294-4314. DOI: 10.3390/toxins7104294

[142] Uhlig S, Eriksen GS, Hofgaard IS, Krska R, Beltrán E, Sulyok M. Faces of a changing climate: Semi-quantitative multi-mycotoxin analysis of grain grown in exceptional climatic conditions in Norway. Toxins. 2013;**5**(10):1682-1697. DOI: 10.3390/toxins5101682

[143] Nesic K, Milicevic D, Nesic V, Ivanovic S. Mycotoxins as one of the foodborne risks most susceptible to climatic change. Procedia Food Science. 2015;**5**:207-210. DOI: 10.1016/j.profoo.2015.09.058

[144] Mitchell NJ, Bowers E, Hurburgh C, Wu F. Potential economic losses to the US corn industry from aflatoxin contamination. Food Additives & Contaminants: Part A: Chemistry, Analysis, Control, Exposure & Risk Assessment. 2016;**33**(3):540-550. DOI: 10.1080/19440049.2016.1138545

[145] Paterson RRM, Lima N. Further mycotoxin effects from climate change. Food Research International. 2011;**44**:2555-2566. DOI: 10.1016/j.foodres.2011.05.038

[146] Battilani P, Toscano P, van der Fels-Klerx HJ, Moretti A, Camardo Leggieri M, Brera C, Rortais A, Goumperis T, Robinson T. Aflatoxin B_1 contamination in maize in Europe increases due to climate change. Scientific Reports. 2016;**6**:24328. DOI: 10.1038/srep24328

[147] Medina A, Rodriguez A, Magan N. Effect of climate change on *Aspergillus flavus* and aflatoxin B_1 production. Frontiers in Microbiology. 2014;**5**:348. DOI: 10.3389/fmicb.2014.00348

[148] Schatzmayr G, Streit E. Global occurrence of mycotoxins in the food and feed chain: Facts and figures. World Mycotoxin Journal. 2013;**6**(3):213-222. DOI: 10.3920/WMJ2013.1572

[149] Streit E, Naehrer K, Rodrigues I, Schatzmayr G. Mycotoxin occurrence in feed and feed raw materials worldwide: Long-term analysis with special focus on Europe and Asia. Journal of the Science of Food and Agriculture. 2013;**93**(12):2871-3136. DOI: 10.1002/jsfa.6225

[150] Rodrigues I, Naehrer K. Prevalence of mycotoxins in feedstuffs and feed surveyed worldwide. Phytopathologia Mediterranea. 2012;**51**(1):175-192

[151] Gizachew D, Szonyi B, Tegegne A, Hanson J, Grace D. Aflatoxin contamination of milk and dairy feeds in the Greater Addis Ababa milk shed, Ethiopia. Food Control. 2016;**59**:773-779. DOI: 10.1016/j.foodcont.2015.06.060

[152] Kamika I, Ngbolua K-te-N, Tekere M. Occurrence of aflatoxin contamination in maize throughout the supply chain in the Democratic Republic of Congo. Food Control. 2016;**69**:292-296. DOI: 10.1016/j.foodcont.2016.05.014

[153] Makau CM, Matofari JW, Muliro PS, Bebe BO. Aflatoxin B_1 and deoxynivalenol contamination of dairy feeds and presence of Aflatoxin M_1 contamination in milk from smallholder dairy systems in Nakuru, Kenya. International Journal of Food Contamination. 2016;**3**(6):1-10. DOI: 10.1186/s40550-016-0033-7

[154] Nishimwe I, Wanjuki I, Karangwa C, Darnell R, Harvey J. An initial characterization of aflatoxin B_1 contamination of maize sold in the principal retail markets of Kigali, Rwanda. Food Control. 2017;**73**(Part B):574-580. DOI: 10.1016/j.foodcont.2016.09.006

[155] Olorunfemi FM, Odebode CA, Ezekiel NC, Krause RW, Njobeh PB, Sulyok M, Krska R. Detection of *Aspergillus* section *Flavi* and aflatoxin in locally formulated fish feeds from South-Western Nigeria. Journal of Aquaculture Research & Development. 2016;**7**(11 (Suppl)):56. DOI: 10.4172/2155-9546.C1.011

[156] Njobeh PB, Dutton MF, Åberg AT, Haggblom P. Estimation of multi-mycotoxin contamination in South African compound feeds. Toxins. 2012;**4**(10):836-848. DOI: 10.3390/toxins4100836

[157] Mwalwayo DS, Thole B. Prevalence of aflatoxin and fumonisins ($B_1 + B_2$) in maize consumed in rural Malawi. Toxicology Reports. 2016;**3**:173-179. DOI: 10.1016/j.toxrep.2016.01.010

[158] Greco MV, Franchi ML, Golba SLR, Pardo AG, Pose GN. Mycotoxins and mycotoxigenic fungi in poultry feed for food-producing animals. The Scientific World Journal. 2014;**2014**:1-9/968215. DOI: 10.1155/2014/968215

[159] Greco M, Pardo A, Pose G. Mycotoxigenic fungi and natural co-occurrence of mycotoxins in rainbow trout (*Oncorhynchus mykiss*) feeds. Toxins. 2015;**7**:4595-4609. DOI: 10.3390/toxins7114595

[160] Oliveira MS, Rocha A, Sulyok M, Krska R, Mallmann CA. Natural mycotoxin contamination of maize (*Zea mays* L.) in the South region of Brazil. Food Control. 2017;**73**(Part B):127-132. DOI: 10.1016/j.foodcont.2016.07.033

[161] de Souza MLM, Sulyok M, Freitas-Silva O, Costa SS, Brabet C, Machinski MJ, Sekiyama BL, Vargas EA, Krska R, Schuhmacher R. Cooccurrence of mycotoxins in maize and poultry feeds from Brazil by liquid chromatography/tandem mass spectrometry. Scientific World Journal. 2013;**2013**:1-9/427369. DOI: 10.1155/2013/427369

[162] Gonçalves-Nunes EMC, Gomes-Pereira MM, Raposo-Costa AP, da Rocha-Rosa CA, Pereyra CM, Calvet RM, Alves-Marques AL, Cardoso-Filho F, Sanches-Muratori MC. Screening of aflatoxin B_1 and mycobiota related to raw materials and finished feed destined for fish. Latin American Journal of Aquatic Research. 2015;**43**(3):595-600. DOI: 10.3856/vol43-issue3-fulltext-22

[163] Maniglia-Merida GC, Ascanio-Evanoff E, Riera JG, Colina-Rivero J, Briceño-Fereira E. del C, Flores-Chona S, Ascanio-Evan A, Arrieta-Mendoza D. Determinación de Aflatoxinas en Alimentos Balanceados para Cerdos en Granjas de los Estados Aragua y Carabobo, Venezuela. Revista Científica, FCV-LUZ. 2015;**25**(3):200-207

[164] Wu L, Li J, Li Y, Li T, He Q, Tang Y, Liu H, Su Y, Yin Y, Liao P. Aflatoxin B_1, zearalenone and deoxynivalenol in feed ingredients and complete feed from different Province in China. Journal of Animal Science and Biotechnology. 2016;**7**(63):1-10. DOI: 10.1186/s40104-016-0122-8

[165] Kotinagu K, Mohanamba T, Rathna Kumari L. Assessment of aflatoxin B_1 in livestock feed and feed ingredients by high-performance thin layer chromatography. Veterinary World. 2015;**8**(12):1396-1399. DOI: 10.14202/vetworld.2015.1396-1399

[166] Anjum MA, Khan SH, Sahota AW, Sardar R. Assessment of Aflatoxin B_1 in commercial poultry feed and feed ingredients. JAPS, Journal of Animal and Plant Sciences. 2012;**22**(2):268-272

Chapter 7

Incidence and Chemical Implications of Aflatoxin in Street-Vended Foods

Michael Dare Asemoloye,
Segun Gbolagade Jonathan, Rashid Saddaf,
Zaffar Habiba, Edward Ehidiabhen Okoawo and
Tunde Sheriffdeen Bello

Additional information is available at the end of the chapter

http://dx.doi.org/10.5772/intechopen.68478

Abstract

It is known that aflatoxin is produced by some *Aspergillus* fungal strains. Incidence of this toxin in food and feeds is as a result of contaminations by these fungal strains. *Aspergillus* strains are common in human environment, and their effects on foods are issues of global concern. First, we need an adequate knowledge of aflatoxin, then we can design proper control and regulatory strategies for its control. In this chapter, we present importance of street-vended foods and their possible health risk on the populace due to possible aflatoxin contamination; we took insights into types and incidences of reported aflatoxin contaminations in street-vended foods and reviewed the chemical nature, structures, and metabolism of aflatoxins and their actions as chemical poisons to human (mycotoxicosis) with deep insight into their toxicology. In addition, we review different environmental factors that may affect aflatoxin production in foods and also considered economic impact of aflatoxin contaminations of foodstuffs. In this chapter, we also discussed different aflatoxin aflatoxin detection methods in foods and examined available or possible regulations to best control its incidence in street-vended food. Adequate understanding of these important information about aflatoxin will form a bedrock for its control in street-vended food.

Keywords: incidence, implication, aflatoxicosis, toxicity, street vended

1. Introduction

Mycotoxins are secondary metabolites of molds (low-molecular-weight organic compounds) that have adverse effects on human, animals, and crops resulting in various illnesses, termed as mycotoxicosis [1]. Mycotoxins are colorless, odorless, and tasteless compounds with diverse characteristic structures and molecular weight, examples are deoxynivalenol (DON), fumonisins, ochratoxins, patulins, zearalenols, trichothecenes, and so on, but the most common of them are the aflatoxins. The name "aflatoxin" is known as toxins that are produced by *Aspergillus* species especially the *Aspergillus flavus* and *Aspergillus parasiticus*; however, other fungi that have recently been reported to be able to produce aflatoxins are *A. nomius, A. pseudotamarii*, and *A. bombycis*. Aflatoxins are difuranocoumarins in nature, and they are of different types such as aflatoxin B_1 (AFB_1), B_2 (AFB_2), G_1 (AFG_1) and G_2 (AFG_2), M_1 (AFM_1), and M_2 (AFM_2), another 18 groups of closely related aflatoxins have been detected in agricultural products and different street-vended foods especially stored grains [2, 3].

Street-vended foods are very important to human health as many people derive pleasure in quick ready-to-be-eaten food without stress especially on the road, streets, and other public places due to their unique flavors and conveniences. Most people that are involved in street food vending are the poor and the lower class people, most of whom are ultimately aimed in getting profit other than meeting people's needs, and some hygienic handlings might be too expensive; they are therefore capable of spreading food-borne diseases if not hygienically handled, and aflatoxins are one of the most frequently reported toxins in street-vended foods.

Aflatoxin contamination of foods can, however, occur at different points in the food chain depending on the time of mold's evasion. It may be produced during preharvest, harvest, drying, or storage period but ultimately depends on the method and handlings, packaging, or transport conditions of the food materials. Once any of the conditions favor the fungal growth, aflatoxins may be secreted. Rice and Ross [4] reported that FAO estimated that approximately 25% of the world's cereal products are contaminated with mycotoxins. Rapid urbanization and population growths increase the labor force, and demand for survival with these street food trades gained its momentum. Majority of the population and labor force belong to the lower class group, and the street food trade is becoming viable informal-sector industry especially for the developing entrepreneurs. Consequently, people residing in aflatoxin-endemic countries are more reported with increased incidence of acute hepatic necrosis, resulting later in cirrhosis, or hepatocellular carcinoma, and this may be increasing due to different cultural systems of food preparation and storage [2, 5–8].

However, when considering different production methods adopted for different street-vended foods, it becomes very important to understand the chemical and health implications of aflatoxin contamination of these foods above stipulated limits and with this, we would be able to design adequate control measures against aflatoxin occurrence in the street-vended foods. Many countries have enacted regulations to prevent mycotoxins in foods due to their effect on human's heath and the world trade, according to the annual report of the Rapid Alert System for Food and Feed (RASFF) [9], mycotoxins were the main hazard in border rejection notifications in the European Union (**Table 1**).

Mycotoxin	2008	2009	2010	2011	2012	Total
Aflatoxins	902	638	649	585	484	3258
Deoxynivalenol (DON)	4	3	2	11	4	24
Fumonisins	2	1	3	4	4	14
Ochratoxin A	20	27	34	35	32	148
Patulin	3	–	–	–	–	3
Zearalenone	2	–	–	–	4	6
Total	933	669	688	635	525	3450

Table 1. Annual reports on mycotoxins from EU countries by Rapid Alert System for Food and Feed (RASFF) [11].

2. Street-vended foods and aflatoxin incidence

Street food vending can make good contribution to the economy of developing countries, for example, the Indian National Policy for Urban Street Vendors/Hawkers reported street vendors of about 2% of the metropolis population [10, 11]. However, street-vended foods are perceived to pose some major public health risks mostly in developing countries, and this is due to unavailability of basic infrastructures and difficulty in controlling large number of food-vending operators. There are several reported incidences of food poisoning due to consumption of street-vended foods. For example, the Shandong Province in China has recorded about 691 outbreaks of food poisoning from vended foods which are responsible for over 49 deaths during the period of 1983–1992 [12–19]. Aflatoxins are mostly reported in stored cereals, legumes, and nuts, and their derivatives (European Food Safety Authority (EFSA) report [20]). Marin et al. [21] stated that different aflatoxin contents were detected in 34,326 food samples from different European Union countries (**Table 2**) and also in some food materials from African and Asian countries; selected examples are presented in **Table 3**.

Insufficient drying and humid storage environmental conditions may result in high mood invasion and concurrent aflatoxin contamination of foodstuffs. Aflatoxin contamination of street-vended foods is of a great concern especially when it occurred above the tolerance limit. Aflatoxin outbreak, such as the Kenya 2004 and 2005 aflatoxin outbreak from locally stored maize [22], may also have economic implications. Therefore, there is a need to recognize the aflatoxin biosynthesis associated with different handlings of street-vended foods as to human tolerance levels of these aflatoxins. Some countries already developed some specific regulations against aflatoxin contaminations with tolerance limit for aflatoxin B1 in foodstuffs ranging between 0 and 30 μg/kg and total aflatoxin contents of 0–50 μg/kg (worldwide regulations for mycotoxins in food and feed in 2003 [23]). However, prevention of disease in street-vended food in many developing countries was difficult due to uncontrollable environmental factors. The street vendors did not have significant knowledge of epidemiology and their safety measures.

Food category	No. samples	No. samples > LOD	Median AFB1/AFT	Mean AFB1/AFT	Maximum AFB1/AFT
Almonds	1766	471 (27%)	0.20/0.28	1.46/1.82	575/579
Brazil nuts	622	271 (43%)	0.20/0.40	22.2/39.6	1897/3337
Hazelnuts	3163	940 (30%)	0.16/0.30	0.95/1.70	200/200
Cashews	336	33 (10%)	0.10/0.20	0.42/0.60	36/39
Peanuts	8929	1830 (20%)	0.10–0.20	1.93/2.69	935/985
Pistachios	4069	1783 (44%)	0.20/0.40	16.8/19.4	2625/2680
Other nuts	1131	158 (14%)	0.10/0.20	1.16/1.41	385/402
Figs	2067	618 (30%)	0.15/0.24	1.36/2.22	130/151
Other dried fruits	1396	114 (8%)	0.10/0.24	0.26/0.51	20/90
Maize	943	136 (14%)	0.12/0.24	0.26/0.41	8/9
Other cereals	3010	207 (7%)	0.20/0.40	0.35/0.51	109/117
Spices	4698	1988 (42%)	0.20/0.40	1.46/1.88	96/96
Baby foods	592	23 (4%)	0.02/0.04	0.07/0.14	1/2
Other foodstuffs	1604	303 (19%)	0.10/0.20	0.53/0.75	99/99

Table 2. Detected aflatoxin contents in food sample (μg/kg) from different EU countries as reported by the European Food Safety Authority [23].

Country	Commodity	Mycotoxin	Level	Source
Saudi Arabia	Peanuts	Aflatoxins	28 μg/kg	[26]
Nigeria	Groundnuts	Aflatoxins	10–176 ppb	[27]
South Africa	Cowpeas	Fumonisins	0.6–25.30 μg/kg	[27]
Iran	Walnuts	Aflatoxins	14.4 ± 8.4 μg/kg	[29]
Iran	Peanut (roasted)	Aflatoxins	17.99 ± 18.70 μg/kg	[29]
Benin Republic	Cowpea	Aflatoxins	3.52 μg/kg	[28]
Nigeria	Yam chips	Aflatoxins	4–18 μg/kg	[25]
Nigeria	Maize	Aflatoxins	3–138 μg/kg	[31]
Nigeria	Shelled melon	Aflatoxins	5–20 μg/kg	[28]
Pakistan	Chili	Aflatoxins	0.1–96.2 μg/kg	[29]
Pakistan	Milk and sweets	Aflatoxin M1	0.05–0.48 μg/kg	[30]

Table 3. Mycotoxins contamination in some foodstuffs in Afro-Asia.

http://dx.doi.org/10.5772/intechopen.68478

3. Structures and aflatoxin metabolisms in humans

Aflatoxins were first isolated from the Turkish X-disease patients that consumed the mold-contaminated food as a result of production of aflatoxins. Later these toxins were also found in *Aspergillus* and few *Penicillium* species. These aflatoxins were present in food and dairy products. On the basis of fluorescence UV light, the aflatoxins were characterized as B_1, B_2, G_1, and G_2, while minor metabolites were also detected on the basis of thin layer chromatography (TLC) and were known as M_1 and M_2. Aflatoxins of different types have been reported in many street-vended foods, the most common ones are Aflatoxin B, G, and M. They are crystalline, soluble in polar solvents (methanol, chloroform, and water). They fluoresce when exposed to UV light. Ammonia and/or hypochloride solutions have been investigated for removal of aflatoxins in food materials as it is believed that the lactone ring can cause aflatoxin alkaline hydrolysis; however, the toxicity of this breakdown is still of a great concern. Aflatoxins B_1 and G_1 are usually converted into B_{2a} and G_{2a}, respectively, during catalytic acid interference leady to oxidize these toxins; this makes them lose their crystalline nature.

Aflatoxin B_1 is one of the potent, mutagenic, and carcinogenic toxins [32–35]. It was present mostly in conjugation with other aflatoxins like B_2, G_2, and G_1. Structurally, it consists of either five rings along with furofuran moiety or five aromatic carbon rings or six lactone rings. Another parasitical aflatoxin was produced from *Aspergillus* species which is known as B_2. It was first isolated and identified from *Aspergillus parasiticus*. Both B1 and B2 emit lights under UV fluorescence. B_1 and B_2 emit blue light [38]. M_1 and M_2 were minor metabolites produced by the B_1 and B_2 aflatoxins. These aflatoxins were formed when cows were fed with a fungal contaminated food. The most important carriers of M_1 and M_2 in cows are maize, cotton seed, ground nut, etc. M_1 and M_2 were produced in the milk of cow; hence, it is also known as cow milk aflatoxins. Structurally aflatoxins also consist of five/six aromatic or lactone rings along with the furofuran moiety but in the junction of carbon furan rings and the hydroxyl group as shown in **Figure 1**.

The chemical characteristics of these aflatoxins are presented in **Table 4**.

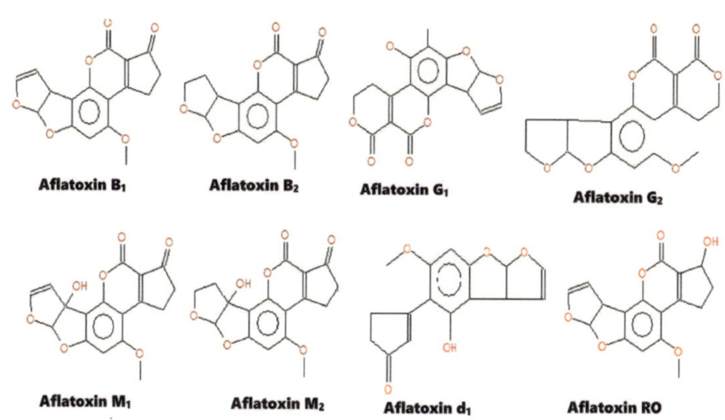

Figure 1. Structure of aflatoxins.

Metabolism of aflatoxins usually form many other metabolites, examples are aflatoxins R_0, RB_1, RB_2, and H. These aflatoxins have hydroxyl group in their ring's carbonyl group, D ring may also be formed in some other cases such as in aflatoxins RB_1 and RB_2 while an opened E ring may be formed in B_3.

Two pathways known for the formation of all these aflatoxins are:

1. Microbial transformation
2. Chemical reduction of sodium borhydride

Aflatoxins are highly toxic and carcinogenic. The disease caused by aflatoxins is known as aflatoxicosis. It mainly targets the liver of humans and animals. Toxicity of aflatoxins depends upon the number of factors such as age, sex, species, and national factors [24]. The aflatoxin metabolic pathways are well studied in animals, but in human, data was very limited. In rats, the DNA/RNA synthesis was inhibited by ingestion of aflatoxins (5 mg/kg). The B1 aflatoxin binds with the N_7 guanine by covalent bonding and formed AFB1-N7-guanine adducts. It results in the transversion of G to T which caused the DNA mutations and carcinogenic. In animals or in humans, the pathway of aflatoxins is described as in **Figure 2**. While in humans (AFB_1-N7-guanine)

Aflatoxin	Molecular formula	Molecular weight	Melting point	UV absorption max (e), nm, methanol	
				265	360–362
AFB1	$C_{17}H_{12}O_6$	312	268–269	12,400	21,800
AFB2	$C_{17}H_{14}O_6$	314	286–289	12,100	24,000
AFG1	$C_{17}H_{12}O_7$	328	244–246	9600	17,700
AFG2	$C_{17}H_{14}O_7$	330	237–240	8200	17,100
AFM1	$C_{17}H_{12}O_7$	328	299	14,150	21,250 (357)
AFM2	$C_{17}H_{14}O_7$	330	293	12,100 (264)	22,900 (357)
AFd1	$C_{16}H_{14}O_5$	286	250–300	11,200	20,120
AFRO	$C_{17}H_{14}O_6$	314	280–290	12,100 (310)	22,900 (320)

Table 4. Chemical characteristic of different aflatoxins.

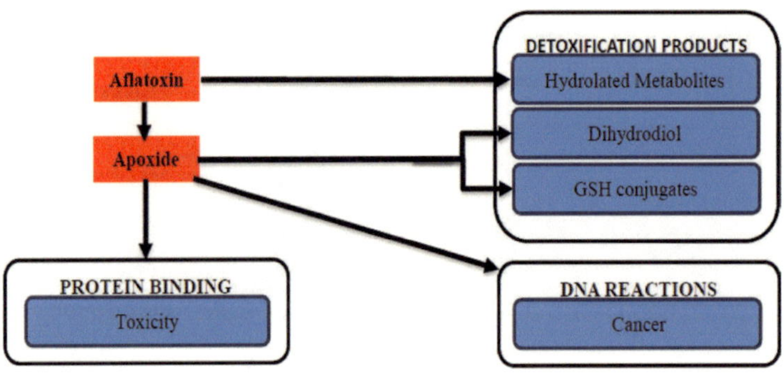

Figure 2. Aflatoxin metabolism in humans.

transverse the nucleotides G to T at the position of codan 249. These suppress the p53 tumor gene. The reactive epoxide formed which hydrolyzed to form AFB_1-8,9-dihydrodiol. The dihydrodiol ionizes and formed a Schiff's base with primary amine groups in the proteins. The AFB_1 also inhibits the phosphodiesterase activity in the kidney, liver, and brain [36].

4. Tolerance, detection, and aflatoxin action as chemical poison

In the 1980s, aflatoxin B1 and sum total aflatoxin tolerance level in foodstuffs range from 0 to 50 μg/kg. Some countries have a zero tolerance which practice the limit of detection depending on the analytical procedures and standards (**Table 3**), for example, Austria and Switzerland have the lowest tolerance limit for aflatoxin B1. And aflatoxin B1 is the most important aflatoxin, based on the occurrence frequency and its toxicity. For detection of many aflatoxins, thin layer chromatography (TLC) or high-performance liquid chromatography (HPLC) is commonly used, but in some cases, mini-column, gas chromatography/mass spectrophotometry and qualitative radioimmunoassay (RIA) methods may be adopted. These methods are frequently coupled with the analytical methods of AOAC or AOAC-derived methods. It may be expected here that the immunoassay detection such as the enzyme-linked immunosorbent assay (ELISA) method will also gain attention, but this methods is presently still undergoing intensive validation processes to be used as a generalized tool for regulatory analysis.

Aflatoxin acts as toxic secondary metabolites, carcinogenic, hepatotoxic, immunotoxic, and teratogenic in humans and many animal species. High ingestion and accumulation of 'aflatoxins' also pose toxic effects on human cell and tissues and even on genes. Aflatoxin poisoning is common in some areas of the world, and different types of this case have been documented. Many cases of aflatoxicosis have also been reported in Africa and Asia, most of which involve the ingestion of contaminated cereals such as maize, rice, or other foods like cassava or dried powdered food materials; some other cases were reported in food products such as pasta or peanut meals. An example of such case is the 1990 Malaysia aflatoxin infection of over 40 adult humans and 13 children deaths associated with the consumption of aflatoxin and borate contamination of noodles. The autopsies of the heart, brain, spleen, liver, kidney, and lung showed damages from aflatoxin interference. Autopsy of brain (cerebrum) specimens from 18 kwashiorkor children and 19 other children who had died from a variety of other diseases in Nigeria showed aflatoxin present in 81% of the cases [37].

5. Aflatoxins and mycotoxicosis (toxicology)

5.1. Effect of fungal toxins on humans

The human body could be exposed to mycotoxins through skin contact or by direct inhalation of spore-borne toxins, and these tend to accumulate within the body organs or tissues during their metabolism [24]. Infections caused by fungi such as *A. flavus* are known as *mycosis*, symptoms of high mycotoxin content in body cells is called *mycotoxicosis*, and diseases or symptoms of aflatoxin poison are referred to *aflatoxicosis*. Therefore, there is a high possibility that aflatoxins might play some important roles in mycotoxicosis. Mycotoxins are potent but silent killer endemic in most third world countries due to poor storage systems, leading to obnoxious levels of mycotoxins in food products [38]. These toxic compounds cause induction of jaundice, cancer, immune suppression, premature puberty in girls, reproductive dysfunction, birth defects, obstruction of liver metabolism, and liver cirrhosis by damaging DNA and sulfhydryl bonds in many enzymes [39, 40]; they also take part in gastrointestinal infection, kwashiorkor, Reye's syndrome, and hepatitis. Aflatoxin B1, for example, may cause chromosomal aberration in human. Generally, when aflatoxins B1, G1, and M1 are accumulated in human tissues, it results in epoxide formation at eight and nine positions in the terminal furan ring and then binds covalently with the nucleic acid.

Diagnostic features of mycotoxicosis:

1. Non-transmissible.
2. Seasonal outbreak.
3. Disruption associated with specific foodstuff.
4. Drug treatments have little or no effect.

5.2. Aflatoxins and aflatoxicosis

Many aflatoxins can cause acute mycotoxicosis (i.e., when the symptoms show within a short period after infection say like 7 days, it may lead to death if not treated on time) and/or chronic mycotoxicosis (i.e., when they appear and persist for a long period of time). In 1993 International Agency for Research on Cancer (IARC) has classified four types of naturally occurring aflatoxins AFs (AFB_1, AFB_2, AFG_1, AFG_2) as the most active substances that cause mutagenicity and carcinogenicity [45]. Aflatoxin AFB_1 is the most prevalent among all types of aflatoxins due to long-term chronic exposure to even very small amount of aflatoxin in food items, and it is an important concern for human health [10]. Its chronic exposure can lead to malnutrition, suppressed immune response, centrilobular necrosis, proliferation of bile duct, hepatic lesions and fatty infiltration of liver, and even hepatomas [41, 42] as shown in **Figure 2**.

Symptoms of mycotoxicosis depend on the quantity or concentration of the toxin, time of exposure, type of the mycotoxin involved, degree of toxin combination, host resistant capacity, physiological status of the host, and so on. Mycotoxin can cause heath effect in contact with the skin and alimentary canal, by inhalation or by other means. The toxins can enter into blood streams and the lymphatic system to inhibit the process of protein synthesis, damage macrophage system, or affect the lung's ability to clear particles or cause immune suppression.

5.3. Hepatocellular carcinoma (HCC)

Carcinogenic chemicals like aflatoxin can bring different modifications in DNA sequence or protein structure which causes DNA adduct formation and finally cancer in effected people. AFB1 is a micro-component of nutrition that causes genetic alteration by inducing adduct formation, leading to DNA strand break. This DNA or oxidative damage can turn out to be hepatocellular carcinoma (HCC) or cancer [43–48].

http://dx.doi.org/10.5772/intechopen.68478

Studies have demonstrated that HCC caused by AFB_1 is due to *p53* gene mutation which occurs as a result of transversion at 249 codon (guanine to thymine). This kind of mutation ends up with the arginine to serine substitution that causes 50% HCC due to AFB1 [45, 46]. The liver is the target organ for AFB_1 metabolism where mechanism of action starts with food ingestion. In the liver, cytochrome-P450 enzyme initiates metabolism of AFB_1 by transformation of AFB_1 to nucleophilic, reactive genotoxic intermediate (aflatoxin $B_{1-8'}$ 9-oxide, AFBO) or hydroxylation and demethylation (as shown in **Figure 3**). When AFBO binds to liver cells, it results in DNA adduct formation, termed as 8, 9, dihydro 8, (N7-guanyl) 9-hydroxy-AFB_1. If this phenomenon extends after DNA replication, adduct reacts with *p53* tumor suppressor gene and causes mutation in it, resulting in HCC. Expression of mutated protein (R249Sp53) may lead to inhibition of apoptosis, inhibition of *p53*-mediated transcription, and liver cell growth [45].

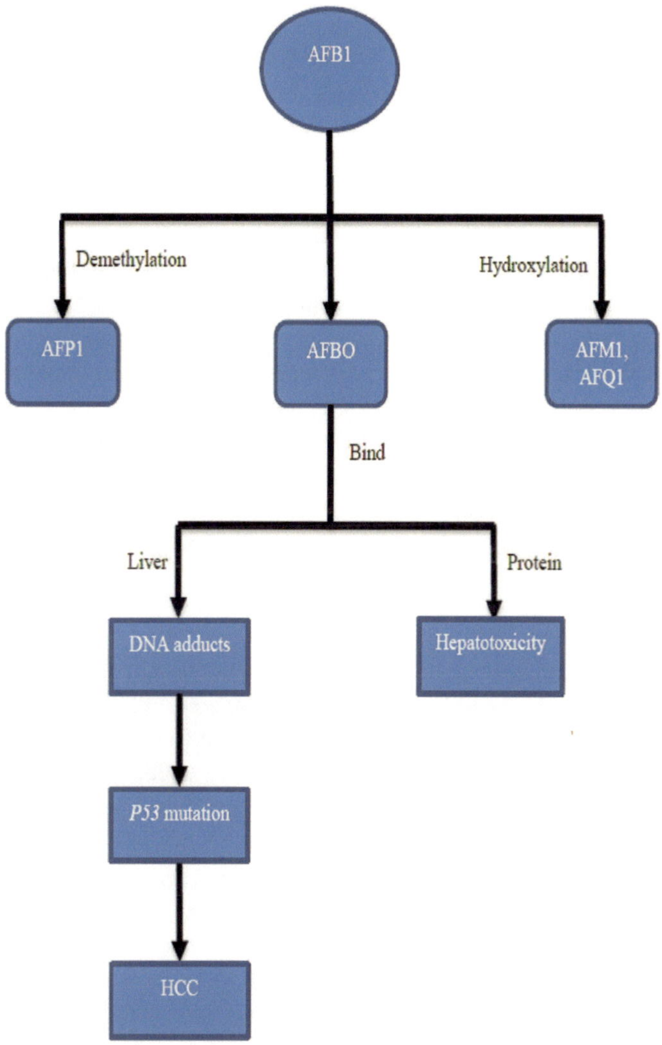

Figure 3. Biotransformation from AFB1 comprises CYP450-mediated reaction resulting in nucleophilic genotoxic reactive intermediate (AFBO), hydroxylation (to AFM_1 and AFQ_1), or demethylation (to AFP_1). AFBO binds to the liver cell and causes mutation in *p53* which leads to HCC. AFBO when binds to protein amino acids causes aflatoxicosis.

5.4. Immunosuppressive action of aflatoxins

Aflatoxins are major factor causing suppression in the immune system which further affects humoral and cellular response. Animal exposure to aflatoxin showed dose-dependent response in percentage of splenic $CD8^+$ T cells, $CD3^-$ $CD8a^+$ cells, natural killer (NK) cells [51, 55, 56]. Reduced expression level of cytokine mRNA in the intestine was observed in a study that might be due to reduced percentage of T cells in broiler. This phenomenon directly affects the immune function of intestinal mucosa [49].

6. Factors affecting the aflatoxin production in street-vended foods

Many fungal strains are genetically capable to produce mycotoxins, but they may not do so until certain conditions are met, and it is known that aflatoxin contents in food vary directly with the processing and storage methods and the associated producer microbe. Physical, chemical, and biological factors can affect the formation of aflatoxins in street-vended food including conditions such as temperature, moisture content, the location of vendors, the storage of utensils, personal hygiene, and method of reheating and storage of food. Availability of nutrients to enhance the fungal growth or energy sources such as sugar and vegetable oil will enhance the toxin production [26], for example, there was a higher aflatoxin produced by *A. flavus* on peanut and cottonseed than rice and sorghum; also, it was reported that some fungi like *Alternaria* and *Fusarium* can cause either preharvest or postharvest contamination of grains. In another study, *P. viridicatum* was reported to produce more stable citrinin and ochratoxin on grain compared to meat [50]. The most important climate factors favoring the aflatoxin production are temperature and moisture; optimum temperature for the growth of aflatoxins ranges from 25 to 37°C. *A. flavus* and *A. parasiticus* produced aflatoxin at 12–41°C, 14% moisture content, and humidity >62% followed by rapid drought [52, 53]; example is the 2004 drought-induced aflatoxicosis breakout in Kenya [49, 54].

The location and conditions set for preparing and storing street food may contribute to the production of fungal toxins. Jonathan et al. [31, 55, 56] reported variations in biodeteriorating fungi and aflatoxin contents of particular foodstuff collected from different locations in Nigeria. Most of the vendors are illiterate and do not have knowledge about the sanitary conditions [57]. A case reported in Africa showed that 85% of food stalls were located near garbage dumping sites [58]. Some storage molds such as *Aspergillus* and *penicillium* can survive environments with very low water content. In postharvest storage of food crops, aflatoxin production may be triggered by prolonged drought-associated elevated temperatures [58]. In some other cases, early harvest is also a good strategy to control mold contaminations [59]. Kaaya et al. [60] observed about 4 times increase in aflatoxin levels during the 3rd week and 7 times more in delay harvest after 4 weeks. However, following early harvest, crops have to be well dried to safe fungi contamination. Rachaputi et al. [61] observed least aflatoxin contents in early groundnut harvest and 27% increase in delayed harvest.

Many traditional ways used for food preparation such as heating and roasting can reduce the aflatoxin contents in foods; ranging values in sausage rolls prepared from different locations in

Nigeria were observed by Jonathan et al. [55] due to different methods adopted for its preparation; Fandohan et al. [62] determined the fates of aflatoxin and fumonisin in different traditional ways of preparing maize and maize foods based on washing, sorting, winnowing, and hulling combined with crushing of the grains; this is also supported by Park [63] and Lopez-Garcia and Park [64]. The presence of other microorganisms either bacteria or fungi may alter elaboration of mycotoxins on food materials, for example, there was reduced aflatoxin production when *A. parasiticus* was grown in the presence of some bacteria, *Streptococcus lactis* and *Lactobacillus casei* [65]. Meanwhile, fungal metabolites such as rubratoxins from *Penicillium purpurogenum*, cerulenin from *Cephalosporium caeruleus*, and *Acrocylindrium oryzae* enhance aflatoxin production even though they repress growth of aflatoxin-producing fungi [66–69].

Aflatoxin-producing fungi can contaminate food crops during cultivation, harvest, transport, storage, or food preparation [70–72]. Iaha et al. [70] reported that insect damage of maize enhances *Fusarium* mycotoxin through the inoculation of fungal spores into the wounds made on the kernels or stalks during feeding [73]. Higher aflatoxin contents were recorded in maize samples stored in high moisture content up till 10 [74–76]. Most of the people involved in street food vending are lower or middle class people who have more concerns for income or quantity than the quality of their products; many of them lack quality information on how to maintain good hygiene. Most street hawkers always believe in improvised cheap usage of materials during the food processing which may contravene the quality of the food product. However, these food products are sold to many people including the high class group of people. The lack of mycotoxin awareness among the street food vendors is one of the factor increasing mycotoxin incidences in foodstuffs. Many contaminated foods are still sold to avert loss of investment; many of the infected foodstuffs are sold at cheaper amount, and people tend to still buy them especially in countries with high poverty level.

7. Economic impact of mycotoxins

Effect of aflatoxin on the economy require good imparts assessment model experts and data sets [80]; this may include income losses due to deaths of livestock, weight loss, reductions in productivity, and the yield of eggs, meat, and milk. It affects aves (poultry) such as ducks, chicken, and turkeys; mostly prime example is that of the 1960 outbreak of "Turkey-X disease" in the United Kingdom caused about 100,000 turkey deaths; it may also lead to loss in fishes, birds, rabbits, dogs, and other mammals. The effect of aflatoxin on economy, storage loss, and generally output loss in animal husbandry due to aflatoxin incidence would be threatened here. Aflatoxin lethal dose (LD-50) is generally between 0.5 and 10 mg/kg of livestock body weight; however, high dose of mycotoxin was responsible for, another case with more economic loss may still occur if mycotoxins are not controlled. There is a need for capacity building to enhance the analysis economic impact of mycotoxin and trade analysis.

Mycotoxin contaminations may also affect other sectors of food production and agriculture. Infected commodities can be rejected from shipments, and some prices may become reduced due to loss in quality; this can devastate export markets especially in developing countries. The consumers may be indirectly affected by increased price/costs of food materials as the sellers/producers levied extra cost due to maintenance against contamination or health risks if they are only able to afford contaminated products which are always cheaper. Farmers may suffer reduced income, feed loss, or low outputs due to aflatoxin/mycotoxin interference most especially on the stored products, and economy impact of mycotoxins stems from high mortality, immunity, weight loss, fertility, and quality of dairy products in livestock.

However, we need to checkmate the mycotoxin preventive cost over economic cost; preventive cost in most cases saves many products as the saying "prevention is better than cure" and yields better production and storage practices. Most African countries such as Gambia, Ghana, Mali, Niger, Nigeria, Senegal, as well as Sudan have reported about 7.5 million US$ cost for mycotoxin control programs due to high-reported mycotoxin incidences on foodstuffs like peanuts and many cereals. In addition, fungi biodeterioration of foodstuffs has caused many countries economic loss due to rejection of exported foodstuffs and affected relationships with trading partners. Implementation of credible food safety controls by exporting countries is needed to foster their smooth exportation of food and agricultural produce. Quality assurance by importing countries has to be routine, and by so doing, the manufactured food will have insignificant mycotoxin levels. The exporting country must be able to comply with this requirement and demonstrate that compliance has been realized. Effectiveness of control measures requires important elements such as administrative structures, resource management, scientific and technical infrastructure, and financing and human capital. Lack of efficient management of resources in many countries has been verified to compromise the credibility of food safety controls. In 2010, foods worth more than US$200,000 were destroyed by regulatory agencies in Nigeria as a result of contamination by mycotoxins.

Aflatoxin (B_1, B_2, G_1, and G_2) level in product (ng/g) can give a mark of quality and can be used as a threshold for distinguishing low-, medium-, and high-quality product. Africa could have accrued an estimate of 67 million dollars annually, but this is lost due to export rejection of its food and agricultural produce contaminated with high levels of mycotoxins. Over the years, the European Union Rapid Alert System has notified Nigeria on its alarming export rejects. Contamination by mycotoxins is owed to hot and humid conditions, soil condition, storage method, crop variety, cultural practices, harvesting procedures, etc.

8. Prevention of aflatoxin contamination in street-vended foods

Adequate practices such as pest and disease control management are needed for safe and healthy food production; this must be judiciously implemented to tackle the menace of aflatoxin spoilage of storage and dairy agricultural produce. Early harvest of agricultural produce may avert aflatoxin/mycotoxin contaminations as the fungi that produce these toxins are usually associated with deterioration of the organic compounds. More also, mycotoxin prevention starts right from the cultivation and rearing of livestock. The use of contaminant-free tools, sterile water for washing, and some other hygienic measures needs to be taken seriously either the products are to be consumed or for sale. Some scientists suggest harvesting at dry period, and storage of agricultural produce in dry environment as wet condition improves the growth of molds [77]. Proper cleaning of harvested products, removal of spoilt products (most especially grains), and proper storage methods are good for mycotoxin control. Good transport and storage mechanisms and frequent check for daily temperature and humidity are good for mycotoxin control.

Many scientists have suggested the use of living organisms to nullify or reduce mycotoxin contamination [78]; this is called *biological control measures*. In this method, the mycotoxin-producing fungi or the pest that can act as vector are targeted. The organisms that kill, feed, or attack the toxin production organisms or vector are introduced. In the United States, biological control measures for mycotoxin have been encouraged for various crops like cotton, peanuts, and maize. In Nigeria, the International Institute for Tropical Agriculture (IITA) has reported much success in biocontrol measures on aflatoxins reported as *Aflasave* measures; this has been tried on many crops [79].

Physical methods such as sorting out bad or infected products from healthy ones, hygienic handling, and maintenance of clean environment can reduce aflatoxin/mycotoxin incidence in street-vended foods or industrial food products [67]. A center control point for testing all street-vended food may be set up by the government to reduce mycotoxin incidence. Regulatory measure on mycotoxin should be strengthened in every country. Food management system point will save a lot from food poisoning [80]. Conduction of test at different steps in food production is also important in big food industries; this may be coupled with the use of supplier's schemes. Hazard and analysis critical control point (HACCP) for pre- and postharvest stage plan has save many commodities from contaminants; this has been used on coconuts and corn in South East Asia and on nuts in South and West Africa; it has also been used on apple juice and pistachio nut in South America [79]. Detailed information on HACCP was reported by FAO [79] and suggested the implementation of HACCP for adequate fungal toxin management.

9. Regulations on fungal toxin (mycotoxin)

Legislation and food inspections are very important for food security. For thousands of years back, food has always been subjected to legislation and inspection as the need was felt for some control on the quality of food materials. This control was intended to safeguard the health of consumers and as well prevent cheating in terms of the composition of food; the principle that food contaminated with a hazardous substance is unfit for human consumption and shall not be sold or offered for sale was not always applied as intended. Before, local and municipal affair/ordinances were used for regulated food quality as there were no advanced scientific methods and tools, but today, advances in science of bacteriology, microscopy, and chemistry have aided official legislations on food. Today, we have legislations that prohibit adulterated or misbranded foodstuffs and incidence of contaminants in food particularly incidence of these contaminants above the tolerance limit in either humans or animals.

Mycotoxins as a food toxic substance have recently been considered in food regulation (**Table 5**). It started shortly after the aflatoxin discovery in the 1960s, and some other specific mycotoxins such as ochratoxin, zearalenone, deoxynivalenol, patulin, and phomopsin were later considered. An attempt to study the whole world's legislation on mycotoxins was made by the International Union of Pure and Applied Chemistry (IUPAC) and the Food and Agriculture Organization which led to an updated paper that was published by Schuller et al. [80] in the

Country	Food material	Tolerance level (μg/kg)		Authority
		B1	B2 + G1 + G2	
Argentine	Sugar-coated nuts	–	5	Ministry of Public Health, Ministry of Agriculture
	Peanuts, peanut products, maize products	5	20	Ministry of Public Health, Ministry of Agriculture
	Infant foods based on cereals and AH foods	2		Ministry of Public Health, Ministry of Agriculture
Australia	All foods except peanut products	3	–	National Health and Medical Research Council, 1982 (recommendation)
Australia	Peanut (products), all foods, except milling and shelled products	15 1	5	Not official Ministry of Public Health
	Milling and shelled products and derived products children foods	2 0.02	5 0.02	Ministry of Public Health Ministry of Public Health
Belgium	All foods	5		Ministry of Public Health and Ministry of Agriculture (not official)
Brazil	Industrially prepared foodstuffs for children from 0 to 2 years and for school meals Imported foodstuffs	5	3 10	Proposed
	Other foodstuffs	15	30	Proposed
Canada	Nuts and nut products		15	Health and Welfare Canada (official)
China	Rice, peanuts, maize, sorghum, beans, wheat barley, oats	50		Ministry of Public Health, Council of Agriculture and Local Authorities (official)
Colombia	Sesame seed		20	Publication official del Instituto Colombiana de normas technicas "Icontec." Norma Colombiana nr. 536, edicion 1981
	Oil seeds (peanuts)		10	Ministry of Public Health (official)
	Cereals, grains (sorghum, millet)		30	Ministry of Public Health (official)
Cuba	Cereals, grains, peanuts		0	Ministry of Agriculture
Czechoslovakia	Cereals, grains, peanuts All foods except infant and children foods	5	10	Ministry of Health (official)
	Infant foods on milk basis (calculated on basis of reconstituted product) Other infant foods and children foods	0.1 1	0.2 2	Ministry of Health (official)
Denmark	Peanuts		10	Ministry of the Environment (official)

Country	Food material	Tolerance level (μg/kg)		Authority
		B1	B2 + G1 + G2	
Dominican Republic	Maize and maize products, peanuts, soya, tomatoes, and products thereof		0	Ministry of Agriculture and Ministry of Public Health
Finland	All foods		5	Decision of the Ministry of Trade and Industry on some food contaminants (762), 1984, Ministry of Trade and Industry; National Board of Trade and Consumer Interests
France	All foods Infant foods Dietary milk Foods		10 5 O-1l^g/100 kilocalories (=0–024 ^g/100 kj)	Conseil Supérieur d'Hygiène Publique de France, séance 25.10.1975 Interests, Ministry of Consumption Arrêté 5.01.1981 Journal Officiel de la Republique Française; 11.01.1981, Ministry of Consumption Arrêté 30.03.1978 Journal Officiel de la Republique Française; 11.01.1981, Ministry of Consumption
Federal Republic Of Germany	Peanuts, peanut products, hazelnuts, walnuts, brazil nuts, pistachio nuts, apricot and peach pits, poppy and sesame seeds, cereals, cereal products, grated coconut, almonds	5	10	Aflatoxinverordnung Bundesgesundheitsblatt IS 3313 of 30.11.1976, Various Not ministries, official depending on the State
German Democratic Republic	All foods	5	10	Circular of Ministry of Health 1970, Ministry of Health
Hong Kong	Infant foods and children foods. Peanuts and peanut products		20	Municipal Services Branch, Goment Secretarial, Hong Kong Government Ministry of Ministry Health of Ordinance Health No. 4 of 25.06.1978
Hungary	All foods	5		Ministry of Health of Ordinance Health No. 4 of 25.06.1978
India	All foods Israel Ireland Italy Grains and nuts, peanuts	30		Ministry of Health and Family Welfare, Department of Health Ministry of Health (official)
Israel	Grains and nuts		20	Ministry of Public Health
Ireland	All foods	5	30	
Italy	Peanuts		50	Ministry of Public Health
Japan	All foods	10		Food Sanitation Investigation Council, April 1974, Ministry of Health and Welfare (official)
Jordan	Almonds, cereals, maize, peanuts, pistachio nuts, pine nuts, rice	15	30	Minister of Finance and Customs Instructions (5/35/8251) 11.03.1981; letter of Minister of Health (48/37/2049) 03.03.1981, Ministry of Health

Country	Food material	Tolerance level (μg/kg)		Authority
		B1	B2 + G1 + G2	
Kenya	Peanuts and other vegetable oils, peanut products		20	Food, Drugs, and Chemical Substances Regulations Kenya Gazette, 01.07.1978, Ministry of Health
Luxembourg	Peanuts and peanut products	5		Règlement Grand-Ducal of 22.09.1978 Loi 25.09.1953, art.7f.ll and 12 Benelux Arrêté M(77) 5-03.05.1977 Ministry of Public Health
Malawi	Peanuts (export)	5		Letter of Malawi Bureau of Standards BS/1/1 of 24.06.1976
Malaysia	All foods		35	Food Regulations 1985
Mauritius	Groundnuts Others	5 5	15 10	Food and Drug (Control of Aflatoxins) Regulations, 1979. Government 1: Notice No. 222 of 19.9.1979
Mexico	All foods		20	Ministry of Official Public Health, Ministry of Agriculture
The Netherlands	Peanuts and peanut products All foods and food ingredient	5 5		Algemeen Besluit (Warenwet) art.3 quinquies. Staatsblad van het Koninkrijk der Nederlanden 46, artikel 1, 19.01.1974, Ministry of Welfare, Public Health and Cultural Affairs Ministry of Welfare, Public Health and Cultural Affairs Official
New Zealand	Peanut butter, shelled nuts and nut portion of products containing nuts, other foods		15	Food Regulations 1984, section 257
Nigeria	All foods Infant foods	20 0		Food and Drug Administration (official) Food and Drug Administration (official)
Norway	Nuts, buck wheat, other foodstuffs	5		Rundskriv IK-1/85 of 08.02.1985
Peru	Maize and peanuts	5		Code of practice
Philippines	Coconut, peanut products (export)	20		Ministry of Health
Poland	All foods	0		Ministry of Public Health
Portugal	Peanuts Infant foods Other foods	25 5 20		Decreto-Lei no.6/83 nr. Diàrio da Républica of 14.01.1983, Ministry of Public Health, Ministry of Agriculture, Ministry of Commerce (official)
Romania	All foods	0		Joint papers of veterinary specialists and doctors of medicine 1978, Ministry of Health Ministry of Agriculture

http://dx.doi.org/10.5772/intechopen.68478

Country	Food material	Tolerance level (μg/kg)		Authority
		B1	B2 + G1 + G2	
Singapore	All foods	0	0	Food Regulations 1974, art. 2, para 3c Government Gazette no. 4959-16.01.1976 Ministry of the Environment
South Africa	All foods	5	10	Government Gazette no. 4959-16.01.1976
Surinam	Peanuts, peanut products, pulses	5		Gouvernementsblad van Suriname nr. 199, 1971, Food Inspection Service (not official)
Sweden	All foods		5	National Food Administration's Ordinance (SLV FS 1983: 1) on Foreign Substances in Food, SLV FS 1985: 16, Swedish National Food Administration
Switzerland (existing regulations) Proposed regulations	Almonds, peanuts, hazelnuts, walnuts, brazil nuts, pistachio, apricot and peach kernels, grated coconut, poppy, sesame, peanut butter, peanut flips, arachis oil (not refined, bottled), pumpkin, kernels, maize, cereals All foods, except maize and cereals	1 2 1	5 5 5	Verordnung über die hygienisch-mikrobiologischen Cantons Anforderungen an Lebensmittel, Gebrauchs-und Verbrauchsgegenstände 817.024 of 14.09.1981
Thailand	All foods		20	Ministry of Public Health Notification no. 98.B.E. 2529: Standard for Food Containing Contaminants, Ministry of Public Health (official)
Union of Socialist Soviet Republics	All foods	5		Methodic Documents Minszdrav USSR 2273-80 of 10.12. 1980 and 4082-86 of 20.03.1986, Ministry of Health and State Agric Industrial Committee of the USSR
The United Kingdom	Nuts and nut products	10		Proposal, MAFF press release 181 of 08.07.1986, Ministry of Agriculture, Fisheries and Food
Yugoslavia	Wheat, maize, rice and other cereals, beans	5	1	Article 57, Federal Register no. 2, 1980, Slüzbeni list Socijalisticka Federation Republika Jugoslavija 2/1980, Federal Committee for Labour, Health and Social Welfare (official) Article 57, Federal Register no. 2, 1980, Slüzbeni list Socijalisticka Federation Republika Jugoslavija 2/1980, Federal Committee for Labour, Health and Social Welfare (official)
Zimbabwe	Groundnuts, maize, sorghum	5	4	Ministry of Agriculture

Table 5. Legal aflatoxin tolerance levels in foodstuffs for humans [80].

1987 (second) mycotoxin joint International annual Conference of FAO/WHO/UNEP. The mycotoxin legislation has since then grow exponentially in various countries. Many countries are known to enforce or propose certain aflatoxin regulations for foodstuffs to continue to increase, and many countries expanded their regulations to specify more types of foodstuffs. As in 1981, the maximum limits for aflatoxins in food (aflatoxin B1 or the sum of aflatoxins B1, B2, G1, and G2) vary from zero detectable to 50 jtg/kg. The strategies which may be employed to limit the establishment of mycotoxins in food should include both on-field and post-field measures. The regulation of a country is to ensure that any food contaminated in an amount that is intolerable from a public health point of view, particularly, at a toxicological level, is not tradable in that country. Contaminant levels are required to be kept as low as possible by proper measures. Regulations are established in many countries to control food contamination so as to protect human health; these regulations may include specific maximum limits for several contaminants for different foods and a reference to the sampling methods and methods of analysis used [81, 82]. Report by FAO in 2003 shows that about 100 countries have existing regulations on mycotoxins in specific foods and feeds; this was about 30% increase over 1995 report.

However, studies have shown that most African countries have adequate mycotoxin regulations due to the fact that many countries in Africa account for highest countries with massive aflatoxin incidence. Morocco had the most detailed regulations on mycotoxins with about 15 nations out of about 99 countries with known mycotoxin regulations in 1995. Nigeria, for example, adopted its regulatory system from the European commission used primarily on export commodities. However, mycotoxin regulatory status quo in African countries still needs to be improved upon especially for effective implementation.

10. Conclusion

In this chapter, it would be understood how historical events should gear enforcement of regulatory limits for foods produced locally in order to maintain a high standard especially in foreign trading. This chapter elucidated the need to devise principal methods of mycotoxin control in food and human. It would be understood that mycotoxins in food commodities are a result of fungi infection. It would reiterate the need for proper cultural practice, sanitation, and good storage procedures, among others, as possible measures.

Author details

Michael Dare Asemoloye[1*], Segun Gbolagade Jonathan[1], Rashid Saddaf[2], Zaffar Habiba[2], Edward Ehidiabhen Okoawo[1] and Tunde Sheriffdeen Bello[1]

*Address all correspondence to: asemoloyemike@gmail.com

1 Mycology/Biotechnology Unit, Department of Botany, University of Ibadan, Ibadan, Nigeria

2 Department of Environmental Sciences, COMSATS Institute of Information Technology, Abbottabad, Pakistan

References

[1] Zain ME. Impact of mycotoxins on humans and animals. Journal of Saudi Chemical Society. 2011;**15**(2):129-144

[2] IARC. Traditional herbal medicines, some mycotoxins, napthalene, and styrene. IARC Monographs on the Evaluation Carcinogenic Risk to Humans. 2002;**82**:1-556

[3] Jonathan SG, Adeniyi MA, Asemoloye MD. Fungal biodeterioration, aflatoxin contamination and nutrient value of 'aadun'. Researcher. 2015;**7**(12):26-32

[4] Rice LG, Ross FB. Methods for detection and quantitation of fumonisins in corn, cereal products and animal excreta. Journal of Food Protection. 1994;**57**(5):36-40

[5] Turner PC, Gong YY, Pourshams A, Jafari E, Routledge MN, Malekzadeh R, Wild CP, Boffetta P, Islami F. A pilot survey for *Fusarium* mycotoxin biomarkers in women from Golestan, Northern Iran. World Mycotoxin Journal 2012;**5**:195-199

[6] Abia WA, Warth B, Sulyok M, Krska R, Tchana AN, Njobeh PB, Dutton MF, Moundipa PF. Determination of multi-mycotoxin occurrence in cereals, nuts and their products in Cameroon by liquid chromatography tandem mass spectrometry (LC-MS/MS). Food Control. 2013;**31**:438-453

[7] Abia WA, Warth B, Sulyok M, Krska R, Tchana AN, Njobeh PB, Turner PC, Kouanfack C, Eyongetah M, Dutton MF, Moundipa PF. Biomonitoring of mycotoxin exposure in Cameroon using a urinary multi-biomarker approach. Food and Chemical Toxicology. 2013;**62**:927-934

[8] Ezekiel CN, Sombie JI. Survey of aflatoxins and fungi in some commercial breakfast cereals and pastas retailed in Ogun State, Nigeria. Nature and Science. 2014;**12**(6):27-32

[9] Rapid Alert System for Food and Feed (RASFF). 2012. Annual Reports 2012. European Commission. http://ec.europa.eu/food/food/rapidalert/rasff_publications_en.htm

[10] Bhowmik SK. Street vendors in Asia: a review. Econ Political Weekly, (2005, May 28-June 4). pp. 2257-2264

[11] Ghosh M, Wahi S, Kumar M, Ganguli A. Prevalence of enterotoxigenic *Staphylococcus aureus* and *Shigella* spp. in some raw street vended Indian foods. International Journal of Environmental Health Research. 2007;**17**(2):151-157

[12] deSousa CP. The impact of food manufacturing practices on food borne diseases. Brazilian Archives of Biology and Technology. 2008;**51**(4):815-823

[13] Kaul M, Agarwal G. Microbial load of common chat products. The Indian Journal of Nutrition and Dietetics. 1988;**25**:197-199

[14] Rodrigue DC, Tauxe RV, Rowe B. International increase in *Salmonella enteritidis*: A new pandemic. Epidemiology and Infection. 1990;**10**:21-27

[15] Desenclos JC, Klontz KC, Wolfe LE, Hoecheri S. The risk of Vibrio illness in the Florida raw oyster eating population. American Journal of Epidemiology. 1991;**134**:290-297

[16] Todd EC. Foodborne disease in *Canada*: A 10-year summary from 1975 to 1984. Journal of Food Protection. 1992;**55**:123-132

[17] Bryan F. Risks of practices, procedures and processes that lead to outbreaks of food borne diseases. Journal of Food Protection. 1988;**51**(8):663-673

[18] Muleta D, Ashnafi M. *Salmonella, Shigella* and growth potential of other food-borne pathogens in Ethiopian street vended foods. East African Medical Journal. 2001;**78**(11):576-580

[19] Rahimi E, Bonyadian M, Rafei M, Kazemeini HR. Occurrence of aflatoxin M1 in raw milk of five dairy species in Ahvaz, Iran. Food and Chemical Toxicology. 2010;**48**(1):129-131

[20] European Food Safety Authority (EFSA). Scientific opinion: Effects on public health of an increase of the levels for aflatoxin total from 4 μg/kg to 10 μg/kg for tree nuts other than almonds hazelnuts and pistachios. Statement of the Panel on Contaminants in the Food Chain. EFSA Journal. 2009;**1168**:1-11

[21] Marin S, Ramos AJ, Cano-Sancho G, Sanchis V. Mycotoxins: Occurrence, toxicology, and exposure assessment. Food and Chemical Toxicology. 2013;**60**:218-237

[22] Ostadrahimi A, Ashrafnejad F, Kazemi A, Sargheini N, Mahdavi R, Farshchian M, Mahluji A. Aflatoxin in raw and salt-roasted nuts (Pistachios, peanuts and walnuts) sold in markets of Tabriz, Iran. Jundishapur Journal of Microbiology. 2014;**7**(1):1-5. DOI: 10.5812/jjm.8751

[23] Kritzinger Q, Aveling TAS, Marasas WFO, Rheeder JP, Van Der Westhuizen L, Shephard G. Mycoflora and fumonisin mycotoxins associated with cowpea (*Vigna unguiculata* L. Walp.) seeds. Journal of Agriculture and Food Chemistry. 2003;**51**:2188-2192

[24] Razzaghi-Abyaneh M, Shams-Ghahfarokhi M, Peng-Kuang C. Aflatoxins: Mechanism of Inhibition by Antagonistic Plants and Bacteria. 2010. www.Intechopen.com/books/... 586-304

[25] Houssou PA, Ahuhuendo BC, Jacobsem M. Natural infection of cowpea by toxigenic fungi and mycotoxins in Benin West Africa. Journal of Stored Products Research. 2009;**45**(1):40-44

[26] Bankole SA, Adebanjo A. Mycotoxins in food in West Africa: Current situation and possibilities of controlling it. African Journal of Biotechnology. 2003;**2**(9):254-263

[27] Bankole SA, Ogunsanwo BM, Mabekoje OO. Natural occurrence of moulds and aflatoxins B1 in melon seeds from markets in Nigeria. Food and Chemical Toxicology. 2004;**42**:1309-1314

[28] Bankole SA, Ogunsanwo BM, Eseigbe DA. Aflatoxins in Nigerian dry-roasted groundnuts. Food Chemistry. 2005;**89**(4):503-506

[29] Paterson RRM. Aflatoxins contamination in chilli samples from Pakistan. Food Control. 2007;**18**:817-820

[30] Asma S, Makhdoom AJ, Youjun D, Erum AH, Sualeha R, Saima N, Muhammad A. A survey of aflatoxin M1 in milk and sweets of Punjab, Pakistan. Food Control. 2012;**26**:235-240

[31] Jonathan SG, Oghodero O, Asemoloye MD. Incidence of molds in treated and untreated drinking water of selected local governments in Ibadan, South-Western Nigeria. Research Reviews in Biosciences. 2016;**11**(4):108-118

[32] MMWR Morbidity and Mortality Weekly Report. 2004 Sep 3;**53**(34):790-793

[33] Barro N, Bello AR, Aly S, Ouattara CM, Ilboudo AJ, Traaore AS. Hygienic status assessment of dish washing waters, utensils, hands and pieces of money from street food processing sites in Ouagadougou (Burkina Faso). African Journal of Biotechnology. 2006;**5**(11):1107-1112

[34] Hussain I, Anwar J. A study on contamination of aflatoxin M1 in raw milk in the Punjab province of Pakistan. Food Control. 2008;**19**(4):393-395

[35] Pier AC, McLoughlin ME, Richard JL, Baetz A, Dahfren RR. Utero transfer of aflatoxin and selected effects in neonatal pigs. In: Lacey J, editor. Trichothecenes and Other Mycotoxins. New York: Wiley; 1985. pp. 495-506

[36] Jonathan HW, Timothy DP, Pauline EJ, Jonathan, KS, Curtis MJ, Deepak A. Human aflatoxicosis in developing countries: A review of toxicology, exposure, potential health consequences and interventions. American Journal of Clinical Nutrition. 2004;**80**:1106-1122

[37] Aflafact Fact Sheet of the Leatherhead Food Research Association, Randalls Road, Leatherhead, Surrey KT22 7RY, England. 2014 June 16: 1-4 www.micotoxinas.com.br

[38] Enyiukwu DN, Awurum AN, Nwaneri JA. Mycotoxins in stored agricultural products: Implications to food safety and health and prospects of plant-derived pesticides as novel approach to their management. Greener Journal of Microbiology and Antimicrobials. 2014;**2**(3):32-48

[39] Reddy KR, Salleh B, Saad B, Abbas HK, Abel CA, Shier WT. An overview of mycotoxin contamination in foods and its implications for human health. Toxin Reviews. 2010;**29**(1):3-26

[40] Pleadin J. Mycotoxins in grains and feed: Contamination and toxic effect in animals. Biotechnology in Animal Husbandry. 2015;**31**(4):441-456

[41] World Health Organization. Evaluation of Certain Food Additives and Contaminants. WHO Technical Report Series. 1999. http://apps.who.int/iris/bitstream/10665/42142/1/WHO_TRS_884.pdf

[42] Christiane LC, Marcia RF, Carla CA, Kemmelmeier C. In vitro activity of Neem oil [*Azadirachta indica* A. Juss (Meliaceae) on *Aspergillus flavus* growth, sporulation, viability of spores, morphology and Aflatoxins B1 and B2 production. Advances in Bioscience and Biotechnology. 2010;**1**(4):292

[43] Jiujiang Y. Current understanding on aflatoxin biosynthesis and future perspective in reducing aflatoxin contamination. Toxins. 2012;**4**:1024-1057

[44] Zia U. Analysis of milk marketing chain—Pakistan. Italian Journal of Animal Sciences. 2007;**6**:1384-1386

[45] Kew MC. Aflatoxins as a cause of hepatocellular carcinoma. Journal of Gastrointestinal and Liver Diseases. 2013;**22**(3):305-310

[46] Jackson PE, Kuang SY, Wang JB, Strickland PT, Muñoz A, Kensler TW, Groopman JD. Prospective detection of codon 249 mutations in plasma of hepatocellular carcinoma patients. Carcinogenesis. 2003;**24**(10):657-1663

[47] Martin J, Dufour JF. Tumor suppressor and hepatocellular carcinoma. World Journal of Gastroenterology. 2008;**14**(11):17-20

[48] Hamid AS, Tesfamariam IG, Zhang Y, Zhang ZG. Aflatoxin B1-induced hepatocellular carcinoma in developing countries: Geographical distribution, mechanism of action and prevention (review). Oncology Letters. 2013;**5**(4):1087-1092

[49] Lewis L, Onsongo M, Njapau H, Rogers HS, Luber G, Kieszak S, Nyamongo J, Backer L, Dahiye AM, Misore A, DeCock K. Aflatoxin contamination of commercial maize products during an outbreak of acute aflatoxicosis in eastern and central Kenya. Environmental Health Perspectives. 2005;**113**:pp. 1763-1767

[50] Obuseh FA, Jolly PE, Kulczycki A, Ehiri J, Waterbor J, Desmond RA, Preko PO, Jiang Y, Piyathilake CJ. Aflatoxin levels, plasma vitamins A and E concentrations, and their association with HIV and hepatitis B virus infections in Ghanaians: A cross-sectional study. Journal of the International AIDS Society. 2011;**14**(1):53 p

[51] Pier AC. The influence of mycotoxins on the immune system. In: Smith JE, Henderson RS, editors. Mycotoxins and Animal Foods. Boca Raton, Florida: CRC Press; 1991. pp. 489-497

[52] Meissonnier GM, Pinton P, Laffitte J, Cossalter AM, Gong YY, Wild CP, Bertin G, Galtier B, Oswald IP. Immunotoxicity of aflatoxin B1: Impairment of the cell-mediated response to vaccine antigen and modulation of cytokine expression. Toxicology and Applied Pharmacology. 2008;**231**(2):142-149

[53] Jiang M, Peng X, Fang J, Cui H, Yu Z, Chen Z. Effects of aflatoxin B1 on T-cell subsets and mRNA expression of cytokines in the intestine of broilers. International Journal of Molecular Sciences. 2015;**16**(4):6945-6959

[54] Nyikal J, Misore A, Nzioka C, Njuguna C, Muchiri E, Njau J, Maingi S, Njoroge J, Mutiso J, Onteri J, Langat A, Kilei IK, Nyamongo J, Ogana G, Muture B, Tukei P, Onyango C, Ochieng W, Tetteh C, Likimani S, Nguku P, Galgalo T, Kibet S, Manya A, Dahiye A, Mwihia J, Mugoya I, Onsongo J, Ngindu A, DeCock KM, Lindblade K, Slutsker L, Amornkul P, Rosen D, Feiken D, Thomas T, Mensah P, Eseko N, Nejjar A, Onsongo M, Kessel F, Njapau H, Park DL, Lewis L, Luber G, Rogers H, Backer L, Rubin C, Gieseker KE, Azziz-Baumgartner E, Chege W, Bowen A. Outbreak of aflatoxin poisoning: Eastern and Central Provinces, Kenya. MMWR. 2004;**53**:790-793

[55] Jonathan SG, Okoawo EE, Asemoloye MD. Fungi and aflatoxin contamination of sausage rolls. International Journal of Scientific and Research. 2016;**4**(5). 99-104

[56] Jonathan SG, Adeniyi MA, Asemoloye MD. Nutrient value, fungal biodeterioration, and aflatoxin contamination of suya spices a novel Nigerian indigenous snacks. Hindawi-Scientifica. 2016;**2016**:4602036

[57] World Health Organization (WHO). Toxicological evaluation of certain food additives and contaminants. The 35th Meeting of the Joint FAO/WHO Expert Committee on Food Additives. Vol. 26. WHO Food Additives; 1990. pp. 143-165

[58] Dawson RJ, Canet C. International activities in street foods. Food Control. 1991;**2**:135-139

[59] Ekanem EO. The street food trade in Africa: Safety and socio-environmental issues. Food Control. 1998;**9**:211-215

[60] Kaaya AN, Kyamuhangire W, Kyamanywa S. Factors affecting aflatoxin contamination of harvested maize in the three agroecological zones of Uganda. Journal of Applied Sciences. 2006;**6**:2401-2407

[61] Rachaputi NR, Wright GC, Kroschi S. Management practices to minimize pre-harvest aflatoxin contamination in Australian groundnuts. Australian Journal of Experimental Agriculture. 2002;**42**:595-605

[62] Fandohan P, Zoumenou D, Hounhouigan DJ, Marasas WF, Wingfield MJ, Hell K. Fate of aflatoxins and fumonisins during the processing of into food products in Benin. International Journal of Food Microbiology. 2005;**98**:249-259

[63] Park DL. Effect of processing on aflatoxin. Advances in Experimental Medicine and Biology. 2002;**504**:173-179

[64] Lopez-Garcia R, Park DL. Effectiveness of post-harvest procedures in management of mycotoxin hazards. In: Mycotoxins in Agriculture and Food Safety. New York: Marcel Dekker; 1998. pp. 407-433

[65] Ominski KH, Marquardt RR, Sinha RN, Abramson D. Ecological aspects of growth and mycotoxin production by storage fungi. In: Mycotoxins in Grain: Compounds Other Than Aflatoxin. St. Paul: Egan Press; 1994. pp. 287-314

[66] Makun HA, Gbodi TA, Akanya HO, Sakalo, AE, Ogbadu HG. Fungi and some mycotoxins contaminating rice (*Oryza sativa*) in Niger state, Nigeria. African Journal of Biotechnology. 2007;**6**(2):99-108

[67] Njobeh PB, Njobeh MF, Dutton AA, Chuturgoon, SH, Koch PA, Steenkamp SD. Identification of novel metabolite and its cytotoxic effect on human lymphocyte cells in comparison to other mycotoxins. International Journal of Biological and Chemical Sciences. 2009;**3**(3):524-531

[68] Abbas HK, Williams WP, Windham GL, Pringle HC, Xie W, Shier WT. Aflatoxin and fumonisin contamination of commercial corn (*Zea mays*) hybrids in Mississippi 2011. Journal of Agricultural and Food Chemistry. 2002;**50**:5246-5254

[69] Maqbool U, Anwar-ul-Haq Ahmed M. ELISA determination of aflatoxin M1 in milk and dairy products in Pakistan. Toxicological and Environmental Chemistry. 2009;**91**(2):241-249

[70] Iaha MH, Barbosa CB, Okada IA, Trucksess MW. Occurrence of aflatoxin M1 in dairy products in Brazil. Food Control. 2011;**22**(12):1971-1974

[71] Fallah AA, Rahnama M, Jafari T, Saei-Dehkordi SS. Seasonal variation of aflatoxin M1 contamination in industrial and traditional Iranian dairy products. Food Control. 2011;**22**:1653-1656

[72] Prandini A, Tansini G, Sigolo S, Filippi L, Laporta M, Piva G. On the occurrence of aflatoxin M1 in milk and dairy products. Food and Chemical Toxicology. 2009;**47**(5):984-991

[73] Turner PC, Hopton RP, White KL, Fisher J, Cade JE, Wild CP. Assessment of deoxynivalenol metabolite profiles in UK adults. Food and Chemical Toxicology. 2011;**49**:132-135

[74] Mutegi CK, Ngugi HK, Hendriks SL, Jones RB. Prevalence and factors associated with aflatoxin contamination of peanuts from western Kenya. International Journal of Food Microbiology. 2009;**130**:27-34

[75] Palumbo JD, O'Keeffe TL, Multiplex PCR Analysis of fumonisin biosynthetic genes in fumonisin-nonproducing *Aspergillus niger* and *A. awamori* strains. Mycologia. 2013;**105** (2):277-284

[76] Murphy PA, Hendrich S, Langren C, Bryant CM. Food mycotoxins: An update. Journal of Food Science. 2006;**71**(5):51-65

[77] Bandyopadhyay R, Kumar M, Leslie JF. Relative of aflatoxin contamination of cereal crops in West Africa. Food Additives and Contaminants. 2007;**24**(10):1109-1114

[78] World Health Organization, Food Safety Unit. (FOS) AFRO Food Safety Newsletter July 2006 Issue No 2; 1-4

[79] FAO/WHO. Dietary Exposure Assessment of Chemicals in Food: Report of a Joint FAO/ WHO Consultation; 2005. http://whqlibdoc.who.int/publications/2008/97892415597470_eng.pdf

[80] Schuller PL, Egmond H, Van P, Stoloff L. Limits and regulations on mycotoxins. Proceedings International Symposium Mycotoxins; 1983. pp. 111-129

[81] Fallah AA, Rahnama M, Jafari T, Saei-Dehkordi SS. Seasonal variation of aflatoxin M1 contamination in industrial and traditional Iranian dairy products. Food Control. 2011;**22**:1653-1656

[82] Egmond HP, Van-Deyll, WE, Paulsch WE, Schuller PL. Het aantonen en bevestigen van de aanwezigheid van aflatoxine B1 in levensmiddelen. De Ware(n) Chemicus. 1980;**10**:99-115

Leaching of Cyanogens and Mycotoxins from Cultivated Cassava into Agricultural Soil: Effects on Groundwater Quality

Elie F. Itoba-Tombo, Seteno K.O. Ntwampe and John B.N. Mudumbi

Additional information is available at the end of the chapter

http://dx.doi.org/10.5772/intechopen.68715

Abstract

Cyanogens and mycotoxins are vital in protecting flora against predation. Nevertheless, their increased concentrations and by-products in agricultural soil could result in produce contamination and decreased crop yield and soil productivity. When exposed to unsuitable weather conditions, agricultural produce such as cassava is susceptible to bacterial and fungal attack, culminating in spoilage, particularly in arid and semi-arid regions, and contributing to cyanogen and mycotoxins loading of the arable land. The movement of cyanogen including mycotoxins in such soil can result in sub-surface and/or groundwater contamination, thus deteriorating the soil's environmental health and negatively affecting wildlife and humans. Persistent cyanogen and mycotoxins loading into agricultural soil changes its physico-chemical characteristics and biotic parameters. These contaminants and their biodegradation by-products can be dispersed from soil's surface and sub-surface to groundwater systems by permeation and percolation through the upper soil layer into underground water reservoirs, which can result in their exposure to humans and wildlife. Thus, an assessment and monitoring of cyanogen and mycotoxins loading impacts on arable land and groundwater in communities with minimal resources should be done. Overall, these toxicants impacts on agricultural soil's biotic community, affect soil's aggregates, functionality and lead to the soil's low productivity, cross-contamination of fresh agricultural produce.

Keywords: agricultural soil, cassava, cyanogen, groundwater, mycotoxin

1. Introduction

Cyanogens have been widely demonstrated to be an important component within the earth's system. These compounds have been reported to have an influential role in the lives of several organisms on earth [1]. Cyanogens are characterised by the presence of two elements: a carbon: nitrogen functional group held together by a triple bond (—C≡N). The simplest form, which is predominant in the environment, is hydrogen cyanide (HCN), with nitriles and cyanogenic glycosides (CGs) being other forms of these compounds [2–5]. Generally, free cyanide originates from both anthropogenic and natural processes [6]. The anthropogenic sources of cyanide range from effluents discharged from municipal wastewater treatment plants, agricultural run-off, mining activities and electroplating industries [7, 8], including the application of some cyanide containing insecticides in the agricultural industry, which culminates in environmental contamination [9]. Cyanides and CGs have also been generated in plants and agricultural produce such as *Manihot esculenta* (cassava), with the waste generated through processing of such produce contributing to the cyanide load into the environment. During cassava harvesting and processing, plant-borne hydrolases result in CGs' conversion into by-products which are released into the soil, although sometimes this is due to rot produce, a consequence of microbial contamination of the produce and wastewater generated for processing of such produce [10, 11].

As a result of produce-facilitated microbial decay due to the availability of pathogenic organisms in soil where the produce grows, mycotoxins are produced. Mycotoxins are fungal secondary metabolites that also have a negative impact on human and animal wellbeing [12–14]. They co-occur with other bacterial toxins in spoiled agricultural produce such as cassava. Previous studies on mycotoxins revealed that these compounds are hazardous to animals and humans. Generally, it has been reported that CGs as well as mycotoxins occur naturally in flora and organisms (fungi) as a result of biosynthesis, with their prevalence being quantifiable in many agricultural products, such as cassava, apples, spinach, apricots, cherries, peaches, plums, quinces, almonds, sorghum, lima beans, corn, yams, chickpeas, cashews and kirsch [15, 16]. Although some microorganisms and plants synthesise these compounds for their survival when exposed to harsh environmental conditions, their cumulative production can contribute to ecological disturbances. Furthermore, various arthropods and invertebrates were also determined to produce cyanogens as a defence mechanism and for a control of mating behaviour [17, 18], although on a minute scale, with research by Jones [19] indicating that plants including microorganisms are known to be major producers of these compounds owing to their physiology. Thus, the presence and loading of these cyanogens and mycotoxins into terrestrial ecosystems are largely overlooked, although they have some negative effects on the physico-chemical and biological properties of soil, particularly arable land as well as the environment in general [10, 20].

Previous studies have stated that cyanogen and mycotoxin loading in agricultural soil can have a serious impact, disturbing the terrestrial ecosystem functionality [10]. Current evidence suggests that most studies on agricultural produce such as cassava, known for its high cyanogen content, have predominantly focused on the production of the crop for nutritional and industrial purposes, with its effects on soil (including the surrounding environment) overlooked [2, 10] Accordingly, minimal research has been completed on cyanogen and mycotoxin loading, including their behaviour and movement in soil that can culminate in groundwater contamination. This is because a large amount of agricultural produce, such as cassava tubers, perishes prior to harvesting for a variety of reasons. Although free cyanide and mycotoxin toxicity is widely reported, their level of toxicity is also influenced by cumulative exposure and the continuation of their release from produce into the environment. Cyanogen and mycotoxin loads and their movement in soil, including their potential to contaminate groundwater, which is used in impoverished communities where cassava is cultivated mostly as a source of protein and starch, are largely under-reported.

The highlights of this review are:

- There are similarities in the movement of cyanogens and mycotoxins, including their degradation by-products in soils due to mass transfer processes influenced by the moisture content in the soil, thus;
- Cyanogen and mycotoxins distort the soil's characteristics with seepage into groundwater systems being of paramount concern, negatively impacting terrestrial, aquatic life and water quality, thus;
- Culminate into prolonged cumulative human and animal exposure.

2. Cyanogen and mycotoxin reduction

Several methods of cyanogen reduction have been proposed and include physical, chemical and biological methods [6, 21]. However, it has been reported that some of these methods require high input costs and sophisticated knowledge and/or training to implement successful strategies for their reduction [4]. Meanwhile, scientists have embarked on intense research and simplify reduction methods for these toxicants in the environment by using techniques which are considered environmentally benign, as such novel ways of reducing both cyanogen and mycotoxin levels in the environment, including in agricultural produce destined for consumption, are generally considered cost effective when compared with long-term outcomes of none implementation of control measures [22–24].

2.1. Biological reduction of cyanogens

The biological reduction of CGs as a source of cyanide, as well as mycotoxins, has gained popularity and has been a huge research focus area [17, 22, 23]. As such, genetically modified cassava cultivars, with a suppressed cytochrome P450 gene (producers of enzymes CYP79D1 and CYP79D2) functionality, may inhibit the infiltration of linamarin as it can be converted to free cyanide from valine [25].

Furthermore, other biological treatments for free cyanide involve microorganisms; these organisms are known to be toxin producers and are organisms, such as *Pseudomonas* sp.,

Nocardia sp., *Flavobacterium* sp., *Bdellovibrio* sp., as well as nitrifiers, such as *Nitrosomonas* sp., *Nitrobacter* sp., *Sphingomonas* sp., *Exophiala* sp., *Bacillus* sp., and fungi such as *Aspergillus* sp. and *Penicillium* sp. [4, 8, 22, 26–28]. Among these microorganisms, *Aspergillus* sp. and *Penicillium* sp. are the most prevalent species able to grow successfully in stringent weather conditions, with some, including *Cunninghamella* sp. being common in soil [29], with the ability to grow on a variety of agricultural produce such as maize, peanuts and tubers [30, 31].

In soil consisting of fungal biocatalysts of different origins, scientific evidence seems to indicate that agricultural produce appears to be susceptible to spoilage due to substrate availability, which results in the proliferation of microbial spoilage organisms [32, 33]. It has also been reported that fruit or produce has trace elements, such as Ca, Na, K and Zn, and low relative molecular weight hydrocarbons, including proteins and moisture, providing conditions which facilitate microbial growth and thus spoilage [34, 35]. Owing to this, some microorganisms produce hydrolases, reducing primary compounds in produce to by-products, furthering physico-chemical changes in the environment in which they are leached [30]. These seem to be the ideal conditions in which cyanide reduction biocatalysts proliferate, i.e. conditions that are nutrient rich as a result of nutrient availability from decaying produce.

Some of the cyanogens are reduced to by-products such as bicarbonate and ammonia. The ammonia formed during the process is further utilised by the microorganisms as a source of nitrogen, supporting increased microbial growth [36, 37]. In the agricultural industry, the reduction of both cyanogens and related compounds is complex, as *in-situ* quantification of such processes is minimally reported. The development of processes and strategies that are environmentally benign; i.e. of biological origin, is gaining popularity due to their simplicity and advantages, as they are considered less harmful, and can be beneficial in the economical management urged for, in the improvement of commercial agro-produce manufacturers [28, 38, 39]. Owing to the exposure to primary and by-products of cyanogen conversion/transformation, some species became tolerant, thus biologically evolve.

For example, Sing et al. [30] successfully isolated a fungus, *Cunninghamella* sp. UMAS SD12 from sawdust, with an ability to biodegrade 51.7% pentachlorophenol (PCP) within 15 days in a controlled static environment. However, more research needs to be conducted to assess direct evolvement of the microbial ecosystem, as other microorganisms that constitute a community, for the betterment of soil, can reduce such soils' viability, and/or result in some organisms producing extracellular secondary metabolites such as mycotoxins.

2.2. Biological reduction of mycotoxins

There are numerous mycotoxins known to contaminate agricultural produce such as cassava. Among these mycotoxins, fumonisin B1 and deoxynivalenol (DON) are common. The biodegradation of fumonisin B1 and deoxynivalenol (DON) can be achieved through their direct conversion using detoxification processes with different pathways [22]. For example, fumonisin biodegradation was observed through the elimination of the tricarballylate side chains and amino groups. The enzymatic hydrolysis of such mycotoxins might involve carboxylesterases and aminotransferases from bacteria such as *Sphingomonas* and *Sphingopyxis* normally found in soil, which have the ability to detoxify recalcitrant persistent organic pollutants (PoPs) such as polycyclic aromatic hydrocarbons (PAHs) [40–43]. Other researchers have reported degradation or detoxification of fumonisin, including by-products, by oxidative deaminase from *Exophiala* sp., a common soil organism [42–44]. *Bacillus* sp., including non-Saccharomyces yeast commonly found in soil, were also suggested to destabilise these mycotoxins' structure, and thus reduce their amino acid functional groups albeit at elevated pH [45].

In most instances, the biodegradation process of most mycotoxins involves a consortium of organisms, which utilises a variety of degradation pathways [42, 44]. Overall, the initial biodegradation stage starts at extracellular level by deamination or facilitation by esterase with the last biodegradation step involving microbial/enzymatic decoupling of the aliphatic chain within the mycotoxin molecule [22]. For example, the first biodegradation steps of DON using *Curtobacterium* sp. and *Eubacterium* sp. were determined to be initiated by the de-epoxidation step which subsequently followed oxidation [22, 46].

3. Toxicity of cyanide as a cyanogen from cassava

3.1. Toxicity of *Manihot esculenta*

Worldwide, cassava is utilised as a primary foodstuff for disadvantaged and needy rural communities of Africa, Asia and South America [23, 47, 48]. Cassava's toxicity is due to cyanogens such as linamarin, lotaustralin and 2-((6-O-(b-D-apiofuranosyl)-b-D-glucopyranosyl) oxy)-2-methylbutanenitrile) that are biologically transformed into hydrogen cyanide [25, 49]. As a result of enzymatic hydrolysis, for which the linamarin from the plant tissue is transformed into acetone cyanohydrin through linamarases [3]. At an increased temperature and pH of >30°C and 5, respectively, conditions associated with arid regions which are suitable for microbial proliferation and thus agricultural produce contamination or spoilage, acetone cyanohydrin is released, resulting in its decomposition into acetone and hydrogen cyanide [3, 25] (**Figure 1**). Several studies have been done on the impact of the cultivar on humans as a result of direct ingestion [24, 48, 50], as cyanide concentration in the tuber is estimated to reach 50 mg/kg [51].

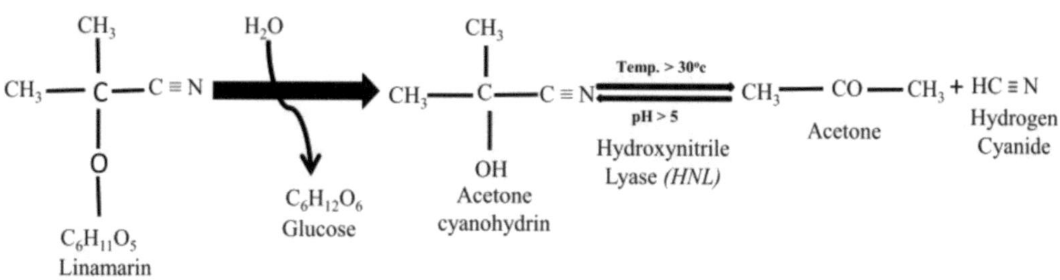

Figure 1. Enzymatic hydrolysis of linamarin to hydrogen cyanide.

Thus, its prolonged consumption may be toxic. However, there is minimal information on hydrogen cyanide loading into irrigable land in which cassava is cultivated. Free metal ions in such soil exposed to hydrogen cyanide can form metallic cyanide complexes under suitable conditions, further prolonging cyanide-based compounds' prevalence in the soil, which might leach into groundwater.

3.2. Production of mycotoxins

Terrestrial ecosystems are populated by a diversity of microorganisms that contribute to and maintain the ecological and biological balance. These organisms contribute to the characteristics of the soil that directly influence soil productivity and crop yield in the agricultural sector [52–54], although some have been shown to exhibit pathogenicity toward mature produce. For example, during the growth and up to the harvest stage of cassava tubers, several pathogenic organisms with mycotoxin-production potential can dominate several other types of bacteria and fungi on the tuber and in cassava-cultivated soils [30]. Some of these organisms are resistant even to the free cyanide in cassava, and with their inherent characteristics, such as their predisposition for survival, they produce mycotoxins such as ochratoxin A, aflatoxins, fumonisin B, pyranonigrin A, tensidol B, funalenone, naphtho-y-pyrones, deoxynivalenol (DON) and malformins [55–57]. Research revealed that exposure to mycotoxins pre/post-harvest and their presence in soil can render the cassava tubers inedible [58, 59], leading to their cumulative and increased levels due to sustained use of pre-recovery land for cultivation to produce an essential food source—a method that will affect the soil's ecology.

3.3. Mycotoxins' effects on soil ecology

Soil ecology is influenced by the biochemical including biotic relationships and physical conditions paramount for its good health [52, 54]. The biochemical aspect of soil used for cultivation is related to its microbial diversity as well as its chemical/pollutant content [53], with the soil's microbial community playing a transformative role with regard to the soil nutrient availability, health and fertility, which enhance the soil's quality, including its productivity [52]. The microbial ecology of any soil facilitates nutrient flow through immobilisation processes, which may result in its bioaugmentation [54, 60, 61], contributing to suitable soil structure that assists in the formation of nutrient-rich aggregates. According to Knudsen [53], soil aggregates are created by microbial activity, albeit at a microscopic level, linking soil particles, while the external polysaccharide tissues of bacterial cells play a role in holding soil aggregates together [52], with subsequent structuring and compaction, parameters influencing the quality of the soil's texture, porosity, aeration, moisture permeability, water circulation and organic matter content [52]. Soil grain cohesion, porosity, permeability and organic matter content are vital for soil quality and fertility, particularly for soil demarcated for sustaining the production of agricultural produce.

All these parameters are indispensable for sustainable use of arable soil for food production and productivity for crop yield [52, 53]. Additionally, soil health can also be affected by surface, subsurface and groundwater supply, including quality.

Soil moisture content is vital for soil functionality as it serves as a water reservoir for the terrestrial ecosystem, playing a major role in the water cycle between surface and subsurface water, thus affecting the quality of groundwater [62, 63]. High mycotoxin loading into the soil may impact its functionality. Thus, the interaction of microorganisms, invertebrates, vertebrates, and planted crops, which eventually leads to the depletion of groundwater quality, leads to sustained leaching or periodic contamination of the water, which can easily lead to human exposure. The disturbance in a terrestrial hydrological movement may have long-term disastrous consequences for surface, subsurface and groundwater supplies [62, 64].

Previous studies on mycotoxin mobility in soil revealed that the movement of these contaminants is influenced by processes such as deposition, decomposition, distribution and accumulation [2, 65], while the compounds' concentration increases with depth. A soil with a high moisture content creates conditions that lead to the furtherance of the contaminants' ability to be transferred, based on processes such as infiltration, percolation and leaching into groundwater [62, 63]. The detoxification bioprocesses and strategies may involve extended periods during which the land is unusable. Furthermore, several studies on the effects of cassava effluents on soil, including microbiota, stated that a high mycotoxin concentration in soil is harmful to these soil microorganisms. Some of these mycotoxins are produced because of inhibitive competition, i.e. organisms will produce these mycotoxins to limit the proliferation of others, particularly under nutrient-depleted conditions.

A study by Knudsen [53] revealed that mycotoxins from cassava are mobile in soil and destroy the resident soil's organisms. Additionally, Okechi et al. [10] showed that the effects of cassava effluents on soil microbial populations revealed a discrepancy in bacterial and fungal populations at different pH levels and soil depths. This indicates that the bacterial populations from the upper layers of soil counts revealed an increase in comparison with those recorded in the lower soil layers, with high concentrations of the mycotoxins observed on the lower soil layers, a process furthered by leaching. Similar total fungal population counts revealed a similar phenomenon with surface, subsurface and deeper soil layers.

3.4. Impacts of hydrovgen cyanide on biochemical and physical properties of agricultural soil

Although the conditions and diversity of habitats contribute to and thus influence the biochemical and physical properties of arable soil [62, 64], a high cyanogen load in soil can have a negative impact on soil microbial populations, with sustained exposure and an increased concentration of cyanogens hindering the microbial activity, and thus the functionality of soil microorganisms, leading to the deformation of the biochemical and physical properties of the soil. A high hydrogen cyanide concentration load in such soil was determined to contribute to an increase in the total oxygen carbon (TOC) and chemical oxygen demand (COD), reducing the ability of *Nitrobacter* sp. to sustain nitration processes [29, 64]. Therefore, an increase in the hydrogen cyanide loading could lead to an imbalance between *Nitrospira* and *Nitrobacter* sp., resulting in a higher count of species with a hydrogen cyanide-resistant ability. The change in the microbial population balance could lead to stunted growth and/ or variations in the growth of a cultivar. This can easily culminate in the dominance of the species, which can be a spoilage organism with free cyanide-resistant characteristics contributing to spoilage patterns/microbial contamination of the produce of interest.

For mitigation strategies in the post-harvesting period, preparation of soil for re-cultivation could lead to inadequate organic matter (OM), variation in total nitrogen (TN) content and availability, which could interfere with soil biochemical and physical properties [64]. Research on the physico-chemical characteristics of cassava-cultivated soil has shown a correlation between continuous cassava cultivation and a decline in the soil's physico-chemical properties (Haplic Acrisols) [64]. Therefore, continuous cultivation of cassava, which normally happens in impoverished communities, could result in a decrease in soil quality, bulk density, organic carbon (OC), OM, trace elements, moisture, infiltration rate, including holding capacity, and aggregate stability. Howeler [66] further reported that the average nutrient removal rate per ton of cassava tuber harvested is equivalent to: N=2.53 (38%), P=0.37 (49%), K=2.75 (56%), Ca=0.44 (16%) and Mg=0.26 (30%). Thus, cyanogen loading indirectly has an impact on C:N ratio, which will result in a pH increase with depth, while OC, nitrogen (N) and OM distortions will be entrenched.

Similarly, Boadi et al. [67] examined the relationship between cyanogen concentration, pH and soil moisture, determining that with an increase in cyanogen values, soil pH increases with moisture content, further supporting the retention of cyanogens at a lower pH. The concentration of cyanogenic compounds was shown to be varied from soil to groundwater and from one site to another [61–63, 68], which suggested that the discrepancies in distribution could be due to the mobility of the contaminant [67].

3.5. Behaviour of cyanogen and mycotoxin in soil

Cyanogen and mycotoxin behaviour in soil, groundwater and the environment is largely controlled by a multitude of chemical reactions and processes. There are similarities and differences between the processes involved for the behaviour of each contaminant, which is largely controlled by conditions the contaminant undergoes when in soil and groundwater. These processes are primarily influenced by numerous biochemical processes and by the compounds' structures, properties and behaviour in the environment. According to Kjeldsen [65], the behaviour of cyanogen and mycotoxins from soil into groundwater is largely influenced by processes such as deposition, dissolution, infiltration, leaching, degradation, transformation and complexation (see **Figure 2**).

Furthermore, human, wildlife exposure and environmental contamination are directly associated with other pathways, such as volatilisation, dermal contact and ingestion of other degradation by-products from the transformation of the primary source due to the transportation pathways facilitated by leaching mechanisms into groundwater [62, 65]. Thus, when not monitored, the environmental prevalence and exposure of these contaminants can be harmful to human health/wildlife. For example, the concentration of leached iron-cyanide complexes in groundwater ranged between 2 and 12 mg/L [59, 69]. The prevalence of such complexes is influenced by the reactivity of free metal ions and free hydrogen cyanide from cassava. These compounds may be transformed (through decomposition) to free cyanide at a later stage, although most are stable with longer half-life, thus they enter an aquifer through processes such as infiltration and leaching.

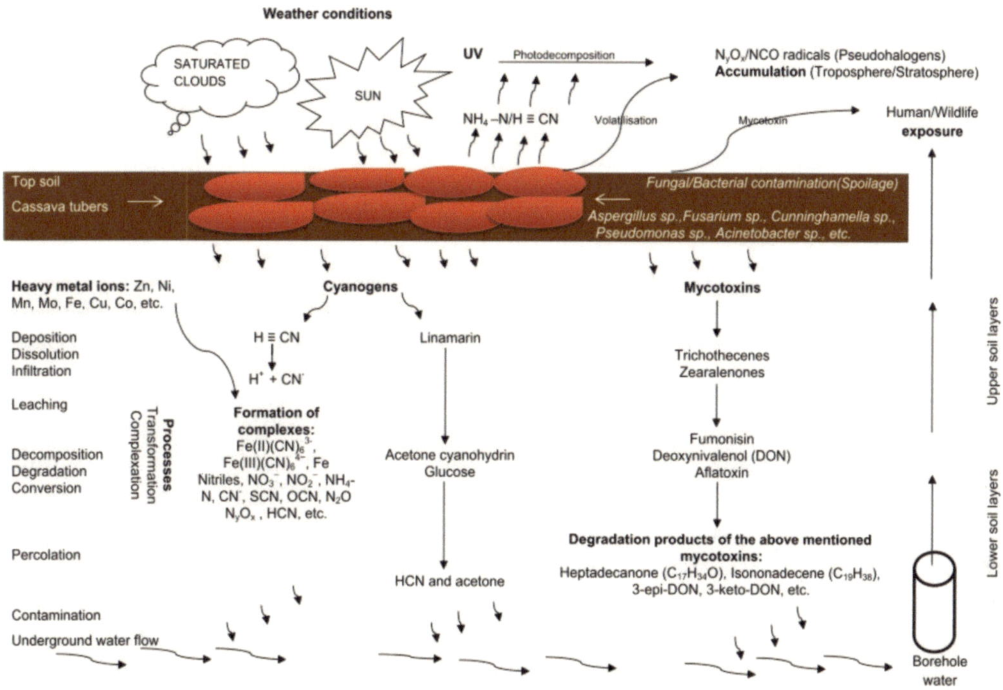

Figure 2. Cassava cyanogen and mycotoxin movement in agricultural soil. Key: NO_3^- (nitrate), NO_2^- (nitrite), Fe(II)(CN)63– (ferrocyanide), Fe(III)(CN)64– (ferricyanide), SCN (thiocyanide), NH_4-N (ammonium nitrogen), CN- (cyanide ion), -OCN (cyanate), N_2O (nitrous oxide) and N_yO_x (nitrogen oxides).

It is also important to point out that the behaviour of contaminant movement in terrestrial or/ and aquatic ecosystems and the environment in general is also influenced by parameters such as wash-off, periodic moisture saturation and time. Based on the stability of each individual contaminant, including its by-products, the mobility can also be spontaneously influenced by the rate of conversion, thus degradation, and can even become volatilised under suitable conditions [65, 70], depending on vapour pressure. Time or length of exposure is a very important aspect, particularly where human exposure is assessed, which is generally neglected or unclear in many recent studies. Similarly, contamination gradients must be established because of groundwater variations in the water flow, as well as the influence of the insolation surrounding the water body that might contribute to acute exposure levels. Furthermore, from produce itself, volatilised compounds can undergo photodecomposition due to UV effects contributing to pseudohalogen accumulation in the troposphere/stratosphere.

3.6. Cyanogen and mycotoxin effects on humans and animals

The focus of this review is that cassava can be toxic when consumed in large quantities owing to its cyanogen content [49]. The prolonged consumption of cassava in different forms can be harmful for humans in particular, owing to inadequacies in post-harvest treatment techniques [8, 21]. For instance, studies on cassava-cyanide effects in humans revealed that a permanent consumption of low-level concentrations of cyanide from poorly processed cassava could result in goitres and Tropical Ataxic Neuropathy (TAN) [24, 59], whereas a high consumption of the produce could result in neurological disorders, such as konzo [23, 50]. Most post-harvest cyanogen removal techniques focus on free-cyanide removal techniques, without accounting for transformed varieties of the cyanogens (see **Figure 2**), such as thiocyanate, etc.

A team of researchers conducting studies on the thiocyanate concentration in urine samples of pupils, who consumed cassava in Mozambique, revealed that a mean concentration of urinary thiocyanate in school children ranged from 225 to 384 mol/L, whereas mean total cyanogen concentrations in processed cassava flour varied with seasons and years from 26 to 186 mg/L [71]. Similarly, a study by Shifrin et al. [59] revealed that mycotoxin can easily be absorbed through dermal contact, ingestion and inhalation. Some mycotoxins are hazardous and are proposed to be carcinogens facilitating mutation in human cells—an effect that can be postulated to suggest their facilitation of cell mutation in humans.

In animals, on the other hand, an increased consumption of tuber debris and waste by-products of produce processing could lead to neuronal disturbances, weight loss and dysfunctional thyroid [23, 50, 72]. Observations reported by Wade et al. [73] on cassava waste in fish, i.e., in the Nile tilapia (*Oreochromis niloticus*), revealed that some cyanogens caused oedema, gill lamellae telangiectasia, gill enlargement, formation of vacuoles and liver cell deterioration. Similar health outcomes for humans and animals observed in acute mycotoxin exposure, including ingestion, were inter alia, weight loss, internal organ bleeding, respiratory diseases (asthma, pneumonia), diarrhoea, liver and kidney cancer and skin irritation [74–76]. Therefore, a large-scale propagation of agricultural produce with cyanogens, which is susceptible to a high concentration of spoilage organisms, particularly mycotoxin producers, requires continuous monitoring to ascertain its quality. Such produce should be free of both cyanogen and mycotoxins, primarily if it is destined for human and animal consumption and/or exposure. In this case, required strategies for the reduction of exposure must be implemented.

4. Conclusion

Cassava, in general, and cassava tubers, in particular, are indispensable for daily self-nourishment of several poor communities worldwide owing to their nutritional value. However, when exposed to environmental processes and bacterial and fungal attacks that can occur prior to harvesting, the produce is susceptible to release cyanogen and mycotoxin compounds that are hazardous to humans, animals and the environment. These contaminants and their by-product mobility into the terrestrial ecosystem are similar and are facilitated by environmental processes such as transformation, complexation, percolation and volatilisation as they can travel from surface and subsurface to groundwater level, which can result in exposure to both animals and humans. The presence of these compounds in arable land can lead to their accumulation, which can negatively affect soil properties, groundwater quality and the environment, thus contributing to a decline in the production of useful produce, such as cassava. Monitoring, particularly in communities that use such arable soil on a continuous basis, can mitigate intoxication of humans and animals, by effectively implementing suitable reduction strategies thus prevent environmental pollution. Therefore, continuous monitoring, quality assurance and a novel *in-situ* biological method (for treatment of the contaminants) are paramount to ensure a healthier agricultural soil, clean surface and groundwater quality.

http://dx.doi.org/10.5772/intechopen.68715

Acknowledgements

This research was funded by the Cape Peninsula University of Technology (CPUT), through the University Research Fund (URF - RK 16). Members of the Bioresource Engineering Research Group (BioERG), Dr Richard Mundembe, Department of Biotechnology (CPUT) and Dr Arnelia N. Paulse, Department of Environmental and Occupational Studies (CPUT) are acknowledged for their support.

Declaration of conflicting interests

The authors declare no potential conflicts of interest with respect to the research, authorship, and/or publication of this manuscript.

Author details

Elie F. Itoba-Tombo*, Seteno K.O. Ntwampe and John B.N. Mudumbi

*Address all correspondence to: itobatomboe@cput.ac.za

Bioresource Engineering Research Group (BioERG), Department of Biotechnology, Cape Peninsula University of Technology, Cape Town, , South Africa

References

[1] Figueira MM, Ciminelli VST, De Andrade MC and Linardi V. Bacterial degradation of metal cyanide complexes. In Jerez CA, Vargas T, Toledo H and Wiertz JE, editors. Biohydrometallurgical Processing: Proceedings of the International Biohydrometallurgical Symposium IBS-95 held in Viña del Mar, Chile, 19-22 November. Santiago: University of Chile; 1995. pp. 333-339

[2] Ubalua AO Cyanogenic glycosides and the fate of cyanide in soil. Australian Journal of Crop Science. 2010;**4**(4):223-238

[3] Montagnac JA, Davis CR and Tanumihardjo SA. Processing techniques to reduce toxicity and antinutrients of cassava for use as a staple food. Comprehensive Reviews in Food Science and Food Safety. 2009;**8**(1):17-27

[4] Baxter J and Cummings S. The impact of bioaugmentation on metal cyanide degradation and soil bacteria community structure. Biodegradation. 2006;**17**(3):207-217

[5] Hidayat A, Zuaraida N, Hanarida I and Damardjati D. Cyanogenic content of cassava root of 179 cultivars grown in Indonesia. Journal of Food Composition and Analysis, 2000;**13**(1):71-82

[6] Cumbana A, Mirione E, Cliff J and Bradbury JH. Reduction of cyanide content of cassava flour in Mozambique by the wetting method. Food Chemistry. 2007;**101**(3):894-897

[7] Yu XZ. Uptake, assimilation and toxicity of cyanogenic compounds in plants: Facts and fiction. International Journal of Environmental Science and Technology. 2015;**12**(2):763-774

[8] Akcil A, Karahan A, Ciftci H and Sagdic O. Biological treatment of cyanide by natural isolated bacteria (*Pseudomonas* sp.). Minerals Engineering. 2003;**16**(7):643-649

[9] Farenhorst A, Andronak LA and McQueen, RDA. Bulk Deposition of Pesticides in a Canadian City: Part 1. Glyphosate and Other Agricultural Pesticides, Water Air Soil Pollution. 2015; **226**(47):1-11

[10] Okechi R, Ihejirika C, Chiegboka N and Ibe I. Evaluation of the effects of cassava mill effluent on the microbial populations and physicochemical parameters at different soil depths. International Journal of Biosciences. 2012;**2**(12):139-145

[11] Siller H and Winter J. Degradation of cyanide in agroindustrial or industrial wastewater in an acidification reactor or in a single-step methane reactor by bacteria enriched from soil and peels of cassava. Applied Microbiology and Botechnology. 1998;**50**(3):384-389

[12] Clarke R, Connolly L, Frizzell C and Elliott CT. Cytotoxic assessment of the regulated, co-existing mycotoxins aflatoxin B, fumonisin B1 and ochratoxin, in single, binary and tertiary mixtures. Toxicon. 2014;**90**:70-81

[13] Alassane-Kpembi I, Kolf-Clauw M, Gauthier T, Abrami R, Abiola FA, Oswald IP and Puel O. New insights into mycotoxin mixtures: The toxicity of low doses of Type B trichothecenes on intestinal epithelial cells is synergistic. Toxicology and Applied Pharmacology. 2013;**272**(1):191-198

[14] Klarić MŠ, Rašić D and Peraica M. Deleterious effects of mycotoxin combinations involving ochratoxin A. Toxins. 2013;**5**(11):1965-1987

[15] Cressey P, Saunders D and Goodman J. Cyanogenic glycosides in plant-based foods available in New Zealand. Food Additives and Contaminants: Part A. 2013;**30**(11):1946-1953

[16] Kuti JO and Konoru HB. Cyanogenic glycosides content in two edible leaves of tree spinach (*Cnidoscolus* spp.). Journal of Food Composition and Analysis. 2006;**19**(6):556-561

[17] Ismaiel AA and Papenbrock J. Mycotoxins: Producing fungi and mechanisms of phytotoxicity. Agriculture. 2015;**5**(3):492-537

[18] Gleadow RM and Woodrow IE. Mini-Review: Constraints on effectiveness of cyanogenic glycosides in herbivore defense. Journal of Chemical Ecology. 2002;**28**(7):1301-1313

[19] Jones DA. Why are so many food plants cyanogenic? Phytochemistry. 1998;**47**(2):155-162

[20] Kim YM, Lee DS, Park C, Park D and Park JM. Effects of free cyanide on microbial communities and biological carbon and nitrogen removal performance in the industrial activated sludge process. Water Research. 2011;**45**(3):1267-1279

[21] Akcil A and Mudder T. Microbial destruction of cyanide wastes in gold mining: Process review. Biotechnology Letters. 2003;**25**(6):445-450

[22] Vanhoutte I, Audenaert K and De Gelder L. Biodegradation of mycotoxins: Tales from known and unexplored worlds. Frontiers in Microbiology. 2016;**7**(561):1-20

[23] Soto-Blanco B and Górniak SL. Toxic effects of prolonged administration of leaves of cassava (Manihot esculenta Crantz) to goats. Experimental and Toxicologic Pathology. 2010; **62**(4):361-366

[24] Siritunga D and Sayre RT. Generation of cyanogen-free transgenic cassava. Planta. 2003;**217**(3):367-373

[25] Andersen MD, Busk PK, Svendsen I and Møller BL. Cytochromes P-450 from cassava (Manihot esculenta Crantz) catalyzing the first steps in the biosynthesis of the cyanogenic glucosides, linamarin and lotaustralin cloning functional expression in *Pichia pastoris*, and substrate specificity of the isolate recombinant enzymes. Journal of Biological Chemistry. 2000;**275**(3):1966-1975

[26] Dash RR, Balomajumder C and Kumar A. Cyanide removal by combined adsorption and biodegradation process. Journal of Environmental Health Science & Engineering. 2006; **3**(2):91-96

[27] Ebbs S. Biological degradation of cyanide compounds. Current Opinion in Biotechnology. 2004;**15**(3):231-236

[28] Barclay M, Hart A, Knowles CJ, Meeussen JC and Tett VA. Biodegradation of metal cyanides by mixed and pure cultures of fungi. Enzyme and Microbial Technology. 1998;**22** (4):223-231

[29] Zhang J, Zeng G, Chen Y, Yu M, Yu Z, Li H, Yu Y and Huang H. Effects of physico-chemical parameters on the bacterial and fungal communities during agricultural waste composting. Bioresource Technology. 2011;**102**(3):2950-2956

[30] Sing NN, Zulkharnain A, Roslan HA, Assim Z and Husaini A. Bioremediation of PCP by Trichoderma and Cunninghamella strains isolated from sawdust. Brazilian Archives of Biology and Technology. 2014;**57**(6):811-820

[31] Adebajo LO, Idowu AA and Adesanya OO. Mycoflora, and mycotoxins production in Nigerian corn and corn-based snacks. Mycopathologia. 1994;**126**(3):183-192

[32] Njumbe Ediage E, Hell K and De Saeger S. A comprehensive study to explore differences in mycotoxin patterns from agro-ecological regions through maize, peanut, and cassava products: A case study, Cameroon. Journal of Agricultural and Food Chemistry. 2014;**62**(20):4789-4797

[33] Munkvold GP and Desjardins AE. Fumonisins in maize: Can we reduce their occurrence? Plant Disease. 1997;**81**(6):556-565

[34] Liang S, Guo X.-Y, Feng NC and Tian QH. Effective removal of heavy metals from aqueous solutions by orange peel xanthate. Transactions of Nonferrous Metals Society of China. 2010;**20**(1):187-191

[35] Sud D, Mahajan G and Kaur M. Agricultural waste material as potential adsorbent for sequestering heavy metal ions from aqueous solutions: A review. Bioresource Technology. 2008;**99**(14):6017-6027

[36] Mrudula S and Anitharaj R. Pectinase production in solid state fermentation by *Aspergillus niger* using orange peel as substrate. Global Journal of Biotechnology and Biochemistry. 2011;**6**(2):64-71

[37] Umsza-Guez MA, Díaz AB, De Ory I Blandino A, Gomes E and Caro I. Xylanase production by *Aspergillus awamori* under solid state fermentation conditions on tomato pomace. Brazilian Journal of Microbiology. 2011;**42**(4):1585-1597

[38] Adams D, Komen J and Pickett T. Biological cyanide degradation. In: Young CA, Tidwell LG and Anderson CG, editors. Cyanide: Social, Industrial and Economic Aspects. Warrendale, PA: The Metals Society; 2001. pp. 203-213

[39] Gouw M, Bozic R, Koopman B and Svoronos SA. Effect of nitrate exposure history on the oxygen/nitrate diauxic growth of *Pseudomonas denitrificans*. Water Research. 2001; **35**(11):2794-2798

[40] Amodu OS, Ojumu TV and Ntwampe SKO. Bioavailability of high molecular weight polycyclic aromatic hydrocarbons using renewable resources. In Petre M, editor. Environmental Biotechnology: New Approaches and Prospective Applications. Rijeka, Croatia: InTech; 2013. pp. 171-194

[41] Heinl S, Hartinger D, Thamhesl M, Schatzmayr G, Moll WD and Grabherr R. An aminotransferase from bacterium ATCC 55552 deaminates hydrolyzed fumonisin B1. Biodegradation. 2011;**22**(1):25-30

[42] Duvick J, Maddox J and Gilliam J. Compositions and methods for fumonisin detoxification. US Patent No 6,538,177 B1. Washington, DC: US Patent and Trademark Office; 2003

[43] Duvick J, Rood T, Maddox J and Gilliam J. Detoxification of mycotoxins in planta as a strategy for improving grain quality and disease resistance: Identification of fumonisin-degrading microbes from maize. In: Kohmoto K and Yoder OC, editors. Molecular Genetics of Host-Specific Toxins in Plant Disease: Proceedings of the 3rd Tottori International Symposium on Host-Specific Toxins, Daisen, Tottori, Japan, 24-29 August 1997. Dordrecht: Springer; 1998. pp. 369-381

[44] Blackwell BA, Gilliam JT, Savard ME, Miller D Jr and Duvick JP. Oxidative deamination of hydrolyzed fuminosin B(1) AP(1)1) by cultures of *Exophiala spinifera*. Natural Toxins. 1999;**7**(1):31-38

[45] Camilo SB, Ono CJ, Ueno Y and Hirooka EY. Anti-Fusarium moniliforme activity and fumonisin biodegradation by corn and silage microflora. Brazilian Archives of Biology and Technology. 2000;**43**(2):159-164

[46] Guan S, He J, Young JC, Zhu H, Li XZ, Ji C and Zhou T. Transformation of trichothecene mycotoxins by microorganisms from fish digesta. Aquaculture. 2009;**290**(3):290-295

[47] Mburu FW, Swaleh S and Njue W. Potential toxic levels of cyanide in cassava (Manihot esculenta Crantz) grown in Kenya. African Journal of Food Science. 2012;**6**(16):416-420

[48] Fasuyi AO. Nutrient composition and processing effects on cassava leaf (*Manihot esculenta,* Crantz) antinutrients. Pakistan Journal of Nutrition. 2005;**4**(1):37-42

[49] Prawat H, Mahidol C, Ruchirawat S, Prawat U, Tuntiwachwuttikul P, Tooptakong U, Taylor WC, Pakawatchai C, Skelton BW and White AH. Cyanogenic and non-cyanogenic glycosides from *Manihot esculenta.* Phytochemistry. 1995;**40**(4):1167-1173

[50] Kamalu BP. The adverse effects of long-term cassava (Manihot esculenta Crantz) consumption. International Journal of Food Sciences and Nutrition. 1995;**46**(1):65-93

[51] Hidayat A, Zuraida N and Hanarida I. The cyanogenic potential of roots and leaves of ninety nine cassava cultivars. Indonesian Journal of Agricultural Science. 2013;**3**(1):25-32

[52] Vineela C, Wani S, Srinivasarao C, Padmaja B and Vittal K. Microbial properties of soils as affected by cropping and nutrient management practices in several long-term manurial experiments in the semi-arid tropics of India. Applied Soil Ecology. 2008;**40**(1):165-173

[53] Knudsen GR. Bacteria, Fungi, and Soil Health. Paper Presented at the Idaho Potato Conference, 18-19 January, Pocatello, ID; 2006

[54] Jenkinson DS and Ladd JN. Microbial biomass in soil: Measurement and turnover. In: Paul EA and Ladd JN, editors. Soil Biochemistry. New York, NY: Marcel Decker; 1981. pp. 415-471

[55] Perrone G, Stea G, Epifani F, Varga J, Frisvad JC and Samson RA. *Aspergillus niger* contains the cryptic phylogenetic species *Aspergillus awamori.* Fungal Biology. 2011;**115**(11): 1138-1150

[56] Varga J, Frisvad JC, Kocsubé S, Brankovics B, Tóth B, Szigeti G and Samson RA. New and revisited species in Aspergillus section Nigri. Studies in Mycology. 2011;**69**(1):1-17

[57] Perrone G, Susca A, Epifani F and Mulè G. AFLP characterization of Southern Europe population of Aspergillus Section Nigri from grapes. International Journal of Food Microbiology. 2006;**111**(S1):S22-S27

[58] Pestka JJ and Smolinski AT. Deoxynivalenol: Toxicology and potential effects on humans. Journal of Toxicology and Environmental Health. 2005;**8**(1):39-69

[59] Shifrin NS, Beck BD, Gauthier TD, Chapnick SD and Goodman G. Chemistry, toxicology, and human health risk of cyanide compounds in soils at former manufactured gas plant sites. Regulatory Toxicology and Pharmacology. 1996;**23**(2):106-116

[60] White DM, Pilon TA and Woolard C. Biological treatment of cyanide containing wastewater. Water Research. 2000;**34**(7):2105-2109

[61] White CS and Markwiese JT. Assessment of the potential for in situ bioremediation of cyanide and nitrate contamination at a heap leach mine in central New Mexico. Soil and Sediment Contamination. 1994;**3**(3):271-283

[62] Chen X and Hu Q. Groundwater influences on soil moisture and surface evaporation. Journal of Hydrology. 2004;**297**(1):285-300

[63] Schaap MG, Leij FJ and Van Genuchten MT. Neural network analysis for hierarchical prediction of soil hydraulic properties. Soil Science Society of America Journal. 1998;**62** (4):847-855

[64] Sat CD, and Deturck P. Cassava soils and nutrient management in South Vietnam. In: Howeler RH, editor. Cassava Breeding, Agronomy and Farmer Participatory Research in Asia: Proceedings of the Fifth Regional Workshop Held in Danzhou, Hainan, China, 3-8 November 1996. Bangkok: CIAT; 1998. pp. 454-470

[65] Kjeldsen P. Behaviour of cyanides in soil and groundwater: A review. Water, Air and Soil Pollution. 1999;**115**(1):279-308

[66] Howeler RH. Cassava mineral nutrition and fertilization. In: Hillocks RJ and Thresh JM, editors. Cassava: Biology, Production and Utilization. Wallingford: CAB International; 2002. pp. 115-147

[67] Boadi NO, Twumasi SK and Ephraim JH. Impact of cyanide utilization in mining on the environment. International Journal of Environmental Research. 2009;**3**(1):101-108

[68] Byers W, Meyers MB and Mooney DE. Analysis of soil from a disused gasworks. Water, Air, and Soil Pollution. 1994;**73**(1):1-9

[69] Meeussen JC, Keizer MG, Van Riemsdijk WH and De Haan FA. Solubility of cyanide in contaminated soils. Journal of Environmental Quality. 1994;**23**(4):785-792

[70] Zidenga T, Leyva-Guerrero E, Moon H, Siritunga D and Sayre R. Extending cassava root shelf life via reduction of reactive oxygen species production. Plant Physiology. 2012;**159** (4):1396-1407

[71] Ernesto M, Cardoso AP, Nicala D, Mirione E, Massaza F, Cliff J, Haque MR and Bradbury JH. Persistent konzo and cyanogen toxicity from cassava in northern Mozambique. Acta Tropica. 2002;**82**(3):357-362

[72] Soto-Blanco B, Maiorka P and Górniak S. Neuropathologic study of long term cyanide administration to goats. Food and Chemical Toxicology. 2002;**40**(11):1693-1698

[73] Wade JW, Omoregie E and Ezenwaka I. Toxicity of cassava (Manihot esculenta Crantz) effluent on the Nile tilapia, *Oreochromis niloticus* (L.) under laboratory conditions. Journal of Aquatic Sciences. 2002;**17**(2):89-94

[74] Afsah-Hejri L, Jinap S, Hajeb P, Radu S and Shakibazadeh S. A review on mycotoxins in food and feed: Malaysia case study. Comprehensive Reviews in Food Science and Food Safety. 2013;**12**(6):629-651

[75] Bhat R, Rai RV and Karim AA. Mycotoxins in food and feed: Present status and future concerns. Comprehensive Reviews in Food Science and Food Safety. 2010;**9**(1):57-81

[76] Bankole S and Mabekoje O. Mycoflora and occurrence of aflatoxin B1 in dried yam chips from markets in Ogun and Oyo States, Nigeria. Mycopathologia. 2004;**157**(1):111-115

Chapter 9

First Report of Mycotoxins in Second Peanuts Crop in Adana and Osmaniye at Harvest, Drying, Prestorage and Storage Periods

Isilay Lavkor, Isil Var, Sevcan Oztemiz and
Manaf AlMatar

Additional information is available at the end of the chapter

http://dx.doi.org/10.5772/intechopen.68210

Abstract

Aflatoxin (AF) and cyclopiazonic acid (CPA) contaminations are very important problems for peanuts and its products. The aim of the study was to detect aflatoxin (types B and G) and cyclopiazonic acid (CPA) occurrence and critical periods of toxin production in peanuts collected from different research areas of Osmaniye and Adana, Turkey, in 2015. Peanut kernels toxin analysis was performed in four different periods during the harvest, drying, prestorage, and storage. Total aflatoxin occurrence in peanut kernels was analyzed by immunoaffinity chromatography-reversed-phase high-performance liquid chromatography (IAC-HPLC) analysis and cyclopiazonic acid occurrence in peanut kernels was analyzed by thin layer chromatography (TLC). Aflatoxin levels in 76 out of 102 contaminated samples were from 0.3 to 1333.42 μg/kg. Cyclopiazonic acid levels in 18 out of 102 peanut samples were from 16.6 to 44.44 μg/kg. An unusual pattern of mycotoxin production (aflatoxin types B and G simultaneously with CPA) was seen in 11 of 102 peanuts samples. Six of nine samples were from the storage period. Aflatoxin contamination during harvesting (64%) and drying (75%) were higher than prestorage (53%). Aflatoxin (93%) and cyclopiazonic acid (30%) were the most produced during storage. The results showed that storage period was significantly important for the presence of two mycotoxins according to the statistical analysis.

Keywords: aflatoxin, cyclopiazonic acid, peanut, drying, harvest, prestorage, storage

INTECH
open science | open minds

1. Introduction

Peanut (*Arachis hypogaea* L., Family: *Fabaceae*) is a rich source of fat, proteins, and vitamins. Peanuts are grown on a large scale in almost all the tropical and subtropical countries, especially in India, China, the USA, and West Africa [1]. Also, peanut is a member of the legume family, an important food and oil crop. It is currently grown on approximately 42 million acres worldwide. It is the third major oilseed of the world after soybean and cotton [2]. Peanut is used for human consumption, oil production, food industries, and animal feeding [3]. It is grown in China (37%), India (20%), Nigeria (6.5%), and the USA (4.1%). Turkey supplies about 0.3% of the world production of peanut [4]. The total production of peanut was 147,537 tons harvested from 377,729 da, with an average yield of 391 kg/da in Turkey in 2015 [5].

Poor agricultural practices and postharvest treatments of peanuts can lead to an infection by mold fungus *Aspergillus flavus* and *Aspergillus parasiticus* releasing the toxic substance aflatoxins (AFs). Contamination may occur when either the grown crops [6] or more badly stored harvests are infested by molds [7, 8]. *A. flavus* also produces other mycotoxins such as cyclopiazonic acid (CPA) and indole-tetramic acid [9]. CPA occurs naturally in peanuts [10–12] and corn [12, 13].

The occurrence of aflatoxins in foods has been also recognized as a potential threat for human health. Aflatoxin, naturally occurring secondary metabolites, are potent hepatotoxic, mutagenic, and carcinogenic toxins, causing serious health hazards in humans and in animals. Aflatoxin B_1, B_2, G_1, and G_2 are found predominantly as the hydroxylated metabolic products of aflatoxins B_1 (AFB_1) and B_2 (AFB_2), respectively [14]. The most toxic aflatoxin known, AFB_1, is cited as a group I carcinogen by the International Agency for Research on Cancer [15, 16].

Another mycotoxin is cyclopiazonic acid (CPA), which causes necrotic foci in internal organs such as the liver and exerts neurotoxic effects [17]. Natural occurrence of CPA has been reported in peanuts, corn, cheese, tomato products, and also meat, eggs, and milk of animals that are fed by contaminated feeds [18]. Incidence of aflatoxigenic *A. flavus* strains was higher in peanuts (69%) than in wheat (13%) or soybeans (5%), while the ratio of CPA producers [12]. Risk of aflatoxin contamination hits top values in such commodities as nuts [19]. The maximum levels of AFB_1 and total aflatoxins allowed in peanut as determined by Commission of the European Communities are 2 and 4 μg/kg, respectively.

Fungal infection of seeds before and after harvest remains a major problem of food safety in most parts of Turkey. Problems associated with this infection include loss of germination, mustiness, moldy smell [20–22], and aflatoxin contamination [23–25]. These problems are, however, dealt with most developed world where a careful commodity screening and improved storage conditions are provided [7, 23, 26]. Though aflatoxins producing fungus are a natural contaminant of peanut and other agricultural commodities, it is aggravated due to poor agricultural practices, harvesting practice, postharvest handling, and storage methods. Some studies undertaken in Turkey in different foods show that aflatoxins levels are substantially higher. One important aspect is traditional harvesting and storage practices [22].

In this study, aflatoxin and cyclopiazonic acid contamination were determined in the second peanut crops at harvest, drying, pre-storage, and the storage periods in Adana and Osmaniye provinces of Turkey in 2015. Aflatoxin is always the most important toxin for peanut because of its toxicity, and the CPA presence in peanuts is also important as it causes necrotic foci in internal organs such as the liver and exerts neurotoxic effects, therefore, in peanuts, aflatoxins and besides, CPA should be investigated.

2. Materials and methods

2.1. Location of the study

In the second crop of peanuts, 102 samples were collected randomly throughout the peanut fields, that is, about 5949.5 da from Adana and Osmaniye provinces during October and November in 2015. Peanut samples were collected from 72 representative fields of 16 different districts of Adana and Osmaniye. Sampling was done according to Bora and Karaca [27] and have been followed in 1% of the survey areas. Also, 30 peanut samples were collected from storage in Adana and Osmaniye provinces. The samples were collected during the following periods of production: preharvest, drying, and prestorage. Also, 30 unshelled peanut samples were collected from the storage.

2.2. Collection of peanut samples

Second crop peanut samples were collected during harvest, drying, prestorage of eliminated soil, and storage periods. Seventy-two crusted peanuts samples were collected at harvest, dried for 7 days, and eliminated soil to prestorage. Each sample has been obtained from different farmer fields. Samples of pods (about 5 kg each) were divided manually and homogeneously to obtain working samples (about 1 kg each) for mycotoxins analyzes. The shells were removed manually. Additionally, 30 unshelled peanut samples were collected from storage. The samples (about 1 kg each) were collected in paper bags for analyzing mycotoxins. All the samples were kept at +4°C [28].

2.3. Analysis of mycotoxins

2.3.1. Determination of AFs with immunoaffinity chromatography-reversed phase high-performance liquid chromatography (IAC-HPLC) analysis

Analysis of aflatoxins was performed using immunoaffinity columns, as described below. Identification and deperiodination of aflatoxins B_1, B_2, G_1, and G_2 in peanut product samples were carried out by high-performance liquid chromatography (HPLC) according to Arzandeh and Jinap [29]. To 50 g of tested sample, 5 g of NaCl and 125 ml of methanol and water (70:30)was mixed in a blender for 2–3 min at high speed. The mixture was filtered through Whatman no. 4 filter paper. Then 30 ml of water was added to a 15 ml of filtrate. About 10 ml of the second filtrate was quantitatively passed through the immunoaffinity column at flow rate of 1 ml/min. The column was washed with 10 ml of water. Aflatoxins were eluted with 1 ml of methanol in an amber vial at flow rate of 1–2 ml/min. The elution step was repeated with 1 ml of water. Thus, Agilent 1100 HPLC was ready for injection. Fluorescence detector (excitation at 360 nm and emission above 440 nm). Mobile phase consisted of methanol/water/acetonitrile (300:600:200, v/v/v) with a flow rate of 1.0 ml/min followed by derivatization with bromine (132 mg/l KBr, 385 μL nitric acid) in C-18 (R-Biopharm Rhône). HPLC column was maintained at a constant temperature ($T = 25°C$). The results were expressed as a μg/kg.

2.3.2. Determination of CPA with thin-layer chromatography (TLC) analysis

The peanut samples were tested for cyclopiazonic acid (CPA) production following the method modified by Somuncuoglu [18]. On each samples, 45 g of the contents of each test plate was macerated in a waring blender with 150 ml of methanol and 2% sodium bicarbonate (7:3). The slurry was twice filtered through a Büchner funnel with Whatman no. 4 filter paper and then concentrated to dryness with a rotary evaporator. The residue was partitioned between 200 ml of dichloromethane-distilled H_2O (1:1), and the dichloromethane layer was extracted three times with a saturated $NaHCO_3$ solution (100 ml). The dichloromethane layer, containing AFs, was rotary evaporated and concentrated under a gentle stream of nitrogen. The aqueous layer, containing CPA, was acidified to pH 2.0 with 0.5 N HCl and extracted two times with 25 ml kloroform (500 ml). The extract was evaporated and concentrated as for CPA.

A total of 45 g of peanuts was used for extraction of CPA. After adding 150 ml methanol and sodium bicarbonate (2%) (7:3), it was stirred in a high-speed mixer for 5 min. The mixture was filtred using Whatman no. 4 filter paper, and then 80 ml was taken from the fitrate obtained. To the filtrate, 30 ml of 0.05 M solution of lead acetate was added and stirred and the precipitate was removed by filtration. 0.5 N HCl reduced the pH to 2 with 50 ml of the filtrate that was extracted twice with 25 ml of chloroform, and the bottom phase was collected. The water in the chloroform phase (lower phase) was removed with 10 g anhydrous sodium sulfate by filtrating through a paper filter. The extract was collected in the flask after extraction and dried at 40°C by rotary vacuum evaporator. The extract in the flask was taken to the tubes with 3-4 ml of chloroform and then the content is dried under nitrogen gas.

The qualitative presence of CPA was determined by thin-layer chromatography (TLC) separation on silica gel plates 60 EM-5721 (20 × 20 cm; Merck). The plates were first dipped in a 2% (wt/wt) solution of oxalic acid in methanol for 10 min, after being heated at 100°C for 1 h and cooled. The plates were spotted with 60–80 μl of the respective extract and developed in the solvent ethyl acetate/2 propanol/sodium hydroxide (50:15:10, v/v/v) for 35–40 min. After this, the plates were being heated at 35–40°C for 1 h and cooled. CPA was viewed after spraying with Ehrlich's reagent (1.0 g of 4-dimethylaminobenzaldehyde in 25 ml of HCl and 75 ml etanol) with subsequent development of a purple color in daylight [30].

The results were calculated applying the formula, and a concentration of CPA in μ/kg = $(S \times Y \times V)/(X \times W)$

where S is the μl aflatoxin CPA standard equal to unknown; Y is the concentration of CPA standard μg/ml; V is the μl of final dilution of sample extract; X is the μl of sample extract spotted to give fluorescent intensity equal to S (CPA standard); and W is the weight of sample in gram of original sample contained in the final extract [18].

2.4. Statistical analysis

To compare the aflatoxin and CPA, periods of harvest, drying, preharvest, and storage results were analyzed using the Kruskal-Wallis one-way analysis on Rank's test (H statistic) and then Mann-Whitney U nonparametric multiple comparison test. All statistical analysis were performed by using SPSS, version 21.0 (IBM Corp., Armonk, NY, USA). Statistical analysis also revealed significant differences among the storage periods.

3. Results and discussion

Of 102 peanut samples analyzed by HPLC, 76 (75%) peanut samples were contaminated with aflatoxins (**Table 1**). High levels of AFs (1333.42 μg/kg) were found in shelled peanuts and unshelled peanuts (1235.15 μg/kg), respectively (**Table 1**). Of the samples analyzed, 32 (31.37%) peanuts samples were above limit as recognized in Turkey (10 μg/kg for AFs) (FAO, 2004) and 27 (26.47%) peanuts samples were above limit as recognized in US Food and Drug Administration (FDA) (20 μg/kg for AFs) [58].

Aflatoxin contamination was determined in 16 (64%) peanut samples of 25 samples collected during harvest. Of these samples, three of them were determined over 10 μg/kg, two of them were over 100 μg/kg, and one of them over 1000 μg/kg (**Table 1**).

Aflatoxin contamination was determined in 24 (75%) peanut samples of 32 samples collected during drying period. Of these samples, four of them were determined over 10 μg/kg, four of them over 100 μg/kg, and one of them over 1000 μg/kg (**Table 1**). Aflatoxin contamination was determined in 8 (53%) peanut samples of the 15 samples collected during prestorage period. Of these samples, two of them were determined over 10 μg/kg, one of them over 100 μg/kg, and one of them over 1000 μg/kg (**Table 1**). A total of 30 peanut samples were taken from the various peanut storages from Adana and Osmaniye provinces. Peanut samples were determined to be infected with aflatoxin levels between 0.18 and 1235.15 μg/kg (**Table 1**). Aflatoxin contamination was determined in 28 (93%) storage samples. Of these samples taken from peanut storage, toxin contamination were determined in five of them over 10 μg/kg, seven of them over 100 μg/kg, and one of them over 1000 μg/kg.

From the 102 peanut samples analyzed by TLC, 18 (17%) peanut samples produced CPA (**Table 2**). Four peanut samples produced CPA in 22.22 μg/kg (16%) during the harvest period. Five peanut samples produced CPA in 22.22 μg/kg (16%) during the drying period (**Table 2**). Nine peanut samples produced CPA in 16.66–44.44 μg/kg (30%) during the storage period (**Table 2**).

Periods of peanuts	Samples (positive samples)	%[b]	Range of AFs (μg/kg)	Samples >10 (μg/kg) (%)[b]	Range of AFs (μg/kg)	Samples >100 (μg/kg) (%)[b]	Range of AFs (μg/kg)	Samples >1000 (μg/kg) (%)[b]	Range of AFs (μg/kg)
Harvest	25 (16)	64	0.03–1333.42	3 (12)	32.45–49.55	2 (8)	137.54–684.30	1 (4)	1333.42
Drying	32 (24)	75	0.06–1106.70	4 (13)	13.93–38.20	4 (13)	105.10–420.77	1 (3)	1106.70
Prestorage	15 (8)	53	0.19–1311.28	2 (13)	11.19–58.89	1 (7)	123.14	1 (7)	1311.28
Storage[a]	30 (28)	93	0.18–1235.15	5 (17)	13.28–37.10	7 (23)	243.25–663.08	1 (3)	1235.15
Total	102 (76)	75	0.03–1333.42	14 (14)	11.93–58.89	14 (14)	105.10–663.08	4 (4)	1106.70–1333.42

[a]Significant differences ($p < 0.05$).

[b]Percentage related to the total number of samples in each period.

Table 1. Occurrence of aflatoxins in peanuts samples ($n = 102$) collected from Adana and Osmaniye provinces in Turkey and analyzed by HPLC.

Periods of peanuts	CPA samples (positive samples)	% [b]	Range of CPA (µg/kg)	Positive CPA samples within AFB (µg/kg)	Positive CPA samples within AFB + AFG (µg/kg)
Harvest	25 (4)	16	22.22	2	2
Drying	32 (5)	16	22.22	2	3
Prestorage	15 (0)	–	–	–	–
Storage[a]	30 (9)	30	16.66–44.44	2	6
Total	102 (18)	17	16.66–44.44		

[a]Significant differences ($p < 0.05$).
[b]Percentage related to the total number of samples in each period.

Table 2. Occurrence of CPA in peanuts samples (n = 102) collected from Adana and Osmaniye provinces in Turkey and analyzed by TLC.

Aflatoxins type B and CPA were detected in 11 of the 102 samples of peanuts, suggesting the possibility of cooccurrence of these toxins. Based on the combination of mycotoxins (AFB+/AFG+/and CPA+/) that were existing in nine of the peanut samples, especially six of them from the storage period can be considered. Only one peanut sample was without contamination of aflatoxin, but it had produced CPA.

Molds may be divided into two main groups, namely the "field fungi" and the "storage fungi." The first contamination is considered to be in the field and during ineligible drying. The reduction occurred in quality due to the mistakes made during the growth period of the peanut plants, exposure to fungus and pest infestation of the fruit, and also when met with climatic conditions such as humidity and temperature; aflatoxin forms of fungi can lead to increased secondary contamination and development [31–33].

Peanuts are considered to be a high-risk product for contamination with aflatoxins since they are frequently contaminated with fungi, particularly *A. flavus* and *A. parasiticus*, and because of long drying times and occurrence of rainy periods after uprooting [34]. Fungi produce carcinogenic aflatoxins. Aflatoxins are highly regulated for both animal feed and food destined for human consumption. Of the naturally occurring aflatoxins, aflatoxin B_1 is the most toxic. *A. flavus* may also produce CPA, which is toxic in a variety of animals and has been implicated in human poisoning. CPA and aflatoxins commonly occur together in contaminated agricultural commodities [35].

CPA is a product of the ubiquitous genera of molds, *Aspergillus* and *Penicillium*. The molds are known to inhabit a number of food sources and may constitute parasitic infections of man and other animals. CPA effects may be masked by concurrent aflatoxicosis; for example, CPA and aflatoxins were isolated from peanut meal related to the Turkey "X" disease that caused the death of over 100,000 turkeys [35]. In this study, CPA was isolated from 18 peanut samples that include harvest, drying, and storage periods. Isolates of *A. flavus* that are able to produce simultaneously aflatoxins type B and CPA were detected in all substrates, suggesting the possibility of co-occurrence of these toxins. CPA occurs naturally in peanuts [10, 11] as a cocontaminant with AFs and may have contributed to the "Turkey X" syndrome in England in 1960 [36].

In general, CPA is produced by *A. flavus* alone or in combination with type B aflatoxins, but not in conjunction with type G aflatoxins. *A. flavus* and *A. parasiticus* are closely related species belonging to the *Aspergillus* section *Flavi*. Both species can produce aflatoxins, but not all isolates of either species do so [37]. Aflatoxins consist of a group of approximately 15–20 related secondary metabolites, although AFB_1 and AFB_2 are produced by *A. flavus*, AFB_1, AFB_2, AFG_1, and AFG_2 are produced by *A. parasiticus* strains. That means in our study, *A. parasiticus* could exist in peanuts during storage, and it would be able to produce CPA as well. Although Vaamonde et al. [12] found that isolates of *A. parasiticus* consistently produce both B and G aflatoxins, they do not produce CPA; according to Oktay and Dinh et al. [16, 37], they could find that *A. parasiticus* are able to produce CPA besides groups B and G.

In the survey areas, farmers do the drying by leaving piles of harvested peanuts on the ground and carry out the mixing with a shovel by shifting the piles. Because of damaged and broken peanuts that hold on to the soil surface and because of lack of ventilation, peanuts become vulnerable to pathogens contained in the surrounding air, so appropriate medium is provided for the development of fungi, such as *Aspergillus*, which infect, primarily, peanuts in the fields and harvest period [38, 39]. Meanwhile, peanuts especially raw, immature, or damaged for any reason with seed coat damage and peanut kernels separated from cotyledons have high potential for production of aflatoxin [40]. As the farmers' peanut drying process under the sun on the soil takes a long time, they can lead to the development of potential producers of aflatoxin fungus. Peanuts are considered to be at high risk of AFs because they are frequently contaminated with *Aspergillus*, especially, aflatoxigenic species. Recently, it has been reported that AFB_1 was detected in 25% of raw peanuts from China, ranging from 0.01 to 720 μg/kg [41]. On the other hand, Juan et al. [42] showed a weak contamination of the analyzed samples of peanuts with AFs (5%). Mphande et al. [43] reported that 78% of raw peanuts from Botswana contained AFs at concentrations ranging from 12 to 329 μg/kg.

The drying stage is very important to reduce attack and damage from insects and fungi. Traditional drying techniques in Turkey involve bare-ground drying and is a major source of fungal contamination. Some farmers do not dry peanuts immediately after harvest. They dry them as a cluster on the ground for a few days waiting for sunshine. They walk on the stacks of peanuts and mix by shovel. Cracks and breaks in peanut pods and testa are caused mainly during shelling by trampling. These practices, coupled with an inefficient and slow drying process under the humid conditions, enhance aflatoxin contamination greatly [38]. When the soil and other materials are removed from the harvested and dried peanuts before entering the storage the amount of aflatoxin possible on the peanut will be reduced. Kacmaz [46] reported that the content of peanut products from order processing contain 10–15% of impurities (stones, earth, garbage, fiber, hernia, etc.) in Osmaniye.

As peanuts come from the field, they are mixed with foreign materials such as rock sediments, moist soil particles, and outer shells of raw peanuts, and they must be removed from pods [45] to avoid forming optimum conditions for the aflatoxin development before entering storage [28, 44, 45].

When determining the aflatoxin contamination compared with second-crop peanuts collected during different periods from the survey areas, it was found that contamination was significantly higher in the period of harvesting (64.00%) and drying (75.00%) than during the prestorage (53.33%) (**Figure 1**). The highest rate of aflatoxin contamination was detected during storage (93%), followed by drying and harvesting (**Figure 1**). Also, CPA was found higher in the period of storage (30.00%) than in harvesting (16.00%) and drying (16.00%) periods (**Figure 2**). The contamination of stored peanuts is the CPA content of peanut during harvesting and drying and the inability to maintain adequately during storage. As a result of statistical analysis it was found that only the storage period was important. Although the storage was statistically significant for samples containing differences in production of the two mycotoxins, and in some instances, a statistical analysis was not evident. Considering the samples taken during harvesting period, aflatoxin contamination continued in the storage conditions, and therefore, it is clear that aflatoxin contamination has increased [24, 28, 38, 39, 45, 47–51]. In 28% of the newly harvested peanuts aflatoxin contamination was found to be 0-5 ppb whereas in 48% of the stored samples it was found to be 0-22 ppb. [39]. According to Ding et al. [41] low AFs contamination is found in peanuts after harvest, but AF levels might be higher during storage and processing. It is therefore, necessary to monitor the AFs contamination status of peanuts during growth, storage, and processing [17].

CPA of kernel samples was detected. The frequency of detection was 60% for the Caiapó, with mean levels ranging from 304.1 to 2583.7 μg/kg, and 74.3% for the 886, with levels ranging from 288.0 to 4918.1 μg/kg [17]. Other studies investigating the production of CPA in peanuts also reported high rates of 89 [52], 93 [53], and 97% [54]. Aflatoxins and CPA were also detected simultaneously in kernel samples (11.4%). The co-occurrence of CPA and AFs has been reported by several investigators [11, 12, 55, 56]. In addition, Smith et al. [57] demonstrated possible synergistic and cumulative effects of the two mycotoxins [17].

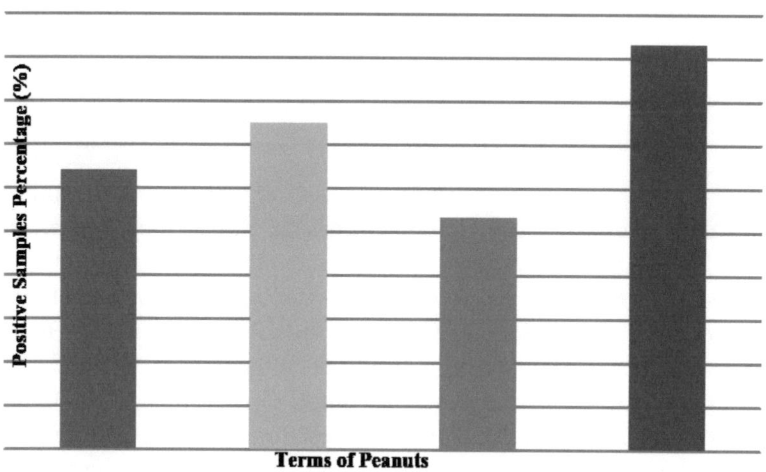

Figure 1. Distribution of the number of AFs found in all positive samples.

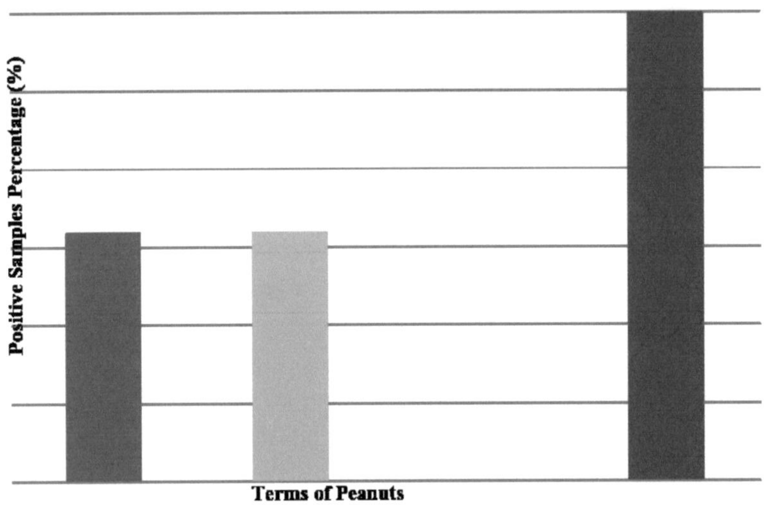

Figure 2. Distribution of the number of CPA found in all positive samples.

The levels of aflatoxin in 27 of the 102 samples analyzed were found to be over 20 μg/kg [58], determined by the FDA. In addition, aflatoxin levels in 32 of the 102 samples were found to be above the legal limit of 10 μg/kg [59] determined in Turkey. In the region where there are major problems in peanut harvest, drying, and storage periods, and as long as measures are not taken, these problems will continue to increase in the future is absolute.

All of these, besides difference in climate conditions, methods of harvesting, drying process, and transferring, leading to mechanical damages of peanuts and inadequate drying after rewetting for dehulling are deperiodinant for the final aflatoxins content. Our results showed that high aflatoxin contamination of 32 of the 102 samples were levels above "recognized" limits in Turkey. So far although aflatoxin is always the most important toxin because of its toxicity, the CPA presence in peanuts is also important by the end of this study.

4. Conclusions

In accordance with results of the study, it was concluded that when deperiodining the aflatoxin contamination compared during different periods, storage period is determined to be higher than the harvesting and drying. So it was concluded that aflatoxin began during the period of harvest, and increased during the drying period. In the period of the prestorage, it was found to decrease as a result of purifying of the soil or other foreign matter partially. Aflatoxin contaminations in peanut samples continue in the storage conditions, and aflatoxin contamination that was detected increased very much. Considering that if the samples were collected at harvest, it can be concluded that the creation of a suitable environment for the production of toxins is inappropriate during drying and storage conditions. In the region, where there are problems in peanut harvesting, drying, and storage periods and as long as

http://dx.doi.org/10.5772/intechopen.68210

measures are not taken, these problems are expected to continue to increase in the future. Also, the sample contaminated with CPA and the simultaneous detection of AFs and CPA highlight the need to investigate factors related to the control and co-occurrence of these toxins in peanuts.

Acknowledgements

This work was financially supported by Tubitak of Turkey (Project no: 115O007). We thank Tubitak for their contributions to Tubitak.

Author details

Isilay Lavkor[1]*, Isil Var[2], Sevcan Oztemiz[3] and Manaf AlMatar[4]

*Address all correspondence to: lavkor@gmail.com

1 Biological Control Institue, Yuregir, Adana, Turkey

2 Department of Food, Faculty of Agriculture, Cukurova University, Adana, Turkey

3 Depertment of Plant Protection, Faculty of Agriculture and Natural Science, Duzce University, Duzce, Turkey

4 Department of Biotechnology, Cukurova University, Adana, Turkey

References

[1] Pitt JI, Hocking AD, Bhudhasamai K, Miscamble BF, Wheeler KA, Tanboon-Ek P. The normal mycoflora of commodities from Thailand. International Journal Food Microbiology. 1993; **20**: 211-226. DOI: 10.1016/0168-1605(93)90166-E.

[2] FAO. FAO achievements in Egypt. 2011. Rome, Italy: Field Program Management Information System.

[3] Embaby EM, Abdel-Galel MM. Detection of fungi and aflatoxins contaminated peanut samples (*Arachis hypogaea* L.). Journal of Agricultural Technology. 2014; **10**: 423-437.

[4] Akdere U, Aslan BH, Saydam IS, Alibekiroglu O, Cınar S. TR63 Region Peanut Sector Report. 2015. Available from: http://www.dogaka.gov.tr/Icerik/Dosya/www.dogaka.gov.tr_620_KB1O20NE_Yerfistigi-Sektor-Raporu-2015.pdf. [Last accessed on 2016 July 22].

[5] TUIK. Summary of agricultural statistics. 2016. Turkish Statistical Institute. Available from: hptt://tuik. [Last accessed on 2016 June 21].

[6] Lillehoj EB, Kwolek WF, Homer ES, Catalano EA, Widstrom NW, Josephson LM, Franz AO. Aflatoxin contamination of preharvest corn: role of *Aspergillus flavus* inoculum and insect damage. Cereal Chemistry. 1980; **57**: 255-257.

[7] Wilson DM, Mubatanhema W, Jurjevic Z. Biology and ecology of mycotoxigenic *Aspergillus* species as related to economic and health concerns. In: DeVries WJ, Trucksess MW, Jackson LS. editors. Mycotoxins and food safety. 2002. NY: Kluwer Academic. pp. 3-17.

[8] Rodríguez Velasco ML, Calonge Delso MM, Ordónez Escudero D. ELISA and HPLC determination of the occurrence of aflatoxin M(1) in raw cow's milk. Food Additives and Contaminants. 2003; **20**: 276-280.

[9] Luk KC, Kobbe B, Townsend JM. Production of cyclopiazonic acid by *Aspergillus flavus* link. Applied and Environmental Microbiology. 1977; **33**: 211-212.

[10] Lansden JA, Davidson JI. Occurrence of cyclopiazonic acid in peanuts. Applied and Environmental Microbiology. 1983; **45**: 766-769.

[11] Ferna'ndez Pinto V, Patriarca A, Locani O, Vaamonde G. Natural co-occurrence of aflatoxin and cyclopiazonic acid grown in Argentina. Food Additives and Contaminants. 2001; **18**: 1017-1020. DOI: 10.1080/02652030110057125.

[12] Vaamonde G, Patriarca A, Pinto VF, Comerio R, Degrossi C. Variability of aflatoxin and cyclopiazonic acid production by *Aspergillus* section *Flavi* from different substrates in Argentina. International Journal of Food Microbiology. 2003; **88**: 79-84. DOI: 10.1016/S0168-1605(03)00101-6.

[13] Gallagher RT, Richard JL, Stahr HM, Cole RJ. Cyclopiazonic acid production by aflatoxigenic and non-aflatoxigenic strains of *Aspergillus flavus*. Mycopathologia. 1978; **66**: 31-36. DOI: 10.1007/BF00429590.

[14] Richard JL. Some major mycotoxins and their mycotoxicoses—An overview. International Journal of Food Microbiology. 2007; **119**: 3-10. DOI: 10.1016/j.ijfoodmicro.2007.07.019.

[15] IARC. Some naturally occurring substances: Food items and constituents, heterocyclic aromatic amines and mycotoxins. IARC monographs on the evaluation of carcinogenic risks to human. IARC Press; 1993.

[16] Oktay HI, Heperkan D, Yelboga E, Karaguler NG. *Aspergillus flavus*—Primary causative agent of aflatoxins in dried figs. Mycotaxon. 2011; **115**: 425-433. DOI: 10.5248/115.425.

[17] Zorzete P, Baquião AC, Atayde DD, Reis TA, Gonçalez E, Corrêa B. Mycobiota, aflatoxins and cyclopiazonic acid in stored peanut cultivars. Food Research International. 2013; **52**: 380-386. DOI:10.1016/j.foodres.2013.03.029.

[18] Somuncuoglu S. Occurrence of cyclopiazonic acid in dried figs [thesis]. Istanbul Technical University; 2007.

[19] Riba A, Matmoura A, Mokrane S, Mathieu F, Sabaou N. Investigations on aflatoxigenic fungi and aflatoxins contamination in some nuts sampled in Algeria. African Journal of Microbiology Research. 2013; **7**: 4974-4980. DOI:10.5897/AJMR2013.5867.

[20] Sauer DB, Meronuck RA, Christensen CM. Microflora. In: Sauer DB. editor. Storage of cereal grains and their products. 1992. St. Paul, MN: American Association of Cereal Chemists. pp. 313-340.

[21] Frisvad JC. Mycotoxins and mycotoxigenic fungi in storage. In: Jayas DS, White NDG, Nuir WE, editors. Stored–Grain ecosystems. 1995. N.Y: Marcel Dekker. pp. 251-288.

[22] Aschalew EL. Aflatoxin content of peanut (*Arachis hypogaea*) in relation to shelling and storage practice of ethiopian farmers [thesis]. 2010. Addis Ababa Unıversıty of Addis Ababa.

[23] McAlpin CE, Wicklow TD, Horn BW. DNA fingerprinting analysis of vegetative compatibility groups in *Aspergillus flavus* from a peanut field in Georgia. Plant Disease. 2002; **86**: 254-258. DOI:10.1094/PDIS.2002.86.3.254.

[24] Bankole SA, Adebanjo A. Mycotoxins in food in West Africa: Current situation and possibilities of controlling it. African Journal of Biotechnology. 2003; **2**: 254-256.

[25] Gachomo EW, Mutitu EW, Kotchoni OS. Diversity of fungal species associated with peanuts in storage and the levels of aflatoxins in infected samples. International Journal of Agricultural Biology. 2004; **6**: 955-959. DOI:1560-8530/2004/06-6-955-959.

[26] Ito Y, Peterson SW, Wicklow TD. *Aspergillus pseudotamarii*, a new aflatoxin producing species in *Aspergillus* section *Flavi*. Mycological Research. 2001; **105**: 233-239. DOI:10.1017/S0953756200003385.

[27] Bora T, Karaca I. Measurement of diseases and pests in crops. 1970. University of Ege. Printing Office, publicians no. 1970; **167**.

[28] Lavkor I. Control of diseases and aflatoxin occurrences with proper cultural and disease management practices in peanut growing [thesis]. 2013. University of Cukuroava.

[29] Arzandeh S, Jinap S. Effect of initial aflatoxin concentration, heating time and roasting temperature on aflatoxin reduction in contaminated peanuts and process optimisation using response surface modelling. International Journal of Food Science and Technology. 2011; **46**: 485-491. DOI:10.1111/j.1365-2621.2010.02514.x.

[30] Sosa A, Kobashigawa E, Galindo J, Bocourt R, Corassin H, Oliverira CAF. Mycotoxicological studies of an *Aspergillus oryzae* strain. Journal of Yeast and Fungal Research. 2013; **4**: 26-32. DOI: 10.5897/JYFR13.0111.

[31] Pitt JI, Hocking AD. Current knowledge of fungi and mycotoxins associated with food commodities in Southeast Asia. Microbiology Australia. 2003; **82**: 173-179.

[32] Cotty P, Garcia RJ. Influences of climate on aflatoxin producing fungi and aflatoxin contamination. International Journal of Food Microbiology. 2007; **119**: 109-115. DOI: 10.1016/j.ijfoodmicro.2007.07.060.

[33] Atayde AD, Reis TA, Godoy IJ, Zorzete P, Reis GM, Corrêa B. Mycobiota and aflatoxins in a peanut variety grown in different regions in the State of São Paulo. Brazil Crop Protection. 2012; **33**: 7-12. DOI: 10.1016/j.cropro.2011.11.013.

[34] Fonseca H. A aflatoxina e o amendoim [Internet]. 2012. Available from: http://www.micotoxinas.com.br/Boletim13.htm. [Last accessed on 2012 Feb 01].

[35] Chang PK, Horn BW, Dorner JW. Clustered genes involved in cyclopiazonic acid production are next to the aflatoxin biosynthesis gene cluster in *Aspergillus flavus*. Fungal Genetics and Biology. 2009; **46**: 176-182. DOI: 10.1016/j.fgb.2008.11.002.

[36] Cole RJ. Etiology of turkey "X" disease in retrospect: A case for the involvement of cyclopiazonic acid. Mycotoxin Research. 1986; **2**: 3-7. DOI: 10.1007/BF03191956.

[37] Dinh HQ, Walker N, Song C, Kobayashi A, Hodges LF. Evaluating the importance of multi-sensory input on memory and the sense of presence in virtual environments. 1999. In: IEEE explore conferencence virtual reality proceedings. Publisher IEEE. pp. 222-228.

[38] Kaaya NA, Warren HL. A review of past and present research on aflatoxin in Uganda. African Journal of Food Agriculture and Nutritional Development. 2005; **5**: 1-18.

[39] Okello DK, Kaaya AN, Bisikwa J, Were M, Oloka H. Management of aflatoxins in peanuts. 2010. Entebbe, Uganda: National Agricultural Research Organisation. [Internet]. Available from: https://www.R-project.org. [Last accessed on 2015 Dec 12].

[40] Bicici M. Disease and aflatoxin problem of peanuts. 2008. Farmers. Brochure. Agricultural extension, communication research and application center of Cukurova University, Adana, Turkey. Available from: http://tyhm.cu.edu.tr/Tr/detay.aspx?pageId=1487. [Last accessed on 2016 July12].

[41] Ding X, Li P, Bai Y, Zhou H. Aflatoxin B_1 in post-harvest peanuts and dietary risk in China. Food Control. 2012; **23**: 143-148. DOI:10.1016/j.foodcont.2011.06.026.

[42] Juan C, Zinedine A, Molto JC, Idrissi L, Manes J. Aflatoxins levels in dried fruits and nuts from Rabat-Sale area, Morocco. Food Control. 2008; **19**: 849-853. DOI:10.1016/j.foodcont.2007.08.010.

[43] Mphande FA, Siame BA, Taylor JE. Fungi, aflatoxins, and cyclopiazonic acid associated with peanut retailing in Botswana. Journal of Food Protection. 2004; **67**: 96-102.

[44] Okello DK, Biruma M, Deom CM. Overview of peanuts research in Uganda: Past, present and future. African Journal of Biotechnology. 2010; **39**: 6448-6459. DOI: 10.5897/AJB09.013.

[45] Lavkor I, Bicici M. Aflatoxin occurrence in peanuts grown in Osmaniye at harvest, post-harvest, drying and pre-storage periods. Journal of Agricultural Sciences. 2015; **21**: 394-405.

[46] Kacmaz A. A study on the technology of peanut processing and the evaluation of the working efficiency of the machines employed to this end [thesis]. 2006. University of Cukurova.

[47] Hell K, Cardwell KF, Setamou M, Poehling HM. The influence of storage practices on aflatoxin contamination in maize in four agroecological zones of Benin, West Africa. Journal of Stored Products Research. 2000; **36**: 365-382. DOI:10.1016/S0022-474X(99)00056-9.

[48] Cardwell KF, Desjardins A, Henry SH, Munkvold G, Robens J. Mycotoxins: The cost of achieving food security and food quality. APS*net* Features. Online. 2001. doi: 10.1094/APSnetFeature-2001-0901. Available from: http://www.apsnet.org/publications/apsnet-features/Pages/Mycotoxins.aspx. [Last accessed on 2016 July 01].

[49] Scheidegger KA, Payne GA. Unlocking the secrets of *Aspergillus flavus* from pathogenicity to functional genomics. Journal of Toxicology. 2003; **22**: 423-459. DOI: 10.1081/TXR-120024100.

[50] CAST. Mycotoxins: risks in plants, animals and human systems. 2003. Task Force Report No: 139. Ames, Iowa, USA: The Council for Agricultural Science and Technology.

[51] Shapira R, Paster N. Control of mycotoxins in storage and techniques for their decontamination. In: Magan N, Olsen M, editors. Mycotoxins in food detection and control. 2004. Cambridge: Woodhead Publishing. pp. 190-223.

[52] Blaney BJ, Kelly MA, Tyler AL, Connole MD. Aflatoxin and cyclopiazonic acid production by Queensland isolates of *Aspergillus flavus* and *Aspergillus parasiticus*. The Australian Journal of Agricultural and Resource Economics. 1989; 40: 395-400.

[53] Horn NW, Greene RL, Sobolcv VS, Dorner JW, Powell JH. Association of morphology and mycotoxin production with vegetative compatibility groups in *Aspergillus flavus*, *A. parasiticus*, and *A. tamarii*. Mycologia. 1996; **88**: 574-587. DOI: 10.2307/3761151.

[54] Resnik SL, González HHL, Pacin AM, Viora M, Caballero GM, Gros EG. Cyclopiazonic acid and aflatoxins production by *Aspergillus flavus* isolated from Argentina corn. Mycotoxin Research. 1996; **12**: 61-66.

[55] Gonçalez E, Nogueira JHC, Fonseca H, Felício JD, Pino FA, Corrêa B. Mycobiota and mycotoxins in Brazilian peanut kernels from sowing to harvest. International Journal of Food Microbiology. 2008; **123**: 184-190. DOI:10.1016/j.ijfoodmicro.2008.01.012.

[56] Heperkan D, Somuncuoglu S, Karbancioglu-Güler F, Mecik N. Natural contamination of cyclopiazonic acid in dried figs and co-occurrence of aflatoxin. Food Control. 2012; **23**: 82-86. DOI:10.1016/j.foodcont.2011.06.015.

[57] Smith EE, Kubena LF, Braithwaite CE, Harvey RB, Phillips TD, Reine AH. Toxicological evaluation of aflatoxin and cyclopiazonic acid in broiler chickens. Poultry Science. 1992; **71**: 1136-1144.

[58] FDA. U.S. Food and Drug Administration. CPG Sec. 570.375 aflatoxin in peanuts and peanut products. 2016. Available from: http://www.fda.gov/iceci/compliancemanuals/compliancepolicyguidancemanual/ucm074598.htm. [Last accessed on 2016 July 21].

[59] Anonymous. Turkish food codex contaminants of regulation. Date 29.12.2011 and numbered 28157 (3rd repeated) the official newspaper. 2011. Available from: http://www.resmigazete.gov.tr/eskiler/2011/12/20111229M3-8-1.pdf. [Last accessed on 2016 July 18].

Chapter 10

The Serum MicroRNA Expression Modified the Genic Toxicity Caused by Aflatoxin B1

Xiao-Ying Huang, Chun-Ying Luo, Xue-Ming Wu, Jin-Guang Yao, Chao Wang, Bing-Chen Huang, Jun Lu, Xing-Zhizi Wang, Tian-Qi Zhang, Qiang Xia and Xi-Dai Long

Additional information is available at the end of the chapter

http://dx.doi.org/10.5772/intechopen.68419

Abstract

The serum microRNAs have been reported as potential biomarkers for hepatocellular carcinoma (HCC); however, their role in genic toxicity related to aflatoxin B1 (AFB1), such as TP53 mutation and DNA damage, has not yet been evaluated. Here, we conducted a hospital-based case-control study, including 558 patients with pathologically diagnosed HCC and positive AFB1 and healthy controls (n = 630) without any evidence of liver diseases. Genic toxicity related to AFB1 was evaluated using the hot-spot mutation at the codon 269 of TP53 gene (TP53M) and AFB1-DNA adducts. Through serum microRNA PCR microarray screening analysis, we observed 10 differentially expressed microRNAs (including miR-7-2-3p, miR-4651, miR-127-3p, miR-192-5p, miR-382-5p, miR-10b-5p, miR-532-3p, miR-16-5p, miR-106b-5p, and miR-4688) among HCC cases with positive AFB1 and controls with positive AFB1. The miR-4651 and miR-382-5p were further identified to be significantly higher in AFB1-positive HCC cases compared to controls. This kind of increasing serum levels was significantly and positively associated with frequency of TP53M and the levels of AFB1-DNA adduct. Furthermore, these microRNAs also modified the prognosis of HCC related to AFB1. These results suggest that the serum levels of microRNAs might be able to modify AFB1-induced genic toxicity, and microRNA-4651 and miR-382-5p, are such potential candidates.

Keywords: serum microRNA, hepatocellular carcinoma, aflatoxin B1, genic toxicity, DNA adduct, TP53M

1. Introduction

Aflatoxin B1 (AFB1) is an important mycotoxin mainly produced by the moulds *Aspergillus parasiticus* and *Aspergillus flavus*. Structurally, it is characterized by fusion of a cyclopentanone ring to the lactone ring of the coumarin moiety [1–3]. Because *A. parasiticus* and *A. flavus* usually multiply under hot and humid conditions, AFB1 is as a contaminant of human food (including core, peanuts, soya sauce, and fermented soy beans) in tropical areas [1–4]. Increasing evidence has shown that AFB1 has three toxicological effects: (a) the attraction of specific organs, especially liver; (b) genotoxicity, mainly inducing the formation of the hot-spot mutation of p53 gene (especially mutation at the codon 249) and AFB1-DNA adducts; and (c) carcinogenicity, primarily causing hepatocellular carcinoma (HCC) [4–11]. Studies have shown that DNA damage induced by AFB1 plays the central role of carcinogenesis of HCC related to AFB1 in the toxic studies [5–11]. Today, this toxin has been classified as a known human carcinogen by the International Agency for Research on Cancer [6, 12]. Therefore, early marker of genic toxicity of AFB1 before carcinogenesis induced by this toxin offers the best chance of prevention for individuals with AFB1 exposure.

Increasing evidence has shown that there is a link between dysregulation of microRNAs and HCC [13–19]. In particular, microRNAs are highly stable in circulation and expression patterns seem to be tissue specific, suggesting that circulating microRNAs may be potentially ideal biomarkers for some diseases including HCC [20–30]. However, information on whether serum microRNAs are correlated with AFB1-related genic toxicity is limited. In this study, we investigated the association between serum microRNAs and the toxicological effects of AFB1 exposure through the analysis of AFB1-DNA adduct amount and TP53 gene mutation frequency.

2. Materials and methods

2.1. Study design and participants

The protocol was approved by the Research Ethics Committee of the Affiliated Hospital of Youjiang Medical University for Nationalities. We conducted a hospital-based case-control study in the Guangxi area to elucidate the association between the serum microRNAs and genic toxicity of AFB1. All cases and controls were residents of the Guangxi Zhuang Autonomous Region from AFB1 exposure areas. All participants were recruited from affiliated hospitals of Guangxi Medical University and Youjiang Medical College for Nationalities and accepted enrollment in this study. All newly diagnosed HCC patients in hospitals affiliated with Youjiang Medical College for Nationalities and Guangxi Medical University from January 2006 to December 2015 were utilized. The inclusion criteria on cases are as follows: (1) cases with histopathology-confirmed HCC; (2) cases understanding the objective of the study and providing informed consent; (3) the ability to complete the necessary investigations and questionnaires; (4) cases with negative HBV markers (HBsAg, HBeAg, anti-HBe, anti-HBc, and HBV DNA) and negative anti-HCV; (5) cases with positive ln (AAA) (positive value: ≥1.00 ln fmol/mg); (6) cases without preoperative chemotherapy, radiotherapy, transarterial chemoembolization, or ablation before collection of blood samples; (7) cases receiving resect treatment (curative or partial resection) or resect treatment plus postoperative adjuvant TACE as initial treatment according to Chinese Manage Criteria of HCC, but not treatment with radiotherapy or chemotherapy before surgical operative treatment; and (8) 5-year follow-up completed with available cancerous tissue specimens and clinical data. The exclusion criteria for cases consisted of: (1) cases with HCC but not confirmed by histopathological examination; (2) cases receiving chemotherapy or radiotherapy treatment before surgical operative treatment; (3) cases with positive HBV markers (HBsAg, HBeAg, anti-HBe, anti-HBc, or HBV DNA) or positive anti-HCV; (4) cases without positive ln (AAA) (positive value: ≥1.00 ln fmol/mg); and (5) cases rejected, dropped out, or lost information.

All controls were recruited from the general health check-up center at the same hospitals during the same period for comparison. The inclusion criteria for controls included (1) controls individually matched to HCC cases based on gender, ethnicity (Han, Zhuang), age (±5 years), time when sampled, and hospital locations, to control the effects of confounders; (2) controls understanding the objective of the study and providing informed consent; (3) the ability to complete the necessary investigations and questionnaires; (4) controls with negative HBV markers (HBsAg, HBeAg, anti-HBe, anti-HBc, and HBV DNA) and negative anti-HCV; (5) controls with positive ln (AAA) (positive value: ≥1.00 ln fmol/mg); (6) controls without liver diseases and other systematic diseases; and (7) controls with persistently normal AST, ALT, and AFP levels. The exclusion criteria for controls consisted of (1) individuals with evidence of liver diseases; (2) individuals with positive HBV markers (HBsAg, HBeAg, anti-HBe, anti-HBc, or HBV DNA) or positive anti-HCV; (3) individuals without positive ln (AAA) (positive value: ≥1.00 ln fmol/mg); and (5) individuals rejected, dropped out, or lost information.

According to aforementioned criteria, a total of 558 cases with HCC and 630 controls, representing 97% of eligible cases and 94% of eligible controls, were interviewed and included in the present study. All patients and controls gave informed consent for participation and were interviewed uniformly before surgery by a well-trained interviewer. The questionnaire used in the interview sought detailed information on general demographic data (including sex, age, ethnicity, dietary and living history, medical history, and family disease history). Demographic information and therapeutic data were collected from medical records in the hospitals by a Youjiang Cancer Institution staff member. At the same time, 4 mL of peripheral blood was obtained for serum analysis of microRNAs. Surgically removed tumor samples of all cases were collected for genic toxicity assay of AFB1.

2.2. Serum preparation

For serum preparation, 5-mL peripheral whole blood was collected from each patient with HCC and control. Samples were centrifuged at 3000 r.p.m. for 10 min under the conditions of 4°C, followed by an additional centrifugation at 12,000 r.p.m. for 15 min to completely remove all remaining cells. The serum samples were aliquoted and stored at −80°C until analysis.

2.3. DNA detraction

Genomic DNA was extracted from HCC tumor tissues using *Genomic DNA Prep Kit* (cat#9K-6-0016, Bio Basic, Inc., Ontario, Canada) as described by standard procedures (Protocol #BS474, Bio Basic, Inc., Ontario, Canada). Briefly, about 15-mg fresh cancerous tissue was transferred to a 1.5-mL microcentrifuge tube, and 300 μL of *Cell Lysis Solution* and 1.5 μL of 20 mg/mL *proteinase K* were added for deparaffinization and digestion at 20°C overnight until the tissue had dissolved. After that, 100 μL of *Protein Precipitation Solution* was added for the cell lysate. The supernatant after centrifuge was transferred to another microcentrifuge tube. Then, DNA was extracted by phenol-chloroform extraction and ethanol precipitation and stored at −20°C until additional analysis.

2.4. Laboratory tests

Fasting venous blood samples were collected from all patients for routine workup, including complete blood picture, liver function tests, prothrombin concentration and prothrombin international normalized ratio, AFP, anti-HCV, HBsAg, and HBc-Ab using commercially available assays.

Because AAA is a stable AFB1 exposure biomarker, the levels of AAA were used to evaluate the AFB1 exposure levels of all subjects [1, 3]. AAA levels in the serum were tested using the comparative enzyme-linked immunosorbent assay as previously published. According to our previous reports with respect to AFB1 exposure, value more than 1.00 ln fmol/mg was considered as positive-AFB1 status [31–34].

2.5. Serum microRNAs expression profiling analysis

In this study, we screened the serum microRNAs using two methods: microRNA array analysis and TaqMan-PCR analysis. For microRNA array analysis, we collected sera from six HCC cases and six sex-, age-, and ethnicity-matched controls without any evidence of liver diseases. We sent sera to Shanghai Oe-Bio-Tech Medical Company (Shanghai, China) for microRNA array detection. Briefly, total RNA from 1-mL serum was extracted with the *PAXgene® Blood RNA Kit* (cat#762174, Qiagen, Duesseldorf, Germany), and RNA quality was evaluated using the analyses of RNA purity and concentration by NanoDrop spectrophotometer and RNA integrity (RIN) by BioAnalyzer 2100. RNA samples would be used for microRNA assay if RIN value was more than 7.2. *RT² First Strand Kit* (cat#330401, Qiagen) was used to synthesize all corresponding cDNA. After that, the amounts of human microRNAs in the serum samples were tested through the real-time PCR (on an Applied Biosystems 7900HT Real-Time PCR System) using RT² Profiler PCR Arrays (cat# PAHS-028ZF-2, Qiagen) in combination with RT² SYBR Green Mastermixes (cat# 330500, Qiagen). The cycle threshold (CT) values were analyzed using the PCR Array Data Analysis Web portal (at www.SABiosciences.com/pcrarraydataanalysis.php). In the present study, a total of 10 candidate microRNAs were chosen for TaqMan-PCR analysis according to the following criteria: more than two-times change between cases and controls, coefficient of variation for CT values <0.05, and high expression ($CT_{average}$ <29 cycles) in patients with HCC.

2.6. TaqMan-PCR assay for candidate microRNAs

The serum levels of 10 candidate microRNAs were tested using quantitative reverse transcription-PCR with TaqMan probe described in our previous reports [17–19]. Briefly, total RNA was extracted from 400-μL serum with 0.2 nM of cel-miR-67 using *PureLink® RNA Mini Kit* (cat#12183018A, Ambion, USA), and corresponding first-strand cDNAs were synthesized using *High Capacity cDNA Reverse Transcription Kit* (cat# 4368814, Invitrogen Grand Island, NY) and *TaqMan MicroRNA Reverse Transcription Kit* (cat#4366596, Applied Biosystems, Carlsbad, CA). After that, TaqMan-PCR analysis was performed using standard protocols on a Bio-Rad iCycler CFX Detection System. The serum levels of candidate microRNAs were assessed using *TaqMan microRNA assays* (cat#4427975, Applied Biosystems) with cel-miR-67 as the endogenous control. PCR reactions were run in a 5-μL final volume containing 1 × *TaqMAN Universal Master Mix II* (cat#4440041, Applied Biosystems), 1 × TaqMan microRNA probes and primers (cat#4427975, Applied Biosystems), and about 15 ng of cDNA. Cycling conditions were 30 s at 95°C for the initial denaturation, and 50 cycles of 15 s at 95°C for denaturation and 1 min at 60°C for annealing. All reactions were conducted in triplicate, and controls (including negative and positive control) were performed for each gene. In this study, the relative amount of candidate microRNAs to cel-miR-67 was calculated as $2^{-\Delta CT}$ method, where $\Delta CT = (CT_{microRNA} - CT_{cel\text{-}miR\text{-}67})$.

2.7. Genic toxicity analysis of AFB1

In the present study, genic toxicity of AFB1 was evaluated using two markers: AFB1-DNA adducts and the hot-spot mutation at the codon 249 of TP53 gene (TP53M) in the cancerous tissues. The amount of AFB1-DNA adducts in HCC cancerous tissues was measured by competitive enzyme-linked immunosorbent assay as described by our previous report [33, 35–39]. Briefly, DNA samples were assayed at 50 ng/well and quantitated relative to AFB1-FAPy standard using monoclonal antibody *6A10*. The percent of inhibition was calculated by comparison with the nonmodified heat-denatured calf thymus DNA control. Each sample was measured in triplicate on the three different assay dates and had a variability of less than 10%.

For TP53M assay, TP53 codon 249 genotypes were genotyped using the TaqMan-PCR on iCycler iQ™ real-time PCR detection system (iQ5, Bio-Rad Laboratories Inc., Hercules, CA, USA). Primers and probes for TaqMan-PCR assay of TP53M are as follows: 5′-TTGGC TCTGA CTGTA CCACC AT-3′ (SY#NSO_533299_001, Applied Biosystems), 5′-TGGAG TCTTC CAGTG TGATG ATG-3′ (SY# NSO_533299_002, Applied Biosystems), 5′-FAM-ACCGG AGTCC CATC-MGB-3′ (SY#431603301-001, Applied Biosystems), and 5′- VIC-AACCG GAGGC CCAT-MGB-3′ (SY#431603301-002, Applied Biosystems) [33, 34]. Each PCR was performed in a total volume of 25 μL containing 1 × Premix Ex TaqTM (catalog # DRR039A, TaKaRa Biotechnology (Dalian) Co., Ltd., Dalian, China), 0.2 μM of each primer, 0.2 μM of each probe, and 50–100 ng of genomic DNA using the running conditions: 95°C for 2 min for the initial denaturation and 50 cycles of 10 s at 95°C and 1 min at 60°C. For quality control, each PCR run included negative and positive controls. Additionally, a random 10% samples were analyzed using repeated genotyping and sequencing methods, and 100% identical genotype were yielded.

2.8. HCC patients following up

Patients with HCC were followed and underwent serial monitoring of chest radiograph, ultrasonography, AFP, and emission computed tomography every 2 months for the first 2 years and semiannually thereafter for detection of recurrence. Tumor recurrence was confirmed by imaging techniques (including chest radiograph, ultrasonography, and emission computed tomography), either intrahepatically or extrahepatically (distant metastases or lymph nodes). A new tumor with increasing AFP but without radiologic evidence was not regarded as recurrence until confirmed by imaging. The last follow-up day was set on August 31, 2015, and the survival status was confirmed via clinic records and patient or family contact. The data of two survival types, recurrence-free survival (RFS) and overall survival (OS), were collected in the present study. The duration of RFS was defined as the date of primary treatment to the date of tumor recurrence or last known date alive, whereas the duration of OS was defined as from the date of primary treatment to the date of death or last known date alive [17, 40, 41].

2.9. Statistical analysis

All statistical analyses were done using the statistical package for social science version 18 (SPSS Institute, Chicago, IL). The differences of age, race, gender, and liver function between groups were compared using Student *t* test and the χ^2 test. The nonparametric Mann-Whitney *U* test was used for comparison of microRNA data ($2^{-\Delta CT}$) from independent samples from two groups as this type data were not normally distributed. Unconditional logistical regression was conducted to estimate odds ratios (ORs) for the association between microRNAs and TP53M along with the 95% confidence intervals (CIs). Kaplan-Meier survival analysis with the log-rank test was used to evaluate the effects of the serum microRNAs levels on HCC prognosis. Risk factors for HCC prognosis were selected using the Cox multivariate regression model with stepwise forward selection based on a likelihood ratio test. Hazard ratios (HRs) and 95% CIs for risk factors were then calculated from a multivariate Cox regression model. All statistical tests were two tailed, and a *P*-value of <0.05 was considered statistically significant.

3. Results

3.1. The characteristics of HCC cases and controls

According to eligibility criteria, we collected 1188 serum samples from patients with HCC and controls (**Table 1**). There were no differences between cases and controls in terms of the distribution of age, sex, race, and smoking and drinking status because these were individually matched. All participants had positive marker of AFB1 exposure but not history of HBV or HCV infection. About 60% of HCC cases featured abnormal liver function.

3.2. Differential expression of serum microRNAs levels between AFB1-positive HCC cases and controls

In this study, we first examined the serum microRNA profiles in cases with AFB1-positive HCC compared to controls with positive AFB1 but without any evidence of liver diseases

Variable	Controls (n = 630)		HCCs (n = 558)		*P**
	n	%	n	%	
Age (y)					0.68
≤49	326	51.7	282	50.5	
>49	304	48.3	276	49.5	
Gender					0.90
Female	192	30.5	172	30.8	
Male	438	69.5	386	69.2	
Race					0.86
Han	331	52.5	296	53.0	
Minority	299	47.5	262	47.0	
Smoking					0.81
Negative	253	40.2	228	40.9	
Positive	377	59.8	330	59.1	
Drinking					0.29
Negative	242	38.4	231	41.4	
Positive	388	61.6	327	58.6	
AFP (ng/L)					-
≤20	630	100.0	216	38.7	
>20	0	0.0	342	61.3	
AST					-
Negative	630	100.0	236	42.3	
Positive	0	0.0	322	57.7	
ALT					-
Negative	630	100.0	328	58.8	
Positive	0	0.0	330	59.1	

*The *P* value indicates the statistical significance for the differences between HCC cases and controls.
Abbreviations: AST, aspartate transaminase; ALT, alanine transaminase; AFP, α-fetoprotein.

Table 1. Clinic characteristics of study subjects.

using microRNA PCR Array and identified 10 significantly different microRNAs (including miR-127-3p, miR-7-2-3p, miR-192-5p, miR-4651, miR-10b-5p, miR-382-5p, miR-16-5p, miR-532-3p, miR-4688, and miR-106b-5p) between cases and controls (**Figure 1**). Next, we further investigated the serum expression profiles of these microRNAs in all participants using TaqMan-PCR technique (**Figure 2**). Mann-Whitney *U* test showed that only miR-4651 and miR-382-5p were increased in HCCs compared to controls ($P < 0.05$).

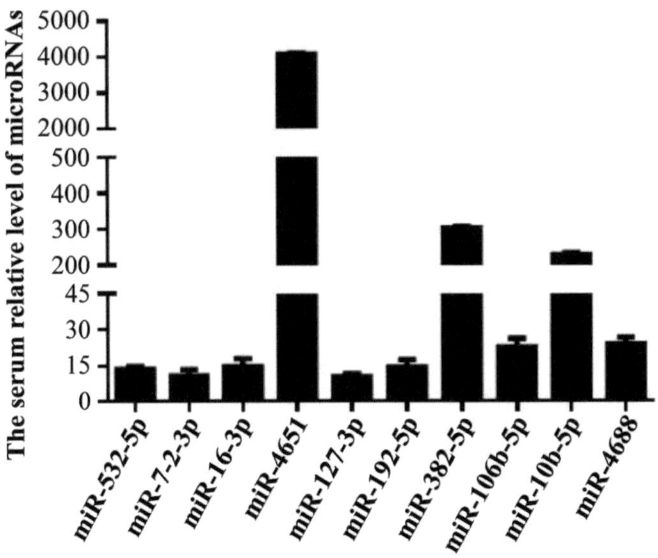

Figure 1. The screening of serum microRNAs for aflatoxin B1 (AFB1)-positive hepatocellular carcinoma (HCC). In this screening analysis, sera samples from six HCC cases and six age-, sex-, and race-matched controls with positive AFB1 exposure but without any evidence of liver tumors were collected, and serum microRNAs were tested using microRNA array analysis. Ten candidate microRNAs were chosen for further analysis according to the fitful criteria (see Section 2).

3.3. The serum miR-4651 and miR-382-5p positively correlated with AFB1-DNA adducts

To investigate whether serum levels of miR-4651 and miR-382-5p were associated with genic toxicity of AFB1, we first explored the correlation between the amount of AFB1-DNA adducts in the cancerous tissues and the serum levels of miR-4651 and miR-382-5p. The increasing serum levels of miR-4651 were found among HCC cases with higher amount of AFB1-DNA adducts in the cancerous tissues (**Figure 3A**). The correlation analysis showed serum miR-4651 level was linearly correlated with the levels of AFB1-DNA adducts, with a linear correlation formula:

$$y = 13.35 + 9.71\chi \quad (1)$$

Where y represents the serum miR-4651 level, and χ represents the amount of AFB1-DNA adducts in the cancerous tissues (mmoL/moL DNA). Similar results were also found in the correlative analysis of miR-382-5p and AFB1-DNA adducts (formula: $y = 2.2 + 0.17\chi$, where y represents the serum miR-382-5p level and χ represents the amount of AFB1-DNA adducts in the cancerous tissues) (**Figure 3B**). Taken together, our data suggested that serum miR-4651 and miR-382-5p expression might be correlated with AFB1-induced DNA damage.

3.4. The serum miR-4651 and miR-382-5p increased risk of TP53M

Because TP53M is the most important molecular signature of AFB1-induced DNA damage [1, 42], we next investigated whether the serum levels of miR-4651 and miR-382-5p modified this mutation in the 558 cancer cases. To analyze, the serum levels of miR-4651 and miR-382-5p

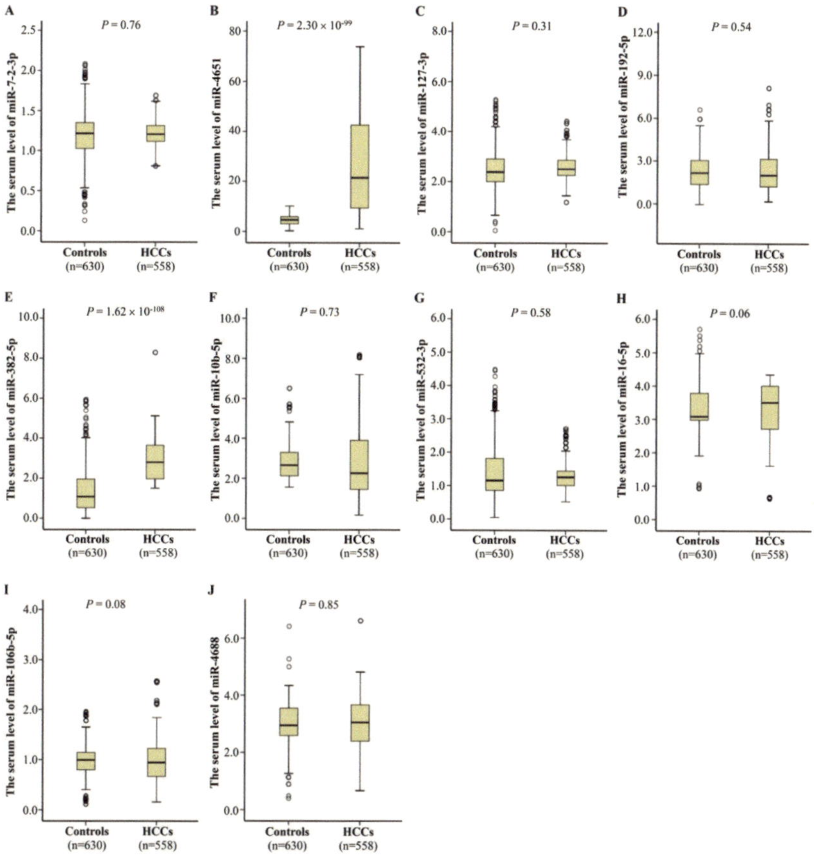

Figure 2. Differentially expressed microRNAs in the serum samples. The differentially expressed microRNAs in the screening analysis, including miR-7-2-3p (**A**), miR-4651 (**B**), miR-127-3p (**C**), miR-192-5p (**D**), miR-382-5p (**E**), miR-10b-5p (**F**), miR-532-3p (**G**), miR-16-5p (**H**), miR-106b-5p (**I**), and miR-4688 (**J**), were further analyzed using TaqMan-PCR method in 558 cases with aflatoxin B1 (AFB1)-positive hepatocellular carcinoma (HCC) and 630 controls with positive AFB1 exposure but without any evidence of liver tumors. The relative levels of microRNA expression were calculated according to $2^{-\Delta CT}$ method (see Section 2). The microRNA data are shown as box plots, with horizontal lines representing the median, the bottom and the top of the boxes representing the 25th and 75th percentiles, respectively. We compared expression data between groups using the Mann-Whitney *U* test.

were divided into two classifications: low (relative expression value ≤20) and high (relative expression value >20) for miR-4651 levels and low (relative expression value ≤3) and high (relative expression value >3) for miR-382-5p, respectively, according to their median relative expression levels. Increasing serum levels of miR-4651 and miR-382-5p increased the frequency of TP53M (**Table 2**); the corresponding risk values were 2.52 (1.65–3.84) and 4.06 (2.72–6.07) for miR-4651 and miR-382-5p, respectively (**Table 2**).

3.5. The serum miR-4651 and miR-382-5p modified the prognosis of AFB1-positive HCC

To study the effects of the serum miR-4651 and miR-382-5p on outcome of patients with AFB1-positive HCC, we analyzed the survival follow-up information of all HCC patients. Results from the Kaplan-Meier survival analysis showed that increasing serum miR-4651

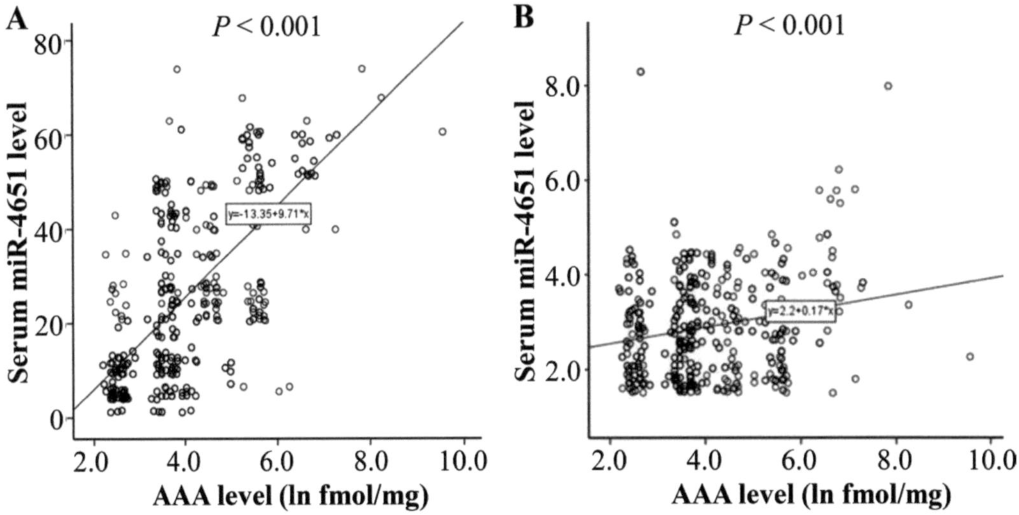

Figure 3. The correlation between the serum levels of miR-4651 and miR-382-5p and the amount of aflatoxin B1 (AFB1)-DNA adducts in cancerous tissues among patients with AFB1-positive hepatocellular carcinoma (HCC) (n = 558). AFB1-DNA adducts and microRNAs were tested using the comparative enzyme-linked immunosorbent assay and TaqMan-PCR techniques, respectively. The serum levels of miR-4651 (**A**) and miR-382-5p (**B**) were linearly associated with the amount of AFB1-DNA adducts.

level significantly correlated with shorter OS and RFS of HCC cases (**Figure 4A and B**). From Cox regression analysis (**Figure 4A and B**), we showed that the miR-4651 is correlated with poor prognosis of HCC (high miR-4651-level risk value, HR = 1.86 and $P = 2.42 \times 10^{-8}$ for OS and 2.28 and 4.32×10^{-9} for RFS, respectively). Survival analysis also exhibited that increasing serum level of miR-382-5p increased death risk (HR = 2.46 and 95% CI = 1.99–3.03) and tumor-recurrence risk (HR = 2.64 and 95% CI = 1.89–3.69) of HCC (**Figure 4C and D**). Taken together, these results indicated that the serum miR-4651 and miR-382-5p are independent of other clinical covariates and suggested its potential as an independent prognostic factor for HCC related to AFB1.

Serum level	**TP53M (–) (n = 174)**		**TP53M (+) (n = 384)**		**OR (95% CI)[a]**	*P*
	n	**%**	**n**	**%**		
miR-4651						
Low	58	33.3	64	16.7	Reference	
High	116	66.7	320	83.3	2.52 (1.65–3.84)	1.70×10^{-4}
miR-382-5p						
Low	86	49.4	75	19.5	Reference	
High	90	51.7	309	80.5	4.06 (2.72–6.07)	8.54×10^{-12}

[a]Adjusted by age, race, and gender.

Table 2. The serum miR-4651 and miR-382-5p levels and TP53M risk.

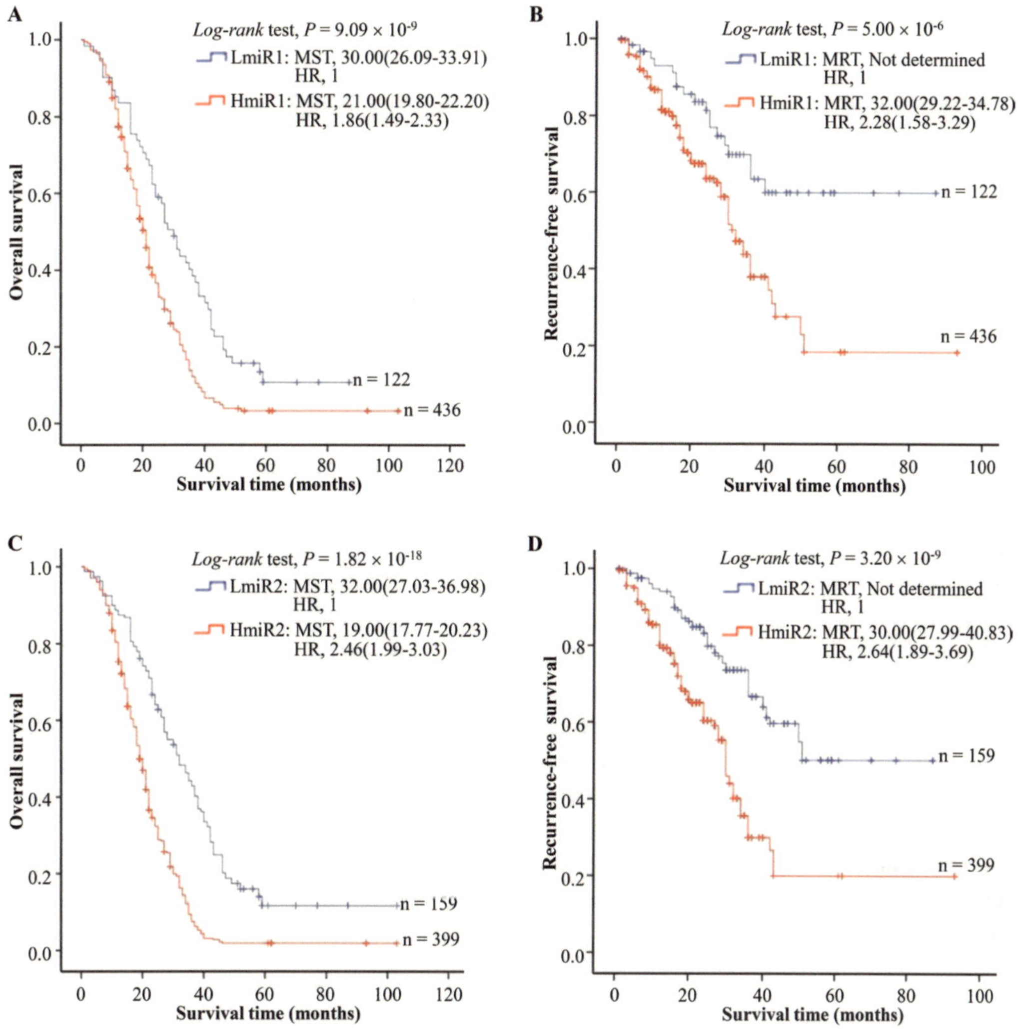

Figure 4. The association between the serum miR-4651 and miR-382-5p and HCC prognosis in 558 aflatoxin B1 (AFB1)-positive cases with hepatocellular carcinoma (HCC). The serum levels of miR-4651 (A and B) and miR-382-5p (C and D) were found to correlate with the overall survival (A and C) or tumor recurrence-free survival (B and D) of HCC. Cumulative hazard function was plotted by Kaplan-Meier's methodology, and *P* value was calculated with two-sided log-rank tests. *Abbreviations*: MST, the median overall survival time; MRT, the median tumor recurrence-free survival time; LmiRN1, low miR-4651 expression; HmiRN1, high miR-4651 expression; LmiRN2, low miR-382-5p expression; HmiRN2, high miR-382-5p; and HR, hazard ratio.

4. Discussion

4.1. The evaluation of toxicological effects of AFB1

A main genic toxicological effect of AFB1 is to induce DNA damage, consisting of AFB1-DNA adducts and the hot-spot mutation of tumor suppressor gene p53 at codon 249 (TP53M) [1, 12].

AFB1 can produce several DNA adducts formation, including 8,9-dihydro-8-(2,6-diamino-4-oxo-3,4-dihydropyrimid-5-yl formamide)-9-hydroxy-AFB1 (AFB1-FAPy) adduct, 8,9-dihydro-8-N^7-guanyl-9-hydroxy-AFB1 (AFB1-N^7-Gua) adduct, and so on. Among these adducts, AFB1-FAPy adduct is a kind of stable imidazole ring-opened form originating from AFB1-N^7-Gua adduct and may display an important role in HCC progress [12]. Furthermore, the accumulation of AFB1-FAPy is nonenzymatic and time-dependent and displays apparent persistence in DNA and potential biological importance in AFB1-related research field [1]. Thus, many researchers in the relative fields regard AFB1-FAPy adduct as a validated biomarker of AFB1 exposure [1]. Increasing evidences have exhibited that the amount of AFB1-FAPy adduct in the liver or placenta tissues are lineally correlated with AFB1 exposure levels and HCC risk, suggesting this adduct should be regarded as a toxicological elucidation biomarker of AFB1 [32]. Our previous studies have shown that the amount of AFB1-DNA adducts in the peripheral blood leukocytes were positively and linearly related to HCC cancerous tissue [3]. These data implied that AFB1-DNA adducts in the peripheral blood leukocytes could be regard as a biomarker for AFB1 exposure as well as adducts in the cancerous tissues. Our following studies exhibited more amount of AFB1-DNA adducts in cancerous tissues than in the peripheral blood. Thus, AFB1-DNA adducts in tumor tissues were furthermore analyzed in the present study.

For the mutations of p53 gene, AFB1 mainly induces the transversion of G → T in the third position at codon 249 of this gene, also called hot-spot mutation at the codon 249 (TP53M). The frequency of TP53M is more persistent biomarker and more directly represents genic toxic effects compared with AFB1-DNA adducts [1, 12]. Our study also showed more than 68.8% (384/558) of patients with AFB1-positive HCC had TP53M in the cancerous tissues. Because of the aforementioned reasons, the genic toxic effects of AFB1 exposure were evaluated through the following two biomarkers: AFB1-DNA adducts amount in HCC cancerous tissues and the frequency of TP53M in this study. Our results also show that these two biomarkers reflected AFB1 exposure information and represented the toxicological capacity of AFB1.

4.2. The serum microRNAs and AFB1-induced genic toxicity

MicroRNAs are a type of small noncoding RNAs and can regulate the translation of protein-coding genes through enhancing protein-coding mRNA degradation or repressing translation of protein-coding mRNA. Their dysregulation affects cell proliferation and differentiation, ultimately resulting to a variety of disorders [43–45]. To date, more than 2000 mature microRNAs have been annotated in the official registry (the MicroRNA Registry). Increasing studies have proved that microRNAs have a crucial role in human carcinogenesis, including hepatocarcinogenesis, via acting as oncogenes or tumor suppressor genes [43–45]. Recent evidence has shown that serum microRNAs are remarkably stable and expression patterns may be tissue specific [20, 21, 24, 27, 28, 30, 46, 47]; thus, they may be important potential candidates for carcinogens and corresponding non-invasive cancer testing. It has been hypothesized that serum microRNAs might correlate with toxicity of carcinogen such as AFB1. Therefore, in the present study, we conducted a case-control study to screen and analyze potential serum microRNAs for genic toxicity testing of AFB1 in a high-AFB1 exposure area, Guangxi area of China [4, 48]. Our results proved that serum microRNAs such as miR-4651 and miR-382-5p were significant and linearly associated with the amounts of AFB1-DNA adducts in the cancerous tissues; moreover, increasing serum levels of these two microRNAs modified the risk of TP53M. Collectively, these results suggest that serum microRNAs might be important biomarkers for predicting genic toxicity of AFB1 and ultimately for preventing HCC induced by AFB1.

In this study, miR-4651 and miR-382-5p were particularly concerned because of its different expression between patients with positive-AFB1 HCC and nontumor controls with positive AFB1. They are encoded by *miR-4651* gene (located at chr7: 75915197–75915269) and miR-382-5p gene (located at chr14: 101520643–101520718), respectively. Until now, it has been not clear whether they act as tumor suppressors or oncogenes. However, we observed that miR-4651 and miR-382-5p had higher expression in the AFB1-positive HCC cases than in non-HCC-harboring individuals, and this increasing expression was further positively associated with poor prognosis of HCC related to AFB1. This implies that they might act as oncogenes or have a similar role of oncogenes through decreasing detoxication and genomic DNA damage repair capacity because the amounts of AFB1-DNA adducts and the frequency of TP53M correlate with HCC risk and prognosis and can reflect the deficiency of detoxication and genomic DNA damage repair capacity. Taken together, these results implied that miR-4651 and miR-382-5p might be useful biomarkers for HCC induced by AFB1 exposure.

This study has several strengths. First, we finished a high-throughput screening analysis for serum microRNAs that exhibited differential levels between cases with positive-AFB1 HCC and healthy controls with positive AFB1. Through this methodology, we not only improved the chance to identify serum biomarkers but also obtained 10 possible AFB1-related microRNAs. Second, only HBV- and HCV-negative cases were included in this study, whereas HBV- or HCV-positive individuals were excluded. This efficiently controlled the effects of other carcinogenetic factors, such as HBV and HCV, and improved correlation analysis of serum microRNAs and genic toxicity of AFB1.

4.3. Limitation

This study had several limitations. First, the increased risk with AFB1 exposure status noted in this study was probably underestimated, because the liver disease itself may affect the metabolism of AFB1 and modify the levels of AFB1-DNA adducts. Second, because the present study is a hospital-based study, potential selection bias might have occurred. Third, in spite of the fact that the status of TP53M was investigated in cases of HCC, other AFB1-related mutations of the TP53 gene were not evaluated. Finally, we did not examine additional functional analysis. Therefore, more functional analyses should be performed based on large samples and a combination of biomarkers and AFB1 exposure.

5. Conclusions

In conclusion, to the best of our knowledge, this is the first report to investigate association between the serum microRNAs and the toxicological effects of AFB1 among Guangxi population from a high AFB1-exposure area. We find that serum levels of miR-4651 and miR-382-5p might increase the amount of AFB1-DNA adducts and the frequency of TP53M, and their dysregulation should contribute to the toxicological effects of AFB1. Given that AFB1 is an important genic agent and a kind of I type carcinogen, our findings might have prevention implications through identifying population with high serum levels of these two microRNAs, once these findings are replicated by other studies based on a larger scale or prospective studies.

Conflicts of interest and source of funding

The authors declare no competing financial interests. This study was supported in part by the National Natural Science Foundation of China (Nos. 81572353, 81372639, 81472243, 81660495, and 81460423), the Innovation Program of Guangxi Municipal Education Department (Nos. 201204LX674 and 201204LX324), the Natural Science Foundation of Guangxi (Nos. 2016GXNSFDA380003, 2015GXNSFAA139223, 2013GXNSFAA019251, 2014GXNSFDA118021, and 2014GXNSFAA118144), Innovation Program of Guangxi Health Department (No. Z2013781), Research Program of Guangxi Clinic Research Center of Hepatobiliary Diseases (No. 2016–11), and Open Research Program from Molecular Immunity Study Room Involving in Acute & Severe Diseases in Guangxi Colleges and Universities (Nos. kfkt20160062 and kfkt20160063).

Acknowledgements

We thank Dr. Hua Huang for molecular biochemical technique; Dr. Yong-Zhi Huang, Qiu-Xiang Liang, Dr. Yun Yi, Dr. Yuan-Feng Zhou, and Yun Yi for sample collection and management. We also thank all members of Department of Medical Test and Infective Control, the Affiliated Hospital of Youjiang Medical University for Nationalities for their help.

Abbreviations

AFB1	Aflatoxin B1
AFB1-FAPy	8,9-Dihydro-8-(2,6-diamino-4-oxo-3,4-dihydropyrimid-5-yl formamide)-9-hydroxy-AFB1
AFB1-N^7-Gua	(8,9-dihydro-8-N^7-guanyl-9-hydroxy-AFB1)
HCC	Hepatocellular carcinoma
HR	Hazard ratio
miR-4651	MicroRNA-4651
miR-382-5p	MicroRNA-382-5p
OR	Odds ratio
PCR	Polymerase chain reaction
TP53M	The hot-spot mutation at codon 249 of TP53 gene

Author details

Xiao-Ying Huang[1†], Chun-Ying Luo[1†], Xue-Ming Wu[1†], Jin-Guang Yao[1†], Chao Wang[2†], Bing-Chen Huang[1†], Jun Lu[3†], Xing-Zhizi Wang[1†], Tian-Qi Zhang[1†], Qiang Xia[3*] and Xi-Dai Long[1,3*]

*Address all correspondence to: sjtulongxd@263.net or xiaqiang@sjtu.edu.cn

1 Department of Pathology, The Affiliated Hospital of Youjiang Medical College for Nationalities (AHYMCN), Baise, China

2 Department of Medicine, AHYMCN, Baise, China

3 Department of Liver Surgery, Ren Ji Hospital, School of Medicine, Shanghai Jiao Tong University, Shanghai, China

†These authors contributed equally to this work.

References

[1] Kensler TW, Roebuck BD, Wogan GN, Groopman JD. Aflatoxin: A 50-year odyssey of mechanistic and translational toxicology. Toxicological Sciences: An Official Journal of the Society of Toxicology. 2011;**120**(Suppl 1):528-548. DOI:10.1093/toxsci/kfq283

[2] Golli-Bennour EE, Kouidhi B, Bouslimi A, Abid-Essefi S, Hassen W, Bacha H. Cytotoxicity and genotoxicity induced by aflatoxin B1, ochratoxin A, and their combination in cultured Vero cells. Journal of Biochemical and Molecular Toxicology. 2010;**24**:42-50. DOI:10.1002/jbt.20310

[3] Long XD, Yao JD, Yang Q, Huang CH, Liao P, Nong LG, Tang YJ, Huang XY, Wang C, Wu XM, Huang BC, Ban FZ, Zeng LX, Ma Y, Zhai B, Zhang JQ, Xue F, Lu CX, Xia Q. Polymorphisms of DNA repair genes and toxicological effects of aflatoxin B1 exposure In: Faulkner AG, editor. Aflatoxins: Food Sources, Occurrence and Toxicological Effects. 1st ed. New York: Nova Science Publishers; 2014. pp. 107-124. DOI:

[4] Sun XD, Su P, Shan H. Mycotoxin contamination of rice in China. Journal of Food Science. 2017;**82**:573-584. DOI: 10.1111/1750-3841.13631

[5] Zhang C, Selvaraj JN, Yang Q, Liu Y. A survey of aflatoxin-producing *Aspergillus* sp. from peanut field soils in four agroecological zones of China. Toxins (Basel). 2017;**9**:40. DOI:10.3390/toxins9010040

[6] Sarma UP. Aspergilli-aflatoxins: health threat. Indian Journal of Clinical Biochemistry. 2017;**32**:1-2. DOI:10.1007/s12291-016-0630-5

[7] Nurul Adilah Z, Mohd RS. Effect of dietary macronutrients on aflatoxicosis: A mini review. The Journal of the Science of Food and Agriculture. 2017. DOI: 10.1002/jsfa.8234

[8] Njoroge SM, Matumba L, Kanenga K, Siambi M, Waliyar F, Maruwo J, Machinjiri N, Monyo ES. Aflatoxin B1 levels in groundnut products from local markets in Zambia. Mycotoxin Research. 2017. DOI: 10.1007/s12550-017-0270-5

[9] Huang L, Duan C, Zhao Y, Gao L, Niu C, Xu J, Li S. Reduction of aflatoxin B1 toxicity by *Lactobacillus plantarum* C88: A potential probiotic strain isolated from Chinese traditional fermented food "Tofu". PLoS One. 2017;**12**:e0170109. DOI:10.1371/journal.pone.0170109

[10] Chen G, Gong YY, Kimanya ME, Shirima CP, Routledge MN. Comparison of urinary aflatoxin M1 and aflatoxin albumin adducts as biomarkers for assessing aflatoxin exposure in Tanzanian children. Biomarkers. 2017;2017:1-6. DOI:10.1080/1354750x.2017.1285960

[11] Kumar P, Mahato DK, Kamle M, Mohanta TK, Kang SG. Aflatoxins: a global concern for food safety. Human health and their management. Frontier Microbiology. 2016;**7**:2170. DOI: 10.3389/fmicb.2016.02170

[12] Wang JS, Groopman JD. DNA damage by mycotoxins. Mutation Research. 1999;**424**:167-181. DOI:

[13] Sun J, Lu H, Wang X, Jin H. MicroRNAs in hepatocellular carcinoma: Regulation, function, and clinical implications. Scientific World Journal. 2013;**2013**:924206. DOI: 10.1155/2013/924206

[14] Persson H, Kvist A, Rego N, Staaf J, Vallon-Christersson J, Luts L, Loman N, Jonsson G, Naya H, Hoglund M, Borg A, Rovira C. Identification of new microRNAs in paired normal and tumor breast tissue suggests a dual role for the ERBB2/Her2 gene. Cancer Research. 2011;**71**:78-86. DOI: 10.1158/0008-5472.CAN-10-1869

[15] Kozomara A, Griffiths-Jones S. miRBase: Integrating microRNA annotation and deep-sequencing data. Nucleic Acids Research. 2011;**39**:D152-D157. DOI:10.1093/nar/gkq1027

[16] Griffiths-Jones S, Grocock RJ, van Dongen S, Bateman A, Enright AJ. miRBase: MicroRNA sequences, targets and gene nomenclature. Nucleic Acids Research. 2006;**34**:D140-D144. DOI: 10.1093/nar/gkj112

[17] Lu YL, Yao JG, Huang XY, Wang C, Wu XM, Xia Q, Long XD. Prognostic significance of miR-1268a expression and its beneficial effects for post-operative adjuvant transarterial chemoembolization in hepatocellular carcinoma. Scientific Reports. 2016;**6**:36104. DOI: 10.1038/srep36104

[18] Liu Y-X, Long X-D, Xi Z-F, Ma Y, Huang X-Y, Yao J-G, Wang C, Xing T-Y, Xia Q. MicroRNA-24 modulates aflatoxin B1-related hepatocellular carcinoma prognosis and tumorigenesis. BioMed Research International. 2014;**2014**:13. DOI: 10.1155/2014/482926

[19] Huang XY, Yao JG, Huang HD, Wang C, Ma Y, Xia Q, Long XD. MicroRNA-429 modulates hepatocellular carcinoma prognosis and tumorigenesis. Gastroenterology Research and Practice. 2013;**2013**:804128. DOI:10.1155/2013/804128

[20] Hung CH, Hu TH, Lu SN, Kuo FY, Chen CH, Wang JH, Huang CM, Lee CM, Lin CY, Yen YH, Chiu YC. Circulating microRNAs as biomarkers for diagnosis of early hepatocellular carcinoma associated with hepatitis B virus. International Journal of Cancer. 2016;**138**:714-720. DOI: 10.1002/ijc.29802

[21] Huang JT, Liu SM, Ma H, Yang Y, Zhang X, Sun H, Zhang X, Xu J, Wang J. Systematic review and meta-analysis: Circulating miRNAs for diagnosis of hepatocellular carcinoma. Journal of Cellular Physiology. 2016;**231**:328-335. DOI: 10.1002/jcp.25135

[22] Zhuang L, Xu L, Wang P, Meng Z. Serum miR-128-2 serves as a prognostic marker for patients with hepatocellular carcinoma. PLoS One. 2015;**10**:e0117274. DOI: 10.1371/journal.pone.0117274

[23] Zhao C, Li Y, Zhang M, Yang Y, Chang L. miR-126 inhibits cell proliferation and induces cell apoptosis of hepatocellular carcinoma cells partially by targeting Sox2. Human Cell. 2015;**28**:91-99. DOI: 10.1007/s13577-014-0105-z

[24] Zhang YC, Xu Z, Zhang TF, Wang YL. Circulating microRNAs as diagnostic and prognostic tools for hepatocellular carcinoma. World Journal of Gastroenterology. 2015;**21**:9853-9862. DOI: 10.3748/wjg.v21.i34.9853

[25] Yu F, Lu Z, Chen B, Dong P, Zheng J. microRNA-150: A promising novel biomarker for hepatitis B virus-related hepatocellular carcinoma. Diagnostic Pathology. 2015;**10**:129. DOI: 10.1186/s13000-015-0369-y

[26] Yin J, Hou P, Wu Z, Wang T, Nie Y. Circulating miR-375 and miR-199a-3p as potential biomarkers for the diagnosis of hepatocellular carcinoma. Tumour Biology. 2015;**36**:4501-4507. DOI: 10.1007/s13277-015-3092-0

[27] Wang Y, Gao Y, Shi W, Zhai D, Rao Q, Jia X, Liu J, Jiao X, Du Z. Profiles of differential expression of circulating microRNAs in hepatitis B virus-positive small hepatocellular carcinoma. Cancer Biomarkers. 2015;**15**:171-180. DOI: 10.3233/CBM-140451

[28] Tsuchiya N, Sawada Y, Endo I, Saito K, Uemura Y, Nakatsura T. Biomarkers for the early diagnosis of hepatocellular carcinoma. World Journal of Gastroenterology. 2015;**21**:10573-10583. DOI: 10.3748/wjg.v21.i37.10573

[29] Tang G, Shen X, Lv K, Wu Y, Bi J, Shen Q. Different normalization strategies might cause inconsistent variation in circulating microRNAs in patients with hepatocellular carcinoma. Medical Science Monitor. 2015;**21**:617-624. DOI: 10.12659/MSM.891028

[30] Tan Y, Lin B, Ye Y, Wen D, Chen L, Zhou X. Differential expression of serum microRNAs in cirrhosis that evolve into hepatocellular carcinoma related to hepatitis B virus. Oncology Reports. 2015;**33**:2863-2870. DOI: 10.3892/or.2015.3924

[31] Yao JG, Huang XY, Long XD. Interaction of DNA repair gene polymorphisms and aflatoxin B1 in the risk of hepatocellular carcinoma. International Journal of Clinical and Experimental Pathology. 2014;**7**:6231-6244

[32] Xia Q, Huang XY, Xue F, Zhang JJ, Zhai B, Kong DC, Wang C, Huang ZQ, Long XD. Genetic polymorphisms of DNA repair genes and DNA repair capacity related to aflatoxin b1 (AFB1)-induced DNA damages. In: Chen C, editor. New Research Directions in DNA Repair. 1st ed. Rijeka: InTech; 2013. pp. 377-412

[33] Long XD, Zhao D, Wang C, Huang XY, Yao JG, Ma Y, Wei ZH, Liu M, Zeng LX, Mo XQ, Zhang JJ, Xue F, Zhai B, Xia Q. Genetic polymorphisms in DNA repair genes XRCC4 and XRCC5 and aflatoxin B1-related hepatocellular carcinoma. Epidemiology. 2013;**24**: 671-681. DOI: 10.1097/EDE.0b013e31829d2744

[34] Long XD, Yao JG, Zeng Z, Ma Y, Huang XY, Wei ZH, Liu M, Zhang JJ, Xue F, Zhai B, Xia Q. Polymorphisms in the coding region of X-ray repair complementing group 4 and aflatoxin B1-related hepatocellular carcinoma. Hepatology. 2013;**58**:171-181. DOI: 10.1002/hep.26311

[35] Long XD, Ma Y, Zhou YF, Ma AM, Fu GH. Polymorphism in xeroderma pigmentosum complementation group C codon 939 and aflatoxin B1-related hepatocellular carcinoma in the Guangxi population. Hepatology. 2010;**52**:1301-1309. DOI: 10.1002/hep.23807

[36] Long XD, Ma Y, Huang YZ, Yi Y, Liang QX, Ma AM, Zeng LP, Fu GH. Genetic polymorphisms in DNA repair genes XPC, XPD, and XRCC4, and susceptibility to Helicobacter pylori infection-related gastric antrum adenocarcinoma in Guangxi population, China. Molecular Carcinogensis. 2010;**49**:611-618. DOI: 10.1002/mc.20630

[37] Long XD, Ma Y, Zhou YF, Yao JG, Ban FZ, Huang YZ, Huang BC. XPD codon 312 and 751 polymorphisms, and AFB1 exposure, and hepatocellular carcinoma risk. BMC Cancer. 2009;**9**:400. DOI: 1471-2407-9-400

[38] Long XD, Ma Y, Qu de Y, Liu YG, Huang ZQ, Huang YZ, Lin ZH, Wei NB, Zhou SC. The polymorphism of XRCC3 codon 241 and AFB1-related hepatocellular carcinoma in Guangxi population, China. Annals of Epidemiology. 2008;**18**:572-578. S1047-2797(08)00058-6 [pii]. DOI: 10.1016/j.annepidem.2008.03.003

[39] Long XD, Ma Y, Huang HD, Yao JG, Qu de Y, Lu YL. Polymorphism of XRCC1 and the frequency of mutation in codon 249 of the p53 gene in hepatocellular carcinoma among Guangxi population, China. Molecular Carcinogensis. 2008;**47**:295-300. DOI: 10.1002/mc.20384

[40] Long XD, Huang XY, Yao JG, Liao P, Tang YJ, Ma Y, Xia Q. Polymorphisms in the precursor microRNAs and aflatoxin B1-related hepatocellular carcinoma. Molecular Carcinogensis. 2016;**55**:1060-1072. DOI: 10.1002/mc.22350

[41] Huang XY, Yao JG, Huang BC, Ma Y, Xia Q, Long XD. Polymorphisms of a disintegrin and metalloproteinase with thrombospondin motifs 5 and aflatoxin B1-related hepatocellular carcinoma. Cancer Epidemiology Biomarkers Prevention. 2016;**25**:334-343. DOI: 10.1158/1055-9965.EPI-15-0774

[42] Qi LN, Bai T, Chen ZS, Wu FX, Chen YY, De Xiang B, Peng T, Han ZG, Li LQ. The p53 mutation spectrum in hepatocellular carcinoma from Guangxi, China: Role of chronic hepatitis B virus infection and aflatoxin B1 exposure. Liver International. 2015;**35**:999-1009. DOI: 10.1111/liv.12460

[43] Zhu Z, Zhang X, Wang G, Zheng H. Role of microRNAs in hepatocellular carcinoma. Hepatitis Monthly. 2014;**14**:e18672. DOI: 10.5812/hepatmon.18672

[44] Ward J, Kanchagar C, Veksler-Lublinsky I, Lee RC, McGill MR, Jaeschke H, Curry SC, Ambros VR. Circulating microRNA profiles in human patients with acetaminophen hepatotoxicity or ischemic hepatitis. Proceedings of the National Academy Science United States of America. 2014;**111**:12169-12174. DOI: 10.1073/pnas.1412608111

[45] Ranganathan K, Sivasankar V. MicroRNAs – Biology and clinical applications. Journal of Oral and Maxillofacial Pathology. 2014;**18**:229-234. DOI: 10.4103/0973-029X.140762

[46] Sohn W, Kim J, Kang SH, Yang SR, Cho JY, Cho HC, Shim SG, Paik YH. Serum exosomal microRNAs as novel biomarkers for hepatocellular carcinoma. Experimental and Molecular Medicine. 2015;**47**:e184. DOI: 10.1038/emm.2015.68

[47] Petrini E, Caviglia GP, Abate ML, Fagoonee S, Smedile A, Pellicano R. MicroRNAs in HBV-related hepatocellular carcinoma: Functions and potential clinical applications. Panminerva Medica. 2015;**57**:201-209.

[48] Long XD, Yao JG, Zeng Z, Huang CH, Huang ZS, Huang YZ, Ban FZ, Huang XY, Yao LM, Fan LD, Fu GH. DNA repair capacity-related to genetic polymorphisms of DNA repair genes and aflatoxin B1-related hepatocellular carcinoma among Chinese population. In: Kruman I, editor. DNA Repair. Rijeka: InTech; 2011. pp. 505-524

Chapter 11

Subcritical Water Extraction and Its Prospects for Aflatoxins Extraction in Biological Materials

Sefater Gbashi, Ntakadzeni Edwin Madala,
Oluwafemi Ayodeji Adebo, Lizelle Piater,
Judith Zanele Phoku and Patrick Berka Njobeh

Additional information is available at the end of the chapter

http://dx.doi.org/10.5772/intechopen.68706

Abstract

Aflatoxins (AFs) are well-known mycotoxins and contaminants of various agricultural commodities globally that are linked to a wide range of adverse health and economic complications. Because of their incessant proliferation and deleterious consequences, it has become mandatory to routinely monitor the levels of these toxins in agricultural products before they go into the market. Essentially, effective analysis is an important component of AFs control, and extraction is a necessary step for their analysis, irrespective of the protocol adopted. Conventional methods for AF extraction are expensive, the processes involved are tedious and utilize large quantities of organic solvents that are environmentally unfriendly. This has necessitated the quest for alternatives that are 'green', cost-effective and easy to perform. In this regard, subcritical water extraction (SWE) is a viable alternative that has proven to be effective in the extraction of other bio-active compounds. This chapter presents a critical appraisal of the principles and dynamics of SWE, and its current applications as a viable tool in the extraction of AFs from various biological matrices. Although further research needs to be performed to enhance its applicability, the adoption of SWE in the extraction of AFs seems very promising and needs to be properly exploited.

Keywords: subcritical water extraction, aflatoxins, temperature, biological materials

1. Introduction

The proliferated contamination of agricultural commodities by AFs has become a serious global concern because of their severe impact on health and the economy [1, 2]. This group of mycotoxins is food contaminants produced by filamentous toxigenic fungal species [3, 4] principally those members within the *Aspergillus* genera. Relative to their contamination of food and feed commodities, approximately 4.5 billion people in the world are at the risk of been chronically exposed to mycotoxins, in particular, AFs [5, 6]. Several reports have implicated AFs as very poisonous human and environmental pollutants [2, 7–10]. In fact, one of the AFs, aflatoxin B_1 (AFB_1), has been recognized as the most potent naturally occurring carcinogen known to man [2]. As a result of their widespread proliferation and associated deleterious effects, there is a growing concern over their intake via consumption of contaminated food and feed by humans and animals alike. This has led to more stringent guidelines and regulatory limits of these toxins, especially with the globalization of the food supply chain, and consequently, necessitating routine surveillance of the prevalence as well as levels of AFs in food and feed [11, 12]. As such, the need for more sensitive and robust analytical methods for the determination of AFs is eminent [12], particularly one that is carried out following the greener route.

Extraction is an important step in AFs analysis. It is inevitable irrespective of the protocol adopted. Although different methods exist for extracting AFs from food and feed such as solvent extraction, solid-phase extraction, and immuno-affinity column (IAC) extraction, there are anxieties over human and environmental health regarding safety in their applications [13, 14]. Conventional techniques also involve labor-intensive and time-consuming procedures [15], requiring relatively large volumes of organic solvents, which are expensive and hazardous [13, 16]. Bearing in mind these concerns associated with extraction of AFs, the design of a greener route that is efficient, cheap, fast and relatively easy to address these challenges is significant. Subcritical water extraction (SWE) seems promising in this regard. Better results, recoveries, and effectiveness have been reported for SWE as compared to other traditional methods for extracting different bioactive compounds [17–19]. In this chapter, a comprehensive review on the implications of AFs and issues with their analysis is presented. The need and potential applicability of SWE in AF analysis are highlighted. Lastly, herein, we demonstrate the basic principle of SWE, underscoring its advantages and disadvantages, and wrapping up the chapter with a discussion on how SWE can be suitable in extracting AFs from biological matrices for routine analysis.

2. Aflatoxins

2.1. Definition and concept of aflatoxins

Aflatoxins are the most perilous and troublesome group of mycotoxins to humans and animals that are generally produced by toxigenic strains of fungi, notably *Aspergillus flavus, A. parasiticus* and *A. niger* [2, 20, 21]. At least 14 different types of AFs are known to exist in nature, however, the major ones of economic and health significance are aflatoxin B_1 (AFB_1),

B_2 (AFB_2), G_1 (AFG_1), G_2 (AFG_2), M_1 (AFM_1) and M_2 (AFM_2). AFM_1 and M_2 are hydroxylated metabolites of AFB_1 and B_2, respectively, bio-transformed by the liver and found in milk, urine, and other body fluids, being less harmful than their precursor toxins [22, 23]. Among the AF group, AFB_1 is considered the most toxic. This one has been established as the most notorious naturally occurring carcinogen [2, 24, 25]. For that reason, it has been classified as a Group 1A human carcinogen [26]. Cereals such as maize are common crops that are contaminated by AFs. Additionally, crops such as oilseeds, including peanuts, different kinds of spices, figs and other dried fruit, are also familiar but most susceptible substrates.

2.2. Physicochemical properties of aflatoxins

Aflatoxins are a group of closely related difuranocoumarin derivatives, with similar structures as they constitute a unique group of naturally occurring heterocyclic compounds that are highly oxygenated [22]. AFs can be broadly classified into two groups based on their chemical structure namely, difurocoumarocyclopentenone series (AFB_1, B_2, M_1, M_2, and other derivatives) and difurocoumarolactone series (AFG_1, G_2, and others) [22]. Their chemistry constitutes highly substituted coumarins containing a fused dihydrofurofuran moiety. The AFBs (*i.e.*, members of the blue fluorescent series) generally feature a fusion of a cyclopentenone ring to the lactone ring of the coumarin moiety, while the AFGs possess a fused lactone ring [27]. AFB_1 and G_1 contain an unsaturated bond on the terminal furan ring at the 8, 9 position. Epoxidation at this position has shown to be essential for their carcinogenicity [28]. The intensity of fluorescence (light) emission differs greatly among the four compounds. This property plays a significant role in their quantification by fluorescence techniques [29]. AFs are also very stable chemical compounds and notoriously difficult to eradicate in food commodities [30, 31]. They are chemically stable during processing and storage, even when heated at quite elevated temperatures such as those achieved during the production of breakfast cereals or baking of bread [31, 32]. This necessitates the avoidance of conditions that favor their production, which is not always feasible in practice [31, 33].

2.3. Exposure and health implications of aflatoxins

The presence of AFs in foods and feeds is problematic as it induces vicious health repercussions in humans and animals when exposed to them. Poisoning from AFs has been reported in different parts of the world, and victims include humans, animals and other non-human primates [34]. Common exposure routes include ingestion of AF contaminated foods and feeds; however, aerosol, parental (placental and breastfeeding) and dermal routes have also been reported [9, 35], but it can be supposed that ingestion is the main source of AF exposure among humans and animals. This group of poisons enters the blood stream and lymphatic system with the liver as targeted organ and damage macrophage systems inhibit protein synthesis and increase sensitivity to opportunistic infections [36]. Exposure to AFs can be chronic or acute, and symptoms and degree of illness depend on the type of AF, concentration, and duration of exposure, as well as species, age, sex, and health status of the exposed individual [37].

Aflatoxicosis refers to poisoning and associated illness resulting from AF exposure [38, 39]. There are numerous cases of aflatoxicosis reported in the literature. In Ibadan, Nigeria, the death of some children who consumed mould-infested *Kulikuli,* was suggested to be due to aflatoxicosis [36]. An outbreak of hepatitis in 1974 in India that killed 100 people and caused ailment in hundreds of others was as a result of AFs via consumption heavily contaminated maize [40]. Incidence of liver cancer and aggravated cases of over 40% of diseases in developing nations including kwashiorkor, growth stunting, and HIV are directly or indirectly associated with dietary AF exposure [5, 27, 41]. One of the most epic episodes of aflatoxicosis reported in human history occurred in rural Kenya, of which 317 cases of illness and 125 deaths were reported [42]. The cause of this outbreak was deciphered to be consumption of maize products heavily contaminated with AFs (several folds above the Kenyan regulatory limit of 20 μg/kg). An outbreak of canine aflatoxicosis occurred in South Africa in 2011 leaving over 220 dogs dead and several others seriously affected after consuming pet food contaminated with high levels of AFs [43]. It is, however, very problematic that aflatoxicosis often remains unrecognized by health workers for an extended period of time, except when a large number of people are affected [9].

2.4. Economic implications of aflatoxins

2.4.1. Economic losses due to aflatoxins contamination of food and feed

The economic significance of AFs is globally illustrious both in the developed and in the underdeveloped nations. In the United States, losses due to AF contamination of maize are estimated at up to 1.68 billion US dollars annually [44, 45]. Globalization of trade has added to the cost and complexity of the situation. For example, adopting the EU standard limit of 4 μg/kg for AFs in peanuts was estimated to cost about 450 million dollars in annual losses on exports [45, 46]. Although estimates on the economic impact of AFs are scarce in the developing countries, based on the literature reports of high levels of mycotoxins found in agricultural commodities in these countries, it is probable that losses consistently far exceed those reported in the United States [45]. To give an instance, in Southeast Asia, the impact of AFs is calculated to a level of 900 million US dollars, of which 500 million are costs directly related to human health effects [45]. In the coastal and eastern regions of Kenya, 2.3 million bags of maize worth over Ksh 3.2 billion (roughly 30 million US dollars) were declared unfit for human consumption by the Ministry of Public Health and Sanitation in 2010 due to high levels of AF contamination [47]. The change in policy by the European Union (EU) is expected to reduce imports on cereals, dried fruits and oilseeds (mainly nuts) by 64%, costing some nine African countries including Egypt, Nigeria and South Africa about 670 million US dollars in trade per year [48]. These economic impacts add to the complexity of the AF malice in developing countries, as they may be compelled to export their best quality produce and sadly retain the poorer commodities for domestic use [47].

2.4.2. Research and surveillance costs of aflatoxins in food and feed

The economic impact of AFs is felt across the entire food and feed supply chains, that is, "from farm to fork." Costs associated with AF management, that is, from sampling and related research expenses, surveillance, mitigation to litigation are also very significant [45, 49]. A study conducted in West Africa estimated annual costs averaging 466 million US dollars from testing, regulatory enforcement, to other quality control measures [50]. In 2000, the USDA's Agricultural Research Service (ARS) instituted a mycotoxin research program worth 17.7 million US dollars primarily geared towards prevention of fungal contamination and toxin production in crops [49]. On average, total value of commercially available test kits for AFs on the market is approximately 10 million US dollars annually, whereas the cost for analysis of AFs alone is placed at 30–50 million US dollars on annual basis [49].

2.5. Regulation of aflatoxins

As elucidated earlier in Section 2.4, AFs constitute a major concern to human health and national economies around the world. Due to the fact that AFs are ubiquitous contaminants and potent carcinogens even at low concentrations, they require stringent regulation to ensure food safety and human health. Different countries have established various limits for AFs in agricultural commodities marketed within their jurisdiction based upon their own perception of risk assessment. At present, over 100 countries have regulations in limiting AFs and other mycotoxins in the food and feed industry [51]. AFs are the most regulated mycotoxins, and 61 countries have regulatory limits of AFB_1 in foodstuffs ranging from 1 to 20 μg/kg, 76 countries have limits up to 35 μg/kg for total AFs in foodstuffs, whereas 21 countries have limits of up to 50 μg/kg for total AFs in animal feeds [52]. In South Africa, regulations exist for total AFs in peanuts intended for further processing (15 μg/kg), in ready-to-eat foodstuffs for humans (10 μg/kg of which AFB_1 is not more than 5 μg/kg), and AFM_1 in diary milk (0.05 μg/L) [53, 54]. The Joint Expert Committee on Food Additives (JECFA) of the Food and Agriculture Organization (FAO) and World Health Organization (WHO) serving as a scientific advisory body to CODEX Alimentarius Commission recommended that levels of AFs in food should be kept As Low As Reasonably Achievable (ALARA) [52]. FAO [52] has published a compendia summarizing worldwide regulations for AFs and other mycotoxins. Many other similar synopses on limits and regulations for AFs are available in literature and could be consulted for additional information [55–57].

2.6. Analysis of aflatoxins in food and feed

Due to the severe effects that AFs elicit in animal and man, several countries and politico-economic unions have placed high priority on the safety of agricultural commodities marketed and consumed within their jurisdiction [51, 52, 58]. Particularly with the globalization of trade, much efforts have been put into mitigation and control of the prevalence of this toxin group in food and feed [59–62]; however, it is apparent that the complete elimination of AFs from foods is an unattainable objective [37]. This has led to various interventions put in place to manage and minimize risk exposure to them [63, 64]. Adequate risk management has been identified as a critical frontline defense in the overall control of AFs in food and feed supplies [36, 63, 65–67]. Any good food safety management program for naturally occurring toxicants [such as Hazard Analysis and Critical Control Point (HACCP)] assumes a holistic approach, involving various phases such as determination of exposure levels, establishment of analytical capabilities, setting and ensuring compliance with regulatory limits, and establishment of surveillance programs [36, 66].

Such critical approaches provide for routine and detailed analysis of every step throughout the food supply chain (from farm to fork) [36, 63]. This positions analysis at the epicenter of AFs management and risk control, which is a global priority [11, 68]. Bearing in mind that decisions relating to regulatory issues or commercial arbitration need to be based on well-defined methods of analysis [66], it is thus vital to ensure that methods for AF analysis are sensitive, efficient and validated against standard guidelines [69, 70]. In AF analysis, the role played by extraction and sample preparation, in general, cannot be overemphasized. Extraction is an inevitable step in AFs analysis no matter the analytical method employed. It has been estimated that up to 70% and perhaps even more of the effort and time that goes into sample analysis comprises the extraction and sample preparation process [71]. Proper design of the extraction process facilitates rapid, efficient and quality analytical results [71].

2.6.1. *Extraction of aflatoxins*

Many efforts have been geared towards developing suitable methods to quantitatively extract and detect AFs in agricultural commodities. For any bioanalytical chemists, the goal is to develop methods with improved sensitivity and selectivity, while at the same time maintaining the credibility of the results, as well as reduce cost and time [72].

2.6.1.1. *Conventional extraction methods for aflatoxins in food and feed*

Different methods have been used for AFs extraction in food and feed. Of these methods, solvent extraction is one of the oldest but still most frequently used method [73]. This method separates analytes based on their relative solubility in two different liquids that are immiscible [74]. One or more solutes contained in a feed solution are transferred to another immiscible solvent, often by rigorously mixing the two immiscible phases, then allowing the two phases to separate [74, 75]. The enriched solvent is called the extract [76]. Common solvents used for solvent extraction include methanol, acetonitrile, chloroform, ethyl acetate, isooctane, ethanol and dichloromethane [4, 73, 77]. The most commonly used solvent extraction approach for AFs is the multi-mycotoxin extraction method of Patterson and Roberts [78]. This method utilizes different organic solvents and reagents such as acetonitrile, isooctane, potassium chloride, dichloromethane and sulfuric acid. It has been widely favored because it selectively extracts several mycotoxins in a single extraction. However, the application of solvent extraction has been greatly limited because it enables the consumption of large quantities of organic solvents, which pose hazards to the environment [16, 79]. Furthermore, solvent extraction often involves long extraction times and laborious procedures with the process extending up to 24 h or more [4, 78]. Moreover, solvents of the required purity tend to be expensive and there are often additional costs with proper disposal of wastes after use [74, 79].

Solid phase extraction (SPE) is another very commonly used extraction method for AFs. It involves the separation of analytes between a liquid mobile phase and a stationary phase contained in a cartridge. Typical materials used at the solid adsorbent phase include ethyl (C2), octyl (C8), octadecyl (C18), cyanopropyl (CN), aminopropyl (NH2), and an ion exchange phase [80]. Non-specific SPE materials are commonly still employed in AF analysis, which is often used for the extraction of more than one mycotoxin [73]. The use of more analyte-specific stationary phases such as immunoaffinity (IA) materials that contain specific antibodies that bind to the analyte of interest is also gaining much attention [70, 81]. Although SPE techniques are relatively simple, have higher specificity and require little quantities of solvents, they are also very expensive and the antibodies are not available for some mycotoxins and products [80].

2.6.1.2. Other methods for aflatoxin extraction

Aside from the extraction methods discussed above, several other methods have been investigated for the extraction of AFs, some of which include quick, easy, cheap, effective, rugged, and safe (QuEChERS) [82, 83], supercritical fluid extraction (SFE) [84], ultrasonic extraction [85], and many others reviewed in the literature [12, 70, 73, 86, 87]. However, as elucidated previously in Section 2.6.1.1, these techniques are fundamentally limited by the use of large volumes of organic solvents, some of which are well known to be toxic and considered as environmental hazards, issues of low recovery, long and laborious procedures and high costs involved amongst others [72]. Moreover, novel advancements in spectrometric analysis of bioactive compounds (e.g., "omics") are pushing the limits of conventional techniques of extraction [72, 88].

Further to this, the adoption of an extraction method strongly depends on the analytical objectives; hence, for AF analysis, methods are required to meet established benchmark standards of the survey, monitory work, legislation and research [89]. It is in line with this that we propose the adoption of SWE as an alternative to conventional extraction methods for AFs, particularly with respect to improved recovery and selectivity, reduced organic solvent consumption and extraction time, at a lower cost. SWE has been in the spotlight as an efficacious and highly promising alternative to traditional techniques of extraction, whose successful applications in the biochemical, pharmaceutical and chemical engineering fields have been well documented in the literature [14, 90–96]. A brief description of this method of extraction is presented in the proceeding sections of this chapter.

3. Subcritical water extraction

3.1. Concept and principle of subcritical water extraction

The term subcritical water refers to liquid water between the boiling point temperature and critical point temperature of water (100–374°C) (**Figure 1**) [14]. Pressure is applied to keep the water in liquid state. Subcritical water extraction (SWE) is a green, cheap and easy-to-adopt extraction technique that utilizes water within its subcritical region as the extraction solvent [14]. The phenomenon behind the extractability of subcritical water is based on the fact that when the temperature of water is raised and the pressure kept sufficient to maintain it in its liquid state (e.g., 250°C and 50 bar), the dielectric constant of water decreases and the hydrogen bond and other intermolecular forces of water weakens, which greatly enhancing its extractability [14, 97].

At atmospheric temperature and pressure (25°C at 1 bar), water has one of the highest dielectric constants amongst non-metallic liquids (ε=80) [98]. However, when the temperature and pressure of water are raised to 250°C and 50 bar, respectively, the dielectric constant falls (ε=27), which is around the range of non-polar solvents such as methanol (ε=33), acetone

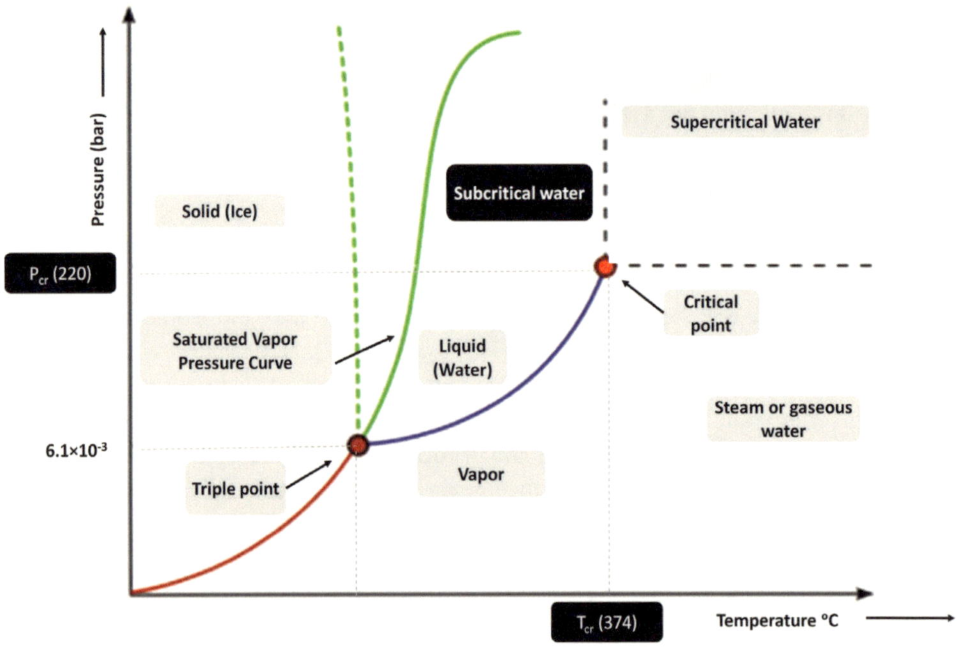

Figure 1. Phase diagram of water as a function of temperature and pressure [14].

(ε=20.7), ethanol (ε=24), and acetonitrile (ε=37) [13, 97]. As a result of the drop in the dielectric constant of subcritical water, its surface tension and viscosity decrease, while its diffusivity increases [13, 14]. As such, water behaves like an organic solvent, dissolving a wide range of low and medium polarity analytes [14]. Interestingly, the extractability and selectivity of subcritical water can be easily maneuvered to extract a range of analytes by simply varying the temperature conditions of the water [90]. Another theoretical explanation on the extractability of subcritical water basis this ability on the fact that, as the temperature of water increases, the average kinetic energy of the molecules of the mixture also increases. This thus disrupts the bonds that exist within and between the molecules, as such, increasing extraction rate.

3.2. Instrumentation of subcritical water extraction

A typical setup of a laboratory scale SWE unit comprises a source of water, temperature retention coil, a solvent pump, an oven and extraction cell, a backpressure valve and a condenser connected to the outlet (**Figure 2**). The grounded sample to be extracted is placed inside the extraction cell, which is located inside the oven. The oven, which usually has an automatic thermostat mechanism, is set to the desired temperature, the backpressure valve is partially locked to maintain the desired pressure and water is pumped at a preset flow rate through the retention coil into the extraction cell. The extraction takes place in the extraction cell as the subcritical water flows through it and mixes with the sample. The hot water extract flows through the condenser and is collected at the outlet [14].

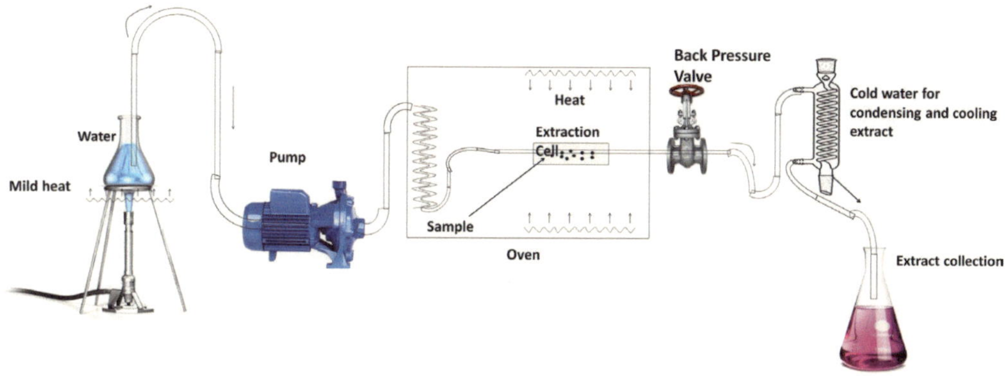

Figure 2. Simple laboratory setup of a PHWE unit [95].

3.3. Operational modes of pressurized hot water extraction

Extraction with subcritical water is performed in two common modes, static mode and flow-through (dynamic) mode. Extraction in the static mode involves retaining the sample in the extraction cell with subcritical water for short periods after which the fluid is allowed to flow out purging the extraction cell and extract collected. It is important to optimize the retention periods to allow for an equilibrium to be reached between the solvent and analyte. The disadvantage of operating in this mode is that within a short time the analyte fluid equilibrium is reached, and hence, no further extraction of the analyte occurs no matter how long the samples are retained in the extraction. On the other hand, extraction in the dynamic mode allows for a continuous flow of fresh fluid through the extraction cell, which reduces or eliminates analyte-fluid equilibrium in a single operation when properly optimized. As such, recovery efficiency is higher in the dynamic mode, although, fluid consumption could be more, resulting in lower energy efficiency compared to the static mode [13, 14]. In a study by Yang *et al.* [99], it was observed that extraction in dynamic mode resulted in the higher recovery of lignin and hemicelluloses from maize stover cellulose than the static mode.

3.4. Factors affecting subcritical water extraction

A number of factors such as temperature, flow rate, pressure, particle size, co-solvents and surfactants affect the performance of SWE. Some of these factors are further described below.

3.4.1. Temperature

The extraction efficiency of SWE is strongly affected by changes in temperature [100]. Generally, extraction efficiency increases with increase in temperature. A higher recovery of total antioxidants was achieved from grape pomace by increasing the extraction temperature [101]. Despite the increase in efficiency by increase in temperature, excess temperatures can result in degradation of thermolabile analytes, hence the need for optimization [100, 102]. The recovery of carvacrol and thymol from *Zataria multiflora* between 100 and 175°C indicated that recoveries increased steadily with the increase in temperature up to 150°C, then a degradation phenomenon followed with a noticeable burning smell [103].

3.4.2. Pressure

The effect of pressure on the extraction efficiency SWE has been described as insignificant [104, 105]. In a study by Shalmashi *et al.* [106], changes in pressure, that is, 20, 30 and 40 bar during SWE did not show any significant effect on the recovery caffeine from tea waste. This is because water is fairly incompressible at temperatures below 300°C, which implies that pressure has very little influence on the physicochemical properties of water, as long as it can maintain in a liquid state [107, 108]. Nevertheless, increased pressure can compromise matrix tissue membranes and force the extraction fluid deep into matrix pores where water at lower pressure may not normally reach [109].

3.4.3. Cosolvents and modifiers

Cosolvents and solvent modifiers are often used to enhance the extractability of SWE. Cosolvents are secondary solvents (usually organic solvents) that are added to subcritical water to enhance its solvation power [95, 110]. The incorporation of methanol during SWE was observed to significantly ($p < 0.05$) increase yield of flavonoids and di-acylated cinnamic acids from *Bidens pilosa* [95, 111]. Solvent modifiers such as salts and other reagents can alter important physicochemical properties of water such as polarity, surface tension, and hydrogen bonding strength, which results in an enhanced extractability [13, 112]. Modifiers can also interact directly with the sample matrix, reducing the activation energy required for analyte desorption and diffusion [14, 113]. Elsewhere [112], it was observed that the solubility of atrazine can be doubled when urea was added to subcritical water, and when ethanol was used, the solubility increased by over 10-folds.

In addition to the above-described factors, other factors that influence the extractability of SWE include solvent flow rate, physicochemical and functional characteristics of the sample matrix and analyte, matrix particle size and geometry of extraction cell [13, 14, 114].

3.5. Advantages and disadvantages of subcritical water extraction

3.5.1. Advantages of subcritical water extraction

The major advantage of SWE is that it is a green (i.e., environmentally friendly) extraction method. The extractant is water, which is non-toxic, non-flammable and renewable. Moreover, water is readily available and cheap, and extraction with it does not generate harmful by-products [90, 115]. In comparison with traditional extraction methods, SWE is less time-consuming and much easier to perform with very few extraction steps, as such, human errors are greatly minimized. When put side-by-side with supercritical fluid extraction (SFE), SWE edges on the basis of being a simple technology, hence, requiring much lower maintenance and engineering cost for equipment [14, 19]. During extraction with subcritical water, the fluid can be maneuvered to selectively extract a range of analytes with different polarities by mere adjusting the temperature of the water, whereas SFE extracts only nonpolar or light-weight compounds [90, 112]. Further to this, SWE is very compatible with various analytical instrumentations because water is colorless and may not interfere with sorts of photodetection such as UV detection or flame ionization detection [14, 116].

3.5.2. Disadvantages of subcritical water extraction

A major setback of SWE is the thermal degradation of some thermolabile analytes at elevated temperatures [117, 118]. When the temperature and pressure of water are extremely high (*i.e.*, above 374°C and 221 bar), there is also the risk that water can become very reactive and could oxidize or catalyze the hydrolysis of some compounds [13]. However, optimization by means of the adoption of a cosolvent or modifier could ameliorate or eliminate these issues [95].

3.6. Application of subcritical water extraction

In the last decade, SWE has been widely investigated for the extraction of various nutritional constituents, organic pollutants, and pharmacoactive compounds from vegetal tissues, food products, soil residues and other ecological biomasses [13, 14, 100, 119]. Free fatty acids and other oils were extracted from spent bleaching earth using SWE [120]. Likewise, it was possible to recover important metabolites from *Moringa oleifera* leaves using SWE [117]. The similar extraction method was used for the recovery of proteins, carbohydrates, and lignans from flaxseed meal [121], catechins and proanthocyanidins from grape seeds [122], flavonoids from aspen knotwood [123] and antioxidants from microalga *Spirulina platensis* [124]. The use of SWE in various applications in different scientific disciplines has been reviewed [90, 100, 107].

3.7. Prospects of subcritical water extraction of aflatoxins

In a recent study [125], we developed and validated an SWE method for the extraction of AFB_1 from maize and subsequently, analysis on high-performance liquid chromatography followed. Results obtained from that study revealed that SWE is suitable for the effective extraction of AFB_1 from maize matrix, with recovery rates ranging from 37 to 128%. Subsequent validation of the optimized method showed acceptable values for accuracy or recovery rate (116%), linearity (%RSD 0.93) and repeatability (%RSD 1.63). It has been stated earlier in Section 2.5 that more countries are enforcing stringent regulations limiting AFs in food and feed, which is increasing demand for their analysis. The efficiency, simplicity, safety and low-cost implications of using SWE are very attractive and compelling in this regard. In comparison with conventional solvent extraction techniques, SWE is very easy to use and requires less time and money [18].

It is known that AFs occur in a diverse manner and can be found deeply deposited inside the food matrices, and as such, their extraction usually requires a process that allows the solvent to penetrate all areas of the matrix to reach hidden toxins trapped in matrix pores [126]. The high pressures involved in SWE seem very suitable in meeting this requirement. Although issues with thermal degradation of some analytes have been a major limiting factor of SWE, it is interesting to know that AFs and most other mycotoxins are relatively thermally stable [31, 62]. Moreover, optimization using cosolvents has been found effective in ameliorating this setback [95, 127]. Accordingly, in our recent study [125] described in the beginning of this section (Section 3.7), we observed a clear positive enhancement on the recovery of AFB_1 by means of a cosolvent (methanol). Based on these observations and other consulted literature reports, it is evident that SWE is a viable alternative to conventional extraction methods for AFs [14, 125].

4. Conclusion

From the literature documents reviewed herein, it has been established that AFs are very potent natural toxins that constitute a significant nuisance to human and animal health as well as the economy. One way to amply combat the prevalence of these toxic substances is by frequent monitoring of their occurrence levels at various critical points along the food supply chain. To this effect, various national and international regulations have been established and are being enforced. Efficient analytical capabilities provide adequate insights on the prevalence of these toxins, which constitute a basis to monitor and where necessary readdress such interventions. This has positioned analysis as a critical element in AF management and control. Extraction is an important step during AF analysis, and hence, improvement in extraction has been a priority in aflatoxicology research. There is a continual quest for efficient extraction methods that are fast, safe and deliver suitable results at reduced cost. SWE meets all these requirements and could make for efficient routine analysis of AFs and other important fungal metabolites in foods and feeds. These observations could stimulate interest and further propel the adoption of SWE in many other applications even beyond the mycotoxicology domain, as well as its scale-up for subsequent industrial applications.

Acknowledgements

This piece of work was financially supported via the Global Excellence and Stature Fellowship of the University of Johannesburg granted to the main author. This chapter was supported in part via the NRF Center of Excellence (CoE) in Food Security co-hosted by the University of Pretoria and the University of the Western Cape in South Africa.

Author details

Sefater Gbashi[1]*, Ntakadzeni Edwin Madala[2], Oluwafemi Ayodeji Adebo[1], Lizelle Piater[2], Judith Zanele Phoku[3] and Patrick Berka Njobeh[1]

*Address all correspondence to: sefatergbashi@gmail.com

1 Department of Biotechnology and Food Technology, Faculty of Science, University of Johannesburg, South Africa

2 Department of Biochemistry, Faculty of Science, University of Johannesburg, South Africa

3 Water and Health Research Centre, Faculty of Health Sciences, University of Johannesburg, South Africa

References

[1] Lizárraga-Paulín EG, Moreno-Martínez E, Miranda-Castro SP. Aflatoxins and their impact on human and animal health: An emerging problem. In: Guevara-Gonzalez RG, editor. Aflatoxins – Biochemistry and Molecular Biology. Rijeka, Croatia: InTech; 2011. pp. 255-282

[2] Makun HA, Dutton MF, Njobeh PB, Gbodi TA, Ogbadu GH. Aflatoxin contamination in foods and feeds: A special focus on Africa. In: Ayman AE, editor. Trends in Vital Food and Control Engineering. Rijeka, Croatia: InTech; 2012. pp. 187-234

[3] Njobeh BP, Dutton FM, Makun HA. Mycotoxins and human health: Significance, prevention and control. In: Mishra AK, Tiwari A, Mishra SB, editors. Smart Biomolecules in Medicine. India: VBRI Press; 2010. pp. 132-177

[4] Chilaka CA, Kock S De, Phoku JZ, Mwanza M, Augustina M, Dutton MF. Fungal and mycotoxin contamination of South African commercial maize. Journal of Food, Agriculture and Environment. 2012;**10**(2):296-303

[5] Turner PC, Collinson AC, Cheung YB, Gong Y, Hall AJ, Prentice AM, et al. Aflatoxin exposure in utero causes growth faltering in Gambian infants. International Journal of Epidemiology. 2007;**36**:1119-1125

[6] Bryden WL. Mycotoxins in the food chain: Human health. Asia Pacific Journal of Clinical Nutrition. 2007;**16**(1):95-101

[7] Zain ME. Impact of mycotoxins on humans and animals. Journal of Saudi Chemical Society. 2011;**15**(2):129-144

[8] Akande KE, Abubakar MM, Adegbola TA, Bogoro SE. Nutritional and health implications of mycotoxins in animal feeds: A review. Pakistan Journal of Nutrition. 2006;**5**(5):398-403

[9] Peraica M, Radic B, Lucic A, Pavlovic M. Toxic effects of mycotoxins in humans. The Bulletin of the World Health Organization. 1999;**77**(9):754-766

[10] Shephard GS. Impact of mycotoxins on human health in developing countries. Food Additives & Contaminants. 2008;**25**(2):146-151

[11] Krska R, Schubert-ullrich P, Molinelli A, Sulyok M, MacDonald S, Crews C. Mycotoxin analysis: An update. Food Additives & Contaminants. 2008;**25**(2):152-163

[12] Rahmani A, Jinap S, Soleimany F. Qualitative and quantitative analysis of mycotoxins. Comprehensive Reviews in Food Science and Food Safety. 2009;**8**:202-251

[13] Teo CC, Tan SN, Yong J, Hew CS, Ong ES. Pressurized hot water extraction (PHWE). Journal of Chromatography A. 2010;**1217**:2484-2494

[14] Gbashi S, Adebo OA, Piater L, Madala NE, Njobeh PB. Subcritical water extraction in biological materials. Separation and Purification Reviews. 2016;**46**:21-34

[15] Choi MPK, Chan KKC, Leung HW, Huie CW. Pressurized liquid extraction of active ingredients (ginsenosides) from medicinal plants using non-ionic surfactant solutions. Journal of Chromatography A. 2003;**983**:153-162

[16] Santana MC, Ferrera ZS, Padrón MET, Rodríguez JJS. Methodologies for the extraction of phenolic compounds from environmental samples: New approaches. Molecules. 2009;**14**:298-320

[17] Hawthorne SB, Grabanski CB, Martin E, Miller DJ. Comparisons of Soxhlet extraction, pressurized liquid extraction, supercritical fluid extraction and subcritical water extraction for environmental solids: Recovery, selectivity and effects on sample matrix. Journal of Chromatography A. 2000;**892**(1-2):421-33

[18] Kubátová A, Miller DJ, Hawthorne SB. Comparison of subcritical water and organic solvents for extracting kava lactones from kava root. Journal of Chromatography A. 2001;**923**(1-2):187-194

[19] Bart JCJ. Additives in Polymers: Industrial Analysis and Applications. West Sussex, England: John Wiley and Sons Ltd, 2005. p. 100

[20] Yang J, Li J, Jiang Y, Duan X, Qu H, Yang B, et al. Natural occurrence, analysis, and prevention of mycotoxins in fruits and their processed product. Critical Reviews in Food Science and Nutrition. 2014;**54**(1):64-83

[21] Mircea C, Poiata A, Tuchilus C, Agoroae L, Butnaru E, Stanescu U. Aflatoxigenic fungi isolated from medicinal herbs. Toxicology Letters. 2008;**180**:32-246

[22] Bbosa GS, Kitya D, Lubega A, Ogwal-Okeng J, Anokbonggo WW, Kyegombe DB. Review of the biological and health effects of aflatoxins on body organs and body systems. In: Razzaghi-Abyaneh M, editor. Aflatoxins – Recent advances and future prospects. Rijeka, Croatia: InTech; 2013. pp. 239-265

[23] Prado G, de Oliveira MS, Lima AS, Moreira APA. Occurrence of aflatoxin M_1 in parmesan cheese consumed in Minas Gerais, Brazil. Ciência e Agrotecnologia. 2008;**32**(6):1906-1911

[24] IARC. IARC Monographs on the Evaluation of Carcinogenic Risks to Humans. Some Traditional Herbal Medicines, Some Mycotoxins, Naphthalene and Styrene. Vol. 82. Lyon, France: IARC; 2002

[25] D'Mello J. Mycotoxins in Cereal Grains, Nuts and Other Plant Products: Food Safety: Contaminants and Toxins. Wallingford, England: CAB International; 2003. pp. 65-90

[26] Park JW, Kim EK, Kim YB. Estimation of the daily exposure of Koreans to aflatoxin B_1 through food consumption. Food Additives & Contaminants. 2004;**21**:70-75

[27] Kensler TW, Roebuck BD, Wogan GN, Groopman JD. Aflatoxin: A 50-year odyssey of mechanistic and translational toxicology. Toxicological Sciences. 2011;**120**(SI):S28-S48

[28] Groopman JD, Kensler TW. Role of metabolism and viruses in aflatoxin-induced liver cancer. Toxicology and Applied Pharmacology. 2005;**206**:131-137

[29] Dhanasekaran D, Shanmugapriya S, Thajuddin N, Panneerselvam A. Aflatoxins and aflatoxicosis in human and animals. In: Guevara-Gonzalez RG, editor. Aflatoxins – Biochemistry and molecular Biology. 1st ed. Rijeka, Croatia: InTech; 2011. pp. 221-254

[30] Bullerman LB, Schroeder LL, Park KY. Formation and control of mycotoxins in food. Journal of Food Protection. 1984;**47**:637-646

[31] Turner NW, Subrahmanyam S, Piletsky SA. Analytical methods for determination of mycotoxins: A review. Analytica Chimica Acta. 2009;**632**:168-180

[32] Kabak B. The fate of mycotoxins during thermal food processing. Journal of the Science of Food and Agriculture. 2009;**89**(4):549-554

[33] Fernández-Cruz ML, Mansilla ML, Tadeo JL. Mycotoxins in fruits and their processed products: Analysis, occurrence and health implications. Journal of Advanced Research. 2010;**1**:113-122

[34] Reddy L. Chemoprotective action of natural products on cultured human epithelial cells exposed to aflatoxin B_1. Durban Institute of Technology, Durban, South Africa; 2005

[35] Boonen J, Malysheva S V, Taevernier L, Diana Di Mavungu J, De Saeger S, De Spiegeleer B. Human skin penetration of selected model mycotoxins. Toxicology. 2012;**301**(1-3):21-32

[36] Atanda O, Makun HA, Ogara IM, Edema M, Idahor KO, Eshiett ME, et al. Fungal and mycotoxin contamination of Nigerian foods and feeds. In: Makun HA, editor. Mycotoxin and Food Safety in Developing Countries. Rijeka, Croatia: InTech; 2013. pp. 1-38

[37] Bennett J, Klich M. Mycotoxins. Clinical Microbiology Reviews. 2003;**16**(3):497-516

[38] Pettersson H. Controlling mycotoxins in animal feed. In: Magan N, Olsen M, editors. Mycotoxins in Food: Detection and Control. Cambridge, England: Woodhead Publishing Ltd. 2004; pp. 269-273

[39] Williams JH, Phillips TD, Jolly PE, Stiles JK, Jolly CM, Aggarwal D. Human aflatoxicosis in developing countries: A review of toxicology, exposure, potential health consequences, and interventions 1, 2, 3. The American Journal of Clinical Nutrition. 2004;**80**:106-122

[40] Montville TJ, Matthews KR. Food Microbiology: An Introduction. 2nd ed. Washington DC, USA: ASM Press. 2008; p. 428

[41] Jolly PE, Shuaib FM, Jiang Y, Preko P, Baidoo J, Stiles JK, et al. Association of high viral load and abnormal liver function with high aflatoxin B_1-albumin adduct levels in HIV-positive Ghanaians: Preliminary observations. Food Additives & Contaminants. Part A, Chemistry, Analysis, Control, Exposure & Risk Assessment. 2011;**28**(9):1224-1234

[42] Lewis L, Onsongo M, Njapau H, Schurz-Rogers H, Luber G, Kieszak S, et al. Aflatoxin contamination of commercial maize products during an outbreak of acute aflatoxicosis in eastern and central Kenya. Environmental Health Perspectives. 2005;**113**(12):1763-1767

[43] Arnot LF, Duncan NM, Coetzer H, Botha CJ. An outbreak of canine aflatoxicosis in Gauteng Province, South Africa. Journal of the South African Veterinary Association. 2012;**83**(1):1-4

[44] Mitchell NJ, Bowers E, Hurburgh C, Wu F. Potential economic losses to the US corn industry from aflatoxin contamination. Food Additives & Contaminants. Part A, Chemistry, Analysis, Control, Exposure & Risk Assessment. 2016;**33**(3):540-550

[45] Schmale DG, Munkvold GP. Mycotoxins in crops: A threat to human and domestic animal health. Plant Health Instructor Journal. 2009;**3**:340-353

[46] Wilson SJ, Otsuki T. What price precaution? European harmonization of aflatoxin regulations and African groundnut exports. European Review of Agricultural Economics. 2001;**28**(3):263-284

[47] Marechera G, Ndwiga J. Farmer perceptions of aflatoxin management strategies in lower Eastern Kenya. Journal of Agricultural Extension and Rural Development. 2014;**6**(12):382-392

[48] Bankole SA, Adebanjo A. Mycotoxins in food in West Africa: Current situation and possibilities of controlling it. African Journal of Biotechnology. 2003;**2**(9):254-263

[49] Robens J, Cardwell K. The costs of mycotoxin management to the USA: Management of aflatoxins in the United States. Journal of Toxicology Toxin Reviews. 2003;**22**(2-3):139-152

[50] CAST. Mycotoxins: Risks in plant, animal, and human systems. Council for Agricultural Science and Technology (CAST). Task Force Report No. 139. Ames, Iowa, USA; 2003

[51] Van Egmond HP, Schothorst RC, Jonker MA. Regulations relating to mycotoxins in food. Analytical and Bioanalytical Chemistry. 2007;**389**:147-157

[52] FAO. Worldwide Regulations for Mycotoxins in Food and Feed in 2003. Food and Agriculture Organization of the United Nations (FAO) Food Quality and Standards Service. Food and Nutrition Paper 81. Rome, Italy: FAO; 2004

[53] MRC. Guidelines on mycotoxin control in South African foodstuffs: From the application of the Hazard Analysis and Critical Control Point (HACCP) system to new national mycotoxin regulations. Medical Research Council Policy Brief. 2009:1-12

[54] Government Gazette. Regulations Governing Tolerances for Fungus-produced Toxins in Foodstuffs. Government Gazette 32409 South Africa: Section 15(1) of the Foodstuffs, Cosmetics and Disinfectants Act, 1972. 2009; p. 6-7

[55] Anukula N, Vangnaib K. Significance of regulation limits in mycotoxin contamination in Asia and risk management programs at the national level. Journal of Food and Drug Analysis. 2013;**21**(3):227-241

[56] Weidenbörner M. Encyclopedia of Food Mycotoxins. 1st ed. Berlin Heidelberg, Germany: Springer-Verlag Berlin Heidelberg; 2001

[57] Guzmán-de-Peña D, Peña-Cabriales JJ. Regulatory considerations of aflatoxin contamination of food in Mexico. Revista Latinoamericana de Microbiología. 2005;**47**(3-4):160-164

[58] Creppy EE. Update of survey, regulation and toxic effects of mycotoxins in Europe. Toxicology Letters. 2002;**127**(1-3):19-28

[59] Suttajit M. Prevention and Control of Mycotoxins [Internet]. Mycotoxin prevention and control in food grains. FAO Agriculture and Consumer Protection; 1989. Available from: http://www.fao.org/docrep/X5036E/x5036E0q.htm [Accessed: February 15, 2017]

[60] Bhat R, Rai R V, Karim AA. Mycotoxins in food and feed: Present status and future concerns. Comprehensive Reviews in Food Science and Food Safety. 2010;**9**(1):57-81

[61] Karlovsky P, Suman M, Berthiller F, De Meester J, Eisenbrand G, Perrin I, et al. Impact of food processing and detoxification treatments on mycotoxin contamination. Mycotoxin Research. 2016;**32**(4):179-205

[62] Bullerman LB, Bianchini A. Stability of mycotoxins during food processing. International Journal of Food Microbiology. 2007;**119**:140-146

[63] Aldred D, Magan N, Olsen M. The use of HACCP in the control of mycotoxins: The case of cereals. In: Magan N, Olsen M, editors. Mycotoxins in Food: Detection and Control. Cambridge, England: Woodhead Publishing Ltd;. 2004. pp. 139-173

[64] Lopez-Garcia R, Park DL, Phillips TD. Integrated mycotoxin management systems. Recent Patents on Food, Nutrition & Agriculture. 1999;**23**:38-48

[65] Toregeani-Mendes KA, Arroteia CC, Kemmelmeier C, Dalpasquale VA, Bando É, Alves AF, et al. Application of hazard analysis critical control points system for the control of aflatoxins in the Brazilian groundnut-based food industry. International Journal of Food Science & Technology. 2011;**46**(12):2611-2618

[66] Gil L, Ruiz P, Font G, Manyes L. An overview of the applications of hazards analysis and critical control point (HACCP) system to mycotoxins. Critical Reviews in Toxicology. 2016;**33**:50-55

[67] FAO/IAEA. Manual on the application of the HACCP system in mycotoxin prevention and control. Joint Food and Agriculture Organization of the United Nations (FAO), and the International Agency for Atomic Energy (IAEA) Training and Reference Centre for Food and Pesticide Control, Rome, Italy. 2001; FAO Food and Nutrition Paper **73**:1-104

[68] Patel P. Mycotoxin analysis: Current and emerging technologies. In: Magan N, Olsen M, editors. Mycotoxins in Food: Detection and Control. Cambridge, England: Woodhead Publishing Limited; 2004. pp. 88-110

[69] Pittet A. Modern methods and trends in mycotoxin analysis. Mitteilungen aus Leb und Hyg. 2005;**96**:424-444

[70] Arroyo-Manzanares N, Huertas-Pérez JF, García-Campaña AM, Gámiz-Gracia L. Mycotoxin analysis: New proposals for sample treatment. Advances in Chemistry. 2014;2014:1-12

[71] Rezaee M, Khalilian F, Pourjavid MR, Seidi S, Chisvert A, Abdel-Rehim M. Extraction and sample preparation. International Journal of Analytical Chemistry. 2015;**2015**:1-2

[72] Augusto F, Hantao LW, Mogollo NGS, Braga SCGN. New materials and trends in sorbents for solid-phase extraction. Trends in Analytical Chemistry. 2013;**43**:14-23

[73] Cigić IK, Prosen H. An overview of conventional and emerging analytical methods for the determination of mycotoxins. International Journal of Molecular Sciences. 2009;**10**(1):62-115

[74] Sapkale GN, Patil SM, Surwase US, Bhatbhage PK. Supercritical fluid extraction. International Journal of Chemical Sciences. 2010;**8**(2):729-743

[75] Zaiz T, Lanez H, Kechida B. Aspen hysys simulation and comparison between organic solvents (sulfolane and DMSO) used for benzene extraction. International Journal of Chemical and Petroleum Sciences. 2013;**2**(1):10-19

[76] Gbashi S. Pressurized hot water extraction (PHWE) and chemometric fingerprinting of phytochemicals from *Bidens pilosa* [thesis]. South Africa: University of Johannesburg, South Africa; 2016

[77] Zabe N, Ayalew K, Powers S. Ethanol extraction method for a rapid test for aflatoxin in corn. Food Additives & Contaminants. 2008;**1001**:297-305

[78] Patterson DSP, Roberts BA. Mycotoxins in animal feedstuffs: Sensitive thin layer chromatographic detection of aflatoxin, ochratoxin A, sterigmatocystin, zearalenone and T-2 toxin. Journal Association of Official Analytical Chemists. 1979;**62**:1265-1267

[79] Tabaraki R, Nateghi A. Optimization of ultrasonic-assisted extraction of natural antioxidants from rice bran using response surface methodology. Ultrasonics Sonochemistry. 2011;**18**(6):1279-1286

[80] Limsuwan S. Development, evaluation and application of methods for mycotoxin analysis [thesis]. Germany. Georg-August-University Göttingen; 2011

[81] Gam L, Tham S, Latiff A. Immunoaffinity extraction and tandem mass spectrometric analysis of human chorionic gonadotropin in doping analysis. Journal of Chromatography B. 2003;**792**(2):187-196

[82] Bardsley J, Oliver M. Determination of multiple mycotoxins in grain using a QuEChERS sample preparation approach and LC-MS/MS detection. Appl Note 21121. 2015;Thermo Fisher Scientific, Runcorn, UK:1-7

[83] Sirhan AY, Tan GH, Al-Shunnaq A, Abdulra'uf L, Wong RCS. QuEChERS-HPLC method for aflatoxin detection of domestic and imported food in Jordan. Journal of Liquid Chromatography & Related Technologies. 2014;**37**(3):321-342

[84] Brown JD. Development of a method for the simultaneous detection of mycotoxins in corn using supercritical fluid extraction (SFE), high performance liquid chromatography (HPLC), and electrospray-atmospheric pressure ionization/mass spectrometry (ES-API/MS) for extraction, separation, and identification [thesis]. USA: The University of Iowa; 1999

[85] Kong WJ, Liu SY, Qiu F, Xiao XH, Yang MH. Simultaneous multi-mycotoxin determination in nutmeg by ultrasound-assisted solid-liquid extraction and immunoaffinity column clean-up coupled with liquid chromatography and on-line post-column photochemical derivatization-fluorescence detection. Analyst. 2013;**138**(9):2729-2739

[86] Turner NW, Bramhmbhatt H, Szabo-Vezse M, Poma A, Coker R, Piletsky SA. Analytical methods for determination of mycotoxins: An update (2009-2014). Analytica Chimica Acta. 2015;**901**:12-33

[87] Anfossi L, Baggiani C, Giovannoli C, Giraudi G. Mycotoxins in food and feed: Extraction, analysis and emerging technologies for rapid and on-field detection. Recent Patents Food, Nutrition & Agriculture. 2010;**2**(2):140-153

[88] Zhang A, Sun H, Wang P, Han Y, Wang X. Modern analytical techniques in metabolomics analysis. Analyst. 2012;**137**(2):293-300

[89] Hussain I. Aflatoxin measurement and analysis. In: Torres-Pacheco I, editor. Aflatoxins – Detection, Measurement and Control. 1st ed. Rijeka, Croatia: InTech; 2011. pp. 129-146

[90] Liang X, Fan Q. Application of sub-critical water extraction in pharmaceutical industry. Journal of Chemical Engineering and Materials Science. 2013;**1**:1-6

[91] Wilson ID, Davis P, Ruane RJ. Supercritical fluid chromatography and extraction of pharmaceuticals. In: Dean JR, editor. Applications of Supercritical Fluids in Industrial Analysis. Dordrecht, Netherlands: Springer Netherlands; 1993. p. 74-103

[92] Ozel MZ, Gogus F, Lewis AC. Subcritical water extraction of essential oils from *Thymbra spicata*. Food Chemistry. 2003;**82**(3):381-386

[93] Gupta A, Naraniwal M, Kothari V. Modern extraction methods for preparation of bioactive plant extracts. International Journal of Applied and Natural Sciences. 2012;**1**(1):8-26

[94] Ibañez E, Kubátová A, Señoráns FJ, Cavero S, Reglero G, Hawthorne SB. Subcritical water extraction of antioxidant compounds from rosemary plants. Journal of Agricultural and Food Chemistry. 2003;**51**(2):375-382

[95] Gbashi S, Njobeh P, Steenkamp P, Tutu H, Madala N. The effect of temperature and methanol-water mixture on pressurized hot water extraction (PHWE) of anti-HIV analogues from *Bidens pilosa*. Chemistry Central Journal. 2016;**10**(37):1-12

[96] Shaddel R, Maskooki A, Haddad-Khodaparast MH, Azadmard-Damirchi S, Mohamadi M, Fathi-Achachlouei B. Optimization of extraction process of bioactive compounds from Bene hull using subcritical water. Food Science and Biotechnology. 2014;**23**(5):1459-1468

[97] Alupului A, Călinescu I, Lavric V. Microwave extraction of active principles from medicinal plants. UPB Scientific Bulletin, Series B: Chemistry and Materials Science. 2012;**74**(2):129-142

[98] Cabane B, Vuilleumier R. The physics of liquid water. Comptes Rendus Geoscience. 2005;**337**:1-23

[99] Yang B, Wyman CE. Effect of xylan and lignin removal by batch and flow through pretreatment on the enzymatic digestibility of corn stover Cellulose. Biotechnology and Bioengineering. 2004;**86**(1):88-95

[100] Asl AH, Khajenoori M. Subcritical water extraction. In: Nakajima H, editor. Mass Transfer – Advances in Sustainable Energy and Environment Oriented Numerical Modeling. Rijeka, Croatia: InTech; 2013. pp. 459-487

[101] Vergara-Salinas JR, Bulnes P, Zúñiga MC, Pérez-Jiménez J, Torres JL, Mateos-Martín ML, et al. Effect of pressurized hot water extraction on antioxidants from grape pomace before and after enological fermentation. Journal of Agricultural and Food Chemistry. 2013;**61**(28):6929-6936

[102] Khoza BS, Dubery IA, Byth-Illing HA, Steenkamp PA, Chimuka L, Madala NE. Optimization of pressurized hot water extraction of flavonoids from *Momordica foetida* using UHPLC-qTOF-MS and multivariate chemometric approaches. Food Analytical Methods. 2015;**2015**:1-10

[103] Khajenoori M, Asl AH, Hormozi F, Eikani MH, Bidgoli HN. Subcritical water extraction of essential oils from Zataria multiflora Boiss. Journal of Food Process Engineering. 2009;**32**(6):804-816

[104] Hawthorne SB, Yang Y, Miller DJ. Extraction of organic pollutants from environmental solids with sub- and supercritical water. Analytical Chemistry. 1994;**66**:2912-2920

[105] Deng CH, Li N, Zhang XM. Rapid determination of essential oil in "*Acorus tatarinowii* Chott" by pressurized hot water extraction followed by solid-phase microextracton and gas chromatography-mass spectrometry. Journal of Chromatography A. 2004;**1059**(1):149-155

[106] Shalmashi A, Abedi M, Golmohammad F, Eikani MH. Isolation of caffeine from tea waste using subcritical water extraction. Journal of Food Process Engineering. 2010;**33**(4):701-711

[107] Ramos L, Kristenson EM, Brinkman UAT. Current use of pressurized liquid extraction and subcritical water extraction in environmental analysis. Journal of Chromatography A. 2002;**975**:3-29

[108] Carr AG, Mammucari R, Foster NR. A review of subcritical water as a solvent and its utilization for the processing of hydrophobic organic compounds. Chemical Engineering Journal. 2011;**172**(1):1-17

[109] Ong ES, Cheong JSH, Goh D. Pressurized hot water extraction of bioactive or marker compounds in botanicals and medicinal plant materials. Journal of Chromatography A. 2006;**1112**(1-2):92-102

[110] Williams HD, Trevaskis NL, Charman SA, Shanker RM, Charman WN, Pouton CW, et al. Strategies to address low drug solubility in discovery and development. Pharmacological Reviews. 2013;**65**:315-449

[111] Gbashi S, Njobeh P, Steenkamp P, Madala N. Pressurized hot water extraction and chemometric fingerprinting of flavonoids from *Bidens pilosa* by UPLC-tandem mass spectrometry. CYTA - J Food. 2016;**15**(2):171-180.

[112] Curren MSS, King JW. Solubility of Triazine pesticides in pure and modified subcritical water. Analytical Chemistry. 2001;**73**:740-745

[113] Plaza M, Turner C. Pressurized hot water extraction of bioactives. TrAC Trends in Analytical Chemistry. 2015;**71**:39-54

[114] Shitu A, Izhar S, Tahir TM. Sub-critical water as a green solvent for production of valuable materials from agricultural waste biomass: A review of recent work. Global Journal of Environmental Science and Management. 2015;**1**(3):255-264

[115] Chemat F, Vian M, Cravotto G. Green extraction of natural products: Concept and principles. International Journal of Molecular Sciences. 2012;**13**:8615-8627

[116] Khoza BS, Gbashi S, Steenkamp PA, Njobeh PB, Madala NE. Identification of hydroxylcinnamoyl tartaric acid esters in *Bidens pilosa* by UPLC-tandem mass spectrometry. South African Journal of Botany. 2016;2016:96-100

[117] Khoza BS, Chimuka L, Mukwevho E, Mukwevho E, Steenkamp PA, Madala NE. The effect of temperature on pressurized hot water extraction of pharmacologically-important metabolites as analyzed by UPLC-qTOF-MS and PCA. Evidence-Based Complementary and Alternative Medicine. 2014;**2014**:1-9. (Article ID 914759)

[118] Moreno E, Reza J, Trejo A. Extraction of polycyclic aromatic hydrocarbons from soil using water under subcritical conditions. Polycyclic Aromatic Compounds. 2007;**27**(4):239-260

[119] Duy L, Sephton MA, Watson JS. Subcritical water extraction of organic matter from sedimentary rocks. Analytica Chimica Acta. 2015;**879**:48-57

[120] Fattah RA, Mostafa NA, Mahmoud MS, Abdelmoez W. Recovery of oil and free fatty acids from spent bleaching earth using sub-critical water technology supported with kinetic and thermodynamic study. Advances in Bioscience and Biotechnology. 2014;**5**:261-272

[121] Ho CHL, Cacace JE, Mazza G. Extraction of lignans, proteins and carbohydrates from flaxseed meal with pressurized low polarity water. Leb Technology. 2007;**40**:1637-1647

[122] García-Marino M, Rivas-Gonzalo J, Ibánez E, García-Marino C. Recovery of catechins and proanthocyanidins from winery by-products using subcritical water extraction. Analytica Chimica Acta. 2006;**563**(1-2):44-50

[123] Hartonen K, Parshintsev J, Sandberg K, Bergelin E, Nisula L, Riekkola ML. Isolation of flavonoids from aspen knotwood by pressurized hot water extraction and comparison with other extraction techniques. Talanta. 2007;**74**(1):32-38

[124] Herrero M, Ibanez E, Senorans J, Cifuentes A. Pressurized liquid extracts from *Spirulina platensis* microalga: Determination of their antioxidant activity and preliminary analysis by micellar electrokinetic chromatography. Journal of Chromatography A. 2004;**1047**:195-203

[125] Gbashi S, Madala N, Adebo OA, Njobeh PB. Development and validation of a pressurized hot water extraction method for the extraction of aflatoxin B_1 in maize. In: 7th Euro Global Summit on Clinical Microbiology and Mycotoxins February 27-28, 2017 Amsterdam, Netherlands. Amsterdam, Netherlands; 2017. p. Full text data not published.

[126] Scudamore KA. Mycotoxins. In: Gilbert J, Senyuva H, editors. Bioactive Compounds in Foods. West Sussex, United Kingdom: Blackwell Publishing Ltd; 2008. pp. 134-162

[127] Curren MSS, King JW. Ethanol-modified subcritical water extraction combined with solid-phase microextraction for determining Atrazine in beef kidney. Journal of Agricultural and Food Chemistry. 2001;**49**:2175-2180

Chapter 12

A Focus on Aflatoxin in Feedstuffs: New Developments in Analysis and Detection, Feed Composition Affecting Toxin Contamination, and Interdisciplinary Approaches to Mitigate It

Fabio Granados-Chinchilla

Additional information is available at the end of the chapter

http://dx.doi.org/10.5772/intechopen.69498

Abstract

Aflatoxins are mold-synthetized secondary metabolites that are capable of causing disease and death in humans and other animals. Aflatoxins hold a prominent place in the discussion on feed safety as are the only mycotoxins with the regulatory framework. Feed ingredients and composition inevitably affect the susceptibility of feed to fungal and toxin contamination. To verify that legal thresholds are being complied, avoiding delivering contaminated feed to animals, and obtain correct prevalence data, analytical methods must be developed which are apt for application on a complex matrix such as animal feed. These methods should include simple screening assays and high-end confirmatory ones. Laboratories without expensive equipment can and should be able to implement methods and to analyze and detect aflatoxins. Aflatoxin contamination is a complex issue that should be assessed interdisciplinarily and farm-to-fork models should be integrated into vigilance. In this chapter, we have devoted some lines to each of the aspects mentioned above focusing on feed aflatoxin contamination.

Keywords: aflatoxins, analytical methods, sample preparation, feed composition, feed safety, farm-to-fork, One Health

1. Introduction

1.1. Aflatoxins

The four major aflatoxins are called B_1, B_2, G_1, and G_2 (**Figure 1**) based on their fluorescence under UV light (blue or green) and retention factors during thin-layer chromatography. AFB_1 has been described as a potent natural carcinogen (classified in group 1; [1]) and is usually the major aflatoxin produced by toxigenic strains. However, other aflatoxins (e.g. AFM_1, B_{2a}, and G_{2a}) have been described, particularly since biotransformation products of the mammalian degradative enzyme metabolism, is based on cytochromes. This biosynthetic pathway is shared by norsolorinic acid, an anthraquinone, and sterigmatocystin (STE), a mutagenic and tumorigenic dihydrofuran toxin. STE is a late metabolite in the aflatoxin pathway and is also produced as a final biosynthetic product by some species such as *Aspergillus, Aspergillus chevalieri, Aspergillus ruber, Aspergillus amstelodami,* and *Aspergillus aureolatus* [2]. The reader is encouraged to consult the papers written by Bbosa and coworkers [3] and Dohnal and coworkers [4] that describe with detail aflatoxin metabolism.

Aflatoxins are difuranocoumarin derivatives produced by a polyketide pathway by many strains of *Aspergillus flavus* and *Aspergillus parasiticus*. Especially, *A. flavus* is a frequent contaminant in agricultural plants and commodities. Other aflatoxin-producing species have been encountered less frequently (**Table 1**). In fact, just in 2011, Varga and coworkers described two new aflatoxin-producing species, *Aspergillus pseudocaelatus* sp. (Argentina) and *Aspergillus pseudonomius* sp. (United States) [5]. Baranyi and coworkers [6] described the phylogenetic association among these strains based on partial calmodulin sequencing. We refer the reader

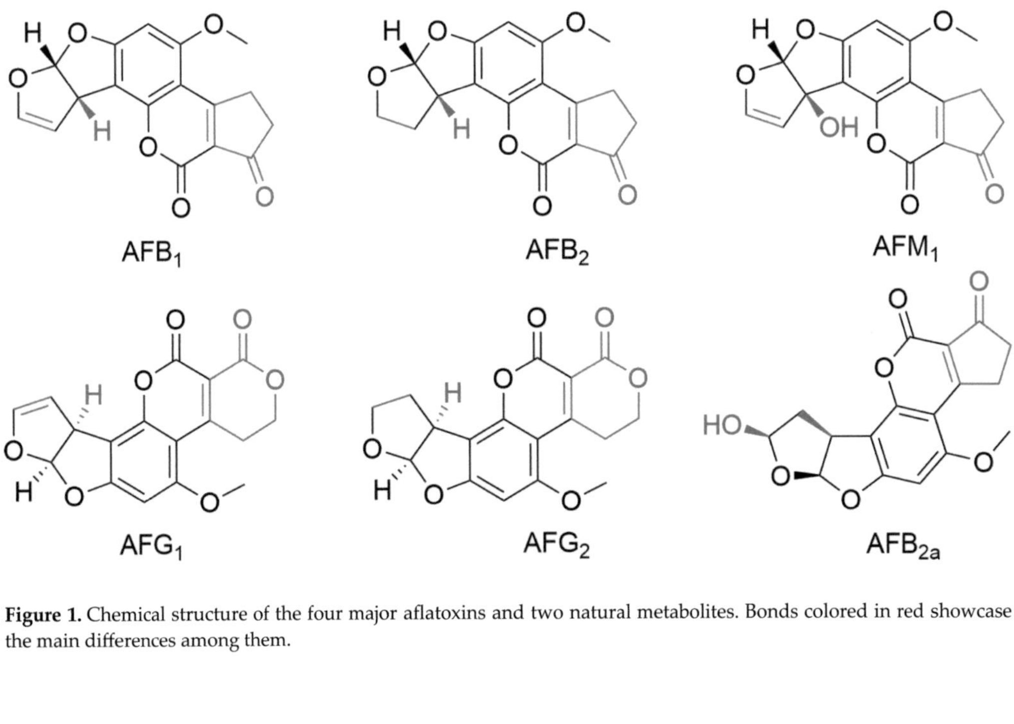

Figure 1. Chemical structure of the four major aflatoxins and two natural metabolites. Bonds colored in red showcase the main differences among them.

Section *Flavi*	Section *Ochraceorosei*	Section *Nidulantes*
Aspergillus arachidicola[2]	*Aspergillus ochraceoroseus*[1]	*Aspergillus astellatus*[1]
Aspergillus bombycis[2]	*Aspergillus rambelli*[1]	*Aspergillus venezuelensis*[1]
Aspergillus minisclerotigenes[2]		
Aspergillus mottae[2]		
Aspergillus nomius[2]		
Aspergillus parvisclerotigenus[2]		
Aspergillus pseudocelatus[2]		
Aspergillus sergii[2]		
Aspergillus pseudonomius[1]		
Aspergillus pseudotamarii[1]		
Aspergillus togoensis[1]		

[1, 2] Blue and green colors represent the type of aflatoxins the strain is capable of producing.

Table 1. Aflatoxigenic fungi species capable of aflatoxin production [124].

to an excellent review by Samson and coworkers [7] regarding phylogeny, identification, and nomenclature of the genus *Aspergillus,* which is composed of more than 339 species.

From the mycological perspective, there are phenotypic and genetic differences in the strains within each aflatoxigenic species and each strain display various toxigenic abilities. For example, *Aspergillus* subgenus *Circumdati* section *Flavi* includes species with usually biseriate conidial heads, in shades of yellow-green to brown, and dark sclerotia [5]. On the other hand, several *afl* genes are involved in the biosynthesis of aflatoxins. Each strain produced toxins differentially (e.g. *A. flavus* and *A. parasiticus* are known to produce aflatoxins B_1 and B_2, aflatoxin fractions the latter only synthesizes G_1 and G_2). Meaning a genotypical difference as well since aflatoxin G producers have integral versions of genes *nad*A and *afl*F [8]. There is, in fact, a battery of molecular tests devoted to the genetical identification of *Aspergillus* section *Flavi* [7, 9]. Several *Aspergillus* strains have been isolated from feeds. For example, Iranian cattle feed [10], poultry feed from South Africa [11], chicken feed from Nigeria [12], and dairy goat feed from Brazil [13].

Several of the species above are important mycotoxin producers including aflatoxins, and like the genetic ability to make aflatoxin, contamination is highly variable. Crops can become contaminated with aflatoxin in the field before harvest, where it is usually associated with drought stress [14]; adding difficulty to this issue, storage conditions may favor mold growth. During storage, usually, the most important variables are the moisture content of the substrate and the relative humidity of the environment [15]. Aflatoxin contamination has been linked to increased mortality in farm animals and, thus, significantly lowering grain value as an animal feed and, thereafter, loss of productivity in the case of food-producing animals [16]. Milk products can also serve as a source of aflatoxin. When cows consume aflatoxin-contaminated feeds, they transform AFB_1 into a hydroxylated form called AFM_1. Cytochrome P_{450} enzymes convert aflatoxins to the reactive 8,9-epoxide form, which is capable of binding to both DNA and proteins [3, 4]. This metabolite is still considered carcinogenic and teratogenic and may affect young and newborn animals and reach human as a final consumption product. Unit operations during milk production usually have little to no effect over the AFM_1. Although we will not explore AFM_1 contamination in detail, we urge the reader to read a very thorough review regarding AFM_1 in bovine milk written by Becker-Algeri and coworkers [17].

Aflatoxin is associated with both toxicity and carcinogenicity in human and animal populations [16]. There are substantial differences in species susceptibility. On the other hand, within a given species, the magnitude of the response is influenced by age, sex, weight, diet, and exposure to infectious agents [16]. The presence of other mycotoxins (most common co-occurrence of AF and ochratoxin A or AF and fumonisins) and pharmacologically active substances may reflect antagonistic, additive, or synergistic effects [18]. Sufficient availability of feed is combined with regulations and continuous surveillance programs to monitor contaminant levels and protect animal populations from significant aflatoxin ingestion. The scarcity of resources (both economic and food supply) may play a role in the use of contaminated feed.

1.2. Feeds and feed ingredients and aflatoxin contamination

Feed is defined as any goods or materials which are consumed by animals and contribute energy and nutrients to the animal's diet [19]. Usually, it is divided into two categories, roughages and compound feed. Roughages comprise diets based on grass, silage, hay, legumes, bagasse and others. Equines and dairy cattle complete rations on occasion are complemented or based on roughages. In Costa Rica, for example, dairy cow diets are composed mainly of forage, including *Cynodon nlemfuensis* Vanderyst, *Pennisetum clandestinum* Hochst, and *Lolium perenne* L. Current global data regarding fungi and mycotoxins in silages have been described [20]. Aflatoxins and *Aspergillus* species have been found to be important especially in corn and sorghum silages.

On the other hand, compound feed is composed primarily of cereals (e.g. rice, wheat, barley, oats, rye, corn, sorghum, and millet), milling by-products (e.g. brans, hulls, pollards), and oil cakes (e.g. palm kernel, soybean, sunflower, rapeseed, peanut, linseed, cottonseed). Other feed ingredients include distillers dried grains. The components above (especially corn and corn by-products) are the most susceptible to aflatoxigenic fungi attack and therefore aflatoxin contamination [13]. The chemical composition, ingredients and nutritional quality of feed inherently influence the capability of fungi to inoculate and even make use of their genetic machinery available to produce aflatoxins within such a substrate. Hence, feed is especially susceptible to aflatoxin contamination.

A few papers have focused on this fact and examined some aspects relating aflatoxin contamination with nutritional analysis. For example, Hashimoto and coworkers analyzed 42 fish feed samples and found no association between pelletized and extruded feed and aflatoxin levels and no nutritional differences between both feed types [21]. However, they did acknowledge an aflatoxin/fumonisin co-occurrence of a 23.8%. Prabakaran and Dhanapal found that natural contamination observed in two Indian regions (220 and 15 $\mu g\ kg^{-1}$) were connected with those areas where feeds were prepared with higher moisture (11.29–11.70 g/100g) and crude fat (4.62–4.64 g/100 g) [22]. Interestingly, the authors also demonstrated that when feed undergo autoclaving and then inoculated with a toxin-producing *A. flavus* strain, higher concentration of toxin was attained when compared with the same non-autoclaved feed. The authors relate this finding to the release of zinc from phytic acid when pressure and temperature are applied, which results in the mineral being available for aflatoxin synthesis [23, 24]. Liu and coworkers found that defatted grains (i.e. soybean, peanut, corn, wheat corn endosperm, and corn germ) showed a significant decrease in aflatoxin concentration when compared to their full-fat counterparts [25]. On the other hand, when the same seeds were treated with corn oil, aflatoxin production capacity was regained by *A. flavus*. Hence, the removal of lipids contains AFB_1 production. The effects of starch, crude protein, soluble sugars (fructose, glucose, sucrose, maltose, raffinose, and stachyose), amino acids (aspartic acid, glutamic acid, glycine, arginine, and alanine), and trace elements (copper, iron, zinc, and manganese) on AFB_1 production and mycelial growth were examined. Maltose, glucose, sucrose, arginine, glutamic acid, aspartic acid, and zinc significantly induced AFB_1 production up to 1.7- to 26.6-fold. Stachyose promoted *A. flavus* growth more so than the other nutrients, playing a pivotal role in grain infection by *A. flavus*. These data provide new insights toward feed protection from contamination. Herzallah modeled aflatoxin carry over from feed to several tissues using a diet based mainly on corn meal (60.5 g/100 g), soybean meal (21.5 g/100 g), and vegetable oil (3.0 g/100 g) [26]. In this case, laying hens were exposed to aflatoxin concentrations up to 965.12 μg kg^{-1} for 6 weeks, and aflatoxin levels were monitored. The author found worrisome tissue levels that rounded up to 0.63 and 2.12 μg kg^{-1} in breast and liver, respectively. A very thorough and sophisticated study [27], applied to dairy cattle and milk, used Monte Carlo simulations to assess different scenarios which contemplated milk yield, feed composition, which considered normal aflatoxin levels found in the feed, and feed ingredients and their inclusion rates. This work reiterates the value of computational modeling to estimate possible contaminant exposure and is unique, as one of the variables used during modeling were aflatoxin levels found in real matrixes (e.g. a naturally contaminated batch of corn, with a maximum concentration of 168 μg kg^{-1}) extracted from the Dutch national surveillance program. The authors modeled diets based on high and low protein. And included relevant feed ingredients in different proportions, e.g. corn (10.24–15.06 g/100 g), soybean meal (14.96–0.23 g/100 g), sunflower seed meal (4.5–3.83 g/100 g), palm kernel (ca. 15 g/100 g), rapeseed meal (7.94–5.54 g/100 g), and corn gluten feed (3.67–1 g/100 g). All these raw materials with a differential potential of aflatoxin contamination. The AFB_1 analysis reflects that corn ingredients exhibit higher aflatoxin levels. In total, five different transfer equations of AFB_1 from feed to AFM_1 in the milk were included, and the results showed that in only 1% of the revised cases, milk toxin levels surpassed the legislative threshold. An increased contamination was found when contaminated feed ingredients were included in the formulation (i.e. contaminated corn), up to 28.5% of the iterations exceeded the threshold. The authors also observed that an increase in the milk production had a minimal effect on these data due to an apparent dilution effect. The same authors conclude that feeding regimes, including the composition of crude fiber and feeding roughages of dairy cows, should be carefully monitored and considered regarding their aflatoxin inclusion potential. Noteworthy, van der Fels-Klerx and Bouzembrak also used a similar approach to estimate the probability of AFB_1 contamination of compound feed for dairy cattle and to limit this contamination [28]. With the results obtained, the authors suggested an optimized feed composition, including a reduction of citrus pulp (10–0 g/100 g), sunflower seed meal (23–1.5 g/100 g), and soybean meal (10–5.1 g/100 g) and an increase in corn ingredients (20.5–29.4 g/100 g), palm kernel (16–22.5 g/100 g), and wheat (2–30 g/100 g), with respect to usual formulations. The authors based their recommended diet on wheat which is relatively inexpensive but may not be available during some seasonal changes or inaccessible to some geographical regions. The authors claimed that 98.8% of the simulated diet would exhibit values below the legal threshold contrary to a 75.6% of cases assumed using a general formulation.

1.3. Aflatoxins, food chain safety, and the One Health approach

The One Health approach highlights the kinship of human, animal, and environmental health and the importance of transdisciplinary and interdisciplinary efforts [29]. Hence, collaborative efforts under this tactic aim to help promote animal and human health. The challenges posed by mycotoxicosis, a foodborne illness that results from consumption of aflatoxin-contaminated food and feed, are more likely to be understood and mitigated through a One Health approach. Shenge and LeJeune reported that it is estimated that a third of global food supplies are contaminated with aflatoxins [30]. This value alone should be cause for concern, as fungal contamination is not only a source of disease for crops but also generate poor harvests, and impact the well-being of animals and humans; creating, even more economic losses when food producing animals are involved. Several aspects of mycotoxins remain unclear, and research is still needed regarding all areas affected by mycotoxins. Although few articles tackle the issue from a holistically standpoint, at least one conference paper presented by Sirma is focused explicitly on using One Health in mycotoxin analysis [31]. On the other hand, Magnussen and Parsi published an article which encompasses a health issue such as hepatocellular carcinoma within the aflatoxin convoluted problematic [32]. More recently, two papers have more specifically considered the issue from the One Health stance. Frazzoli and coworkers contemplated aflatoxin contamination anticipating environment, animal, and human interaction, and the feed and food link with emphasis to the carryover that occurs from the presence of aflatoxins in a feed to milk (i.e. AFM_1) during the entire dairy chain [33]. On the other hand, Ogodo and Ugbogu considered the presence of aflatoxins in food industry, management and its relationship with hepatocellular carcinoma, linking a public health issue again with an agronomical one [34]. The latest effort in integrating the totality of the food chain is in the form of "MyToolBox", a European Commission funded initiative joining knowledge from different sectors to improve risk management, reduce crop losses and its impacts, and provide safe options to treat toxin-contaminated batches [35]. The final objective of the initiative is to offer recommendations and practical measures to the end users along the food and feed chain in a web-based platform.

2. Current methods for the analysis of aflatoxins in feedstuffs

2.1. Relevance of aflatoxin accurate determination

Food safety relies on the capability of laboratories to screen, detect, quantify, and confirm the presence of aflatoxins in different staple foods. Multiple methods have been designed over the years, the authors refer the reader to a good starting point to familiarize with the general principles and mechanisms involved in the main techniques used for aflatoxin analysis [36, 37]. For an ampler view regarding the recent developments in techniques for the detection of aflatoxins, we urge the reader toward the paper written by Yao and coworkers [38]. Herein, we will limit the discussion to techniques applied for the aflatoxin determination in feed, feed ingredients, and some selected related matrices as they lead the food chain.

Accurate mycotoxin analysis is paramount for feed and feed ingredients safety evaluation and epidemiology. Animal feed is at the beginning of the food chain, and any in-feed contaminants may reach the final consumer through food matrixes, such as eggs or meat products [39]. However, few methods are specialized for feeds, which can be noted by the sheer number of official methods for this kind of matrices. For example, AOAC only has three assays: 975.36 (Romer mini column), 989.06 (ELISA), and 2003.02 (liquid chromatography [LC]). Contrasting to the amount of approaches and principles available for other staple foods [40, 41].

2.2. Sampling and some sample preparation highlights

2.2.1. Feed sampling

Aflatoxin sampling is especially complicated since mold growth (and hence toxin distribution), in feed and grain, may not be homogenous. For example, not all the conditions for the production of toxins will be met in the totality of a silo; a storage grain system will reduce the toxin production with a_w values below 0.70 [42]. Errors in sampling methodology carry costs intrinsically. Assuming only a section of a feed batch is contaminated, the composite and homogenized sample is vital. Failure in detecting the mycotoxin will generate adverse effects on farm animals which will be fed with said foodstuff. Sampling directly and only from a "hot zone" will unchain legal events that usually ends in the elimination of a whole feed batch, which is costly for feed manufacturers or importers/exporters [43]. On the other hand, research has demonstrated that the bulk of the variability in mycotoxin analysis comes from sampling [44]. Some papers have focused specifically on aflatoxin sampling. For example, Mallmann and coworkers sampled eight lots of corn using two different plans: manual, using sampling spear for kernels; and automatic, using a continuous flow to collect corn meal [43]. The authors concluded that automatic sampling introduces less analytical variation and it is more accurate than manual sampling. In contrast, Herrman and coworkers sampled Texan grain elevator establishments and determined that while sampling contributes to variability in measuring aflatoxin in grain, aflatoxin analysis using commercially available test kits was a major contributor to variation in aflatoxin test results among commercial food handlers [45].

Several authorities have issued sampling guidelines. The American Association of Feed Control Officials (AAFCO) recommendations for mycotoxin test object collection is detailed in Feed Inspector's Manual for the member States [46]. Similarly, the European Commission has emitted the 2006/401/EC which lay down the sampling methods and analysis for the official control of the levels of mycotoxins in foodstuffs [47]. Food and Agriculture Organization of the United Nations (FAO) have developed a mycotoxin sampling tool (http://www.fstools.org/mycotoxins/). Berthiller and coworkers detailed other sampling and analysis methods that include other mycotoxins [48]. Lee and coworkers designed a statistically derived risk-based sampling plan for surveillance sample assignments of chemical and biological hazards using binomial probability distribution [49]. The authors found that the number of feed samples that exceeded legal thresholds for target analytes (aflatoxins, fumonisins,

Salmonella, and dioxins) in the validation data were lower than those of the average 3-year data in most feed products.

2.2.2. Commercially available tools for aflatoxin analysis

Several companies have devoted their efforts in the fabrication of versatile products useful to isolate aflatoxins from samples. Different researchers have applied diverse approaches to obtain a clean extract, especially to inject into LC systems. For example, R-biopharm AG has two various methods available for the analysis of aflatoxins: RIDASCREEN® for total AF and AFB_1 single analysis and AFLAPREP® an immunoaffinity column which delivers limits of detection as low as 0.007 ng AF mL^{-1} extract, and based on our laboratory experience, sample clean-up is capable and good recoveries are obtained. Noteworthy, R-biopharm also has an immunoaffinity column for sterigmatocystin.

Other laboratories have recently applied these concrete columns to monitor aflatoxin in food and feed [50, 51]. Multiple columns (e.g. AOF MS-PREP®) based on this same principle are also available. However, care must be taken as usually recoveries vary with regard with that of the single toxin column as competition for active sites may arise causing lower recoveries. Romer Labs® has also developed a wide range of detection and sample treatment techniques, including AgraStrip®, which are rapid, ready-to-use (qualitative total aflatoxin or quantitative WATEX) lateral flow devices ideal for on-site or surface testing with a limit of detection of 3.31 μg kg^{-1}. AgraQuant® Aflatoxin/Aflatoxin B_1 which are ELISA tests with sensitivities ranging from 1 to 3 μg kg^{-1} and FluoroQuant® quantitative fluorometric tests based on a solid-phase or immunoaffinity column clean-up. Romer Labs® has two immunochemical columns: Aflastar™ FIT and Aflastar™ R, the latter have been used in our laboratory and applied to feed aflatoxin monitoring with excellent results. Vicam has similar products ranging from strip tests (Afla-V, Afla-V aqua which has removed the use of hazardous organic solvents) to LC clean-up immunoaffinity columns (Afla B, AOZ HPLC [high-performance liquid chromatography]) approached vary from quick response, qualitative to quantitative. Immunoaffinity columns are a very attractive option for sample clean-up and concentration; however, it is important to consider that this approach not only has inherent drawbacks [52, 53] but also may increase laboratory analysis costs considerably.

Although the use of immunoaffinity approaches is appealing, LC-MS or LC-MS-MS techniques usually require chemically based solid phase extraction as several structurally different analytes are analyzed simultaneously. For example, as early as in 2006, Garon and coworkers developed an HPCL-MS/ESI^+ approach to analyzing up to 11 mycotoxins (including AFB_1) in corn silage using an Oasis® HLB cartridges and eluting with a mixture of methyl *tert*-butyl ether/methanol (9:1) [54]. This matrix deserves particular attention since many ruminants' diets are—at least partially—based on silages and forages. Finally, solid phase extraction sorbents based on molecularly imprinted polymers (AFFINIMIP®) have also been developed and are commercially available for mycotoxin analysis. On the other hand, Pickering Laboratories mainly offer two different technologies to enhance aflatoxin sensitivity: i. the Pinnacle PCX derivatization instrument, which is used as a second pulse-free pump, and reaction system and can be coupled to an LC before the fluorescence detector (FLD). The system pumps (with a predetermined flow) a reagent, e.g. iodine/iodide or a pyridinium hydrobromide perbromide (PBPB) solution. And ii. a photochemical reactor equipped with a 254 nm lamp and a knitted reactor coil (UVE™) (which transforms aflatoxins into stable fluorescent hydroxylated counterparts, e.g. AFB1 is converted into AFB_{2a}). The latter approach was used by Soleimany and coworkers which developed a RP-HPLC multiple toxin analysis for cereals with the use of a photodiode array and fluorescence detectors and a photochemical reactor for enhanced detection [55]. The authors found the limits of detection for AFB_1/AFG_1 and AFB_2/AFG_2 to be 0.025 and 0.012 ng g^{-1}, respectively.

Likewise, only a few methods have been described elsewhere, e.g. Shakir Khayoon and coworkers detailed an assay for the determination of aflatoxins in animal feeds and ingredients by LC with multifunctional column clean-up [56]. Biotage® Isolute Multimode® Columns were used to assess aflatoxins successfully. These particular columns have three mechanisms of action: strong cation exchange ($R\text{-}SO_3^-\ H^+$), hydrophobic-based retention [$-(CH_2)_{17}CH_3$], and weak anion exchange [$(CH_2)_3N^+(CH_3)_3Cl^-$]. Based on structural analysis of aflatoxins, not all these mechanisms play a role during their extraction. The authors report great results, i.e. the sensitivity of 0.10 and 0.06 ng g^{-1} for AFG_1/AFG_2 and AFB_1/AFB_2, respectively. Acetonitrile:water (9:1) mixture gave satisfactory recoveries for all aflatoxins (>85%).

2.2.3. *Recent approaches for the extraction of aflatoxins from feeds*

Depending on the method and analytical instrumentation chosen for aflatoxin analysis, the extraction step can become a limiting stage of the overall assay. For example, liquid chromatography coupled with extensive treatment to obtain clean extracts before injection. MS-based approaches have an inherent advantage over classic ones. The detector can differentiate between two different mass/charge units even if chromatographic signals are overlapped. Hence, less intensive and straightforward sample preparation techniques, such as QuEChERS (quick, easy, cheap, effective, rugged and safe), dispersive liquid-liquid microextraction, or "dilute-and-shoot", are employed [57].

Dzuman and coworkers optimized a QuEChERS method for the determination of 56 *Fusarium, Alternaria, Penicillium, Aspergillus,* and *Claviceps* mycotoxins in animal feeds by UHPLC-MS/MS [58]. The authors demonstrated that the pH of extraction solvents was the most critical factor during the preparative step. Silages represent an attractive matrix because the organic acids produced by fermentation acidify and buffer any aqueous media, such conditions, if not considered, may interfere with solvent extraction, chemical or immunological sorbents interaction and may change injection micro-conditions affecting retention times. The same authors applied a dispersive SPE using C_{18} sorbent to avoid coextraction of triacylglycerols and thus prolonging the life of the analytical column. León and coworkers also used QuEChERS to assess 77 banned veterinary drugs, mycotoxins, ergot alkaloids and plant toxins, and a post-target screening for 425 substances, including pesticides and environmental contaminants in feed [59]. Although not specifically in feed, Sirhan and coworkers developed and applied an QuEChERS-based method that included as samples, seeds ($n = 51$), nuts ($n = 78$), and several cereals ($n = 274/669$ samples), that have been also used as feed ingredients (e.g. peanuts, sunflower, almond), and could very easily be applied to other matrices [60]. The authors compared their method to a classic fluorimetric one and found the former to be superior in precision and less biased.

Dispersive liquid-liquid microextraction was applied by Campone and coworkers to the determination of aflatoxins in cereals such as corn, rice, and wheat [61]. Chloroform was selected as transfer solvent, whereas a methanol:water (8:2) was selected as an extraction mixture and a 2.5 enrichment factor was reported. Afzali and coworkers developed a method using dispersive liquid-liquid microextraction for the preconcentration of ultratrace amounts of AFB_1, AFB_2, AFG_1, and AFG_2; the authors validated several parameters as extraction solvent (chloroform), disperser solvent (acetonitrile), sample pH, and centrifugation time finally settling for a two-step approach [62]. Lai and coworkers used a microextraction method to concentrate 1.25 times aflatoxin B_1, B_2, and ochratoxin A with acetonitrile/water/acetic acid mixture as extraction solvent and chloroform as a disperser in rice samples [63]. Noteworthy, it is usual to these microextraction methods to be coupled with immunoaffinity column extraction as an additional step or to compare performance results among methods. Amirkhizi and coworkers used a dispersive liquid-liquid microextraction as a clean-up method before the quantitation of AFB_1 in eggs ($n = 150$) and chicken livers ($n = 50$) obtaining incidences of 72% and 58%, respectively [64]. A review by Spietelun and coworkers treat, in general, miniaturized analytical pretreatment options (e.g. single-drop microextraction, hollow fiber liquid-phase microextraction, dispersive liquid-liquid microextraction) with emphasis in green chemistry [65]. In fact, Zhao and coworkers used ionic liquid-based dispersive liquid-liquid microextraction specifically on feeds, obtaining enrichment factors from 22 to 25 for aflatoxins [66].

Two multi-mycotoxin methods, a dilute-and-shoot LC-MS/MS method and a method based on multi-toxin immunoaffinity columns before LC-MS/MS, were used for the determination of mycotoxins in corn samples, which included integral and moldy grains, harvested in South Africa [67]. Arroyo-Manzanares and coworkers used acetonitrile as an extraction solvent for a "dilute-and-shoot" method for the determination of AFs in animal feed in combination with matrix-matched calibration [68].

Although less complex sample clean ups are very attractive to offer a swift response on a relatively low budget, care must be taken as high matrix interference (when injecting crude extracts) represent a limitation, so some sample treatment methods are usually a requirement. New approaches are continually being developed such as the method selected by Ates and coworkers which injected extracts directly into an automated turbulent flow sample clean-up system, coupled to an LC-HRMS (high-resolution mass spectrometry [Orbitrap]) system to screen up to 600 fungal metabolites to generate feed contaminant profiles [69]. On the other hand, Fabregat-Cabello and coworkers used multi-level external calibration using isotopically labeled internal standards, multiple and single level standard addition, one point isotopic internal calibration and isotope pattern deconvolution to compensate sample extracts, such as those from a feed, that demonstrate powerful matrix effects [70].

On the other hand, Hu and coworkers simplified immunoaffinity column analysis reducing sample extraction and toxin purification to one step and using microbeads coupled with monoclonal antibodies against AFB_1, AFB_2, AFG_1, AFG_2, zearalenone, ochratoxin A, STE, and T-2 toxin [71]. Eighty feed samples were successfully tested using this tactic. Zhao and coworkers described a method for analyzing 30 different mycotoxins (e.g. aflatoxins, ochratoxin A, trichothecenes, zearalenone, fumonisins, and citrinin) in animal feed, animal tissue, and milk [72]. The authors compared three extraction mixtures, different SPE cartridges, including Oasis HLB®, an amino cartridge, Oasis MAX®, and MycoSep® 226 multifunctional cartridge, and sorbents, including C_{18}, chitin, carbon nanotubes, and florisil. The reader is referred to the review by Arroyo-Manzanares and coworkers who cite new techniques in sample preparation for mycotoxins [73].

3. Immunoaffinity-based techniques for aflatoxin detection

Other technologies have helped perform easier and faster toxin analysis. Though, they are limited as to the amount of information that can be drawn from a sample. Recently, the International Crops Research Institute for the Semi-Arid Tropics (ICRISAT) developed the first portable low-cost (up to 2 USD) device designed for rapid detection of aflatoxins. On the other hand, this technology seems to need little to no technical experience to use. This development means that feed producers and farmers may take decisions on location saving unnecessary exposure to toxins and limit economic loss. Another cost-effective approach is based on a lateral flow device (immunodipstick) to assess as little as 5 μg kg^{-1} of AFB_1 in swine feed. We consider this type of approach to be considerably useful to assess aflatoxin cross-contamination in surfaces at feed manufactories, farms, or even dedicated laboratories. Lee and coworkers developed a semi-quantitative one dot lateral flow immunoassay for AFB_1 using a smartphone as a reading system with a sensitivity of 5 μg kg^{-1} [74]. The authors applied this method to whole corn and feed with great results.

With the widespread use of immunochemical based techniques, the development of new toxin-specific monoclonal antibodies with a very high selectivity are in need. Zhang and coworkers reported a new AFB_1 monoclonal antibody (MAb) 3G1 obtained by immunizing Balb/c mice with aflatoxin B_{2a}-Bovine serum albumin [75]. The approach rendered a highly sensitive immunochromatographic assay, a detection limit of 1 ng mL^{-1}, showed no cross-reactivity with other aflatoxins and avoided providing false-positive results. The authors included during validation among other matrices, feedstuffs. Several conjugates and antibodies have been commercially developed for sample preparative purposes. Recently, ImmuneChem® has developed AFB_1, and AFM_1 bovine serum albumin and horseradish peroxidase immobilized antigens for anti-aflatoxin antibody assays. Rabbit and mouse antibodies-based sorbents are also available and can be utilized for detection and quantization of food-borne AFB_1. The standard application of these antibodies is in ELISA test. The usage of monoclonal ELISA test was introduced to research practice early on assessing aflatoxin concentrations in the feed. For example, Banerjee and Shetty applied this technique to poultry feed [76]. Recent approaches have incorporated improvements on ELISA tests. For example, Rossi et al. developed an indirect competitive enzyme-linked immunosorbent assay (ic-ELISA) based on an anti-aflatoxin B_1 monoclonal antibody [77]. The authors reported that the method was validated for aflatoxin screening in poultry feed samples obtaining detection limits and recoveries of 1.25 ng g^{-1} and 98% for broiler feed and 1.41 ng g^{-1} and 102% for laying hen, respectively, on both accounts. The method was also compared with HPLC results, and the authors found a high correlation with HPLC of 0.97 (broiler feed) and 0.98 (laying hen feed). Another research group developed an indirect competitive electrochemical ELISA for the determination of AFB_1 in barley. The method used disposable screen-printed carbon electrodes and anti-AFB_1 monoclonal antibodies (MAb) for immunosensor development. Cross-reactivity of AFG_1 was found, and the authors demonstrated that the coated electrodes could be used for up to 1 month after their preparation when stored at 4°C. The limit of detection was found to be 90 pg mL^{-1}, which translates to 0.36 μg kg^{-1}.

Gold colloid strip tests have also become available for some matrixes and are somewhat popular. For example, Ateko Masinde and coworkers developed a colloidal gold-based immunochromatographic strip which they applied to the analysis of corn and rice [78]. In our context, these matrixes are relevant since both are common feed ingredients. More recently, Sun and coworkers developed a green method using anti-AFB_1 antibody-coated gold colloids as probes in plant oils [79]. Noteworthy, the extraction is attained using water as a solvent. The authors reported a successful visual detection under 5 min with a sensitivity of 1.5 mg kg^{-1}. The methods above are interesting since no professional training needs to be involved in applying them efficiently and can be used in the field.

4. Chromatography coupled with fluorescence detection-based methods for aflatoxin derivatization

AOAC HPLC-based assays are for aflatoxins are scarce. The one method available is 2003.02, which is designed for the determination of AFB_1 in cattle feed. Although it can easily be used to quantitate each AF fraction and other feeds. This last method uses post-column derivatization (a standard approach for AF HPLC-based methods to enhance sensitivity; [80]) using a R-biopharm's KOBRA® CELL, which principle is based on the electrochemical *in situ* Br_2 formation (from potassium bromide) and hence the formation of fluorescent AF derivatives. Similar methods have been reported earlier in the literature [81]. An additional approach is the use of PBPB as another derivatizing agent. Manetta and coworkers already used this method to quantitate a chemically related compound, AFM_1, in milk and cheese [82]. Interestingly, Woodman and Zweigenbaum compared the use of PBPB with the derivatization obtained with a KOBRA® CELL and reported better results using the former [83]. Remarkably, Ramirez-Galicia and coworkers described that AFB_1 suffered fluorescence enhancement when forming AFB_1:β-cyclodextrin inclusion complexes [84]. Hence, β-cyclodextrin could very well be a novel reagent for derivatization.

In our laboratory, we have implemented an accredited assay (according to ISO/IEC 17025 requirements) based on the derivatization of AF using an aqueous I^-/I_2 solution with excellent results ([85], see **Figure 2**). We base our method on the fact that iodine/iodide is less oxidizing than other agents and easier to manipulate. The drawback of this approach is that high temperatures (95°C) must be used to obtain AF derivates swiftly (using a 0.14 mL reaction loop); this is not the case for bromine.

Noteworthy, at 365 and 455 nm as excitation and emission wavelengths, respectively, AFB_2 and AFG_2 show natural fluorescence when no derivatization is used, while the signal for

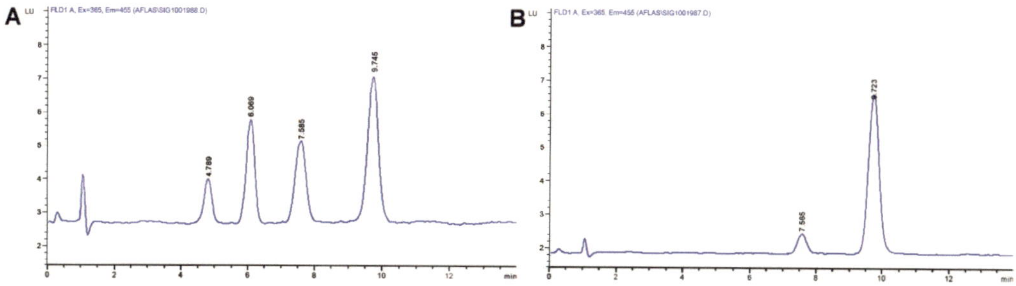

Figure 2. RP-HPLC analysis of aflatoxins using iodine/iodide-based derivatization. (A) 40 μg L^{-1} standard in methanol and 10 μL injection AF fractions in order of elution: AFG_2(4.789), AFG_1(6.069), AFB_2(7.585), and AFB_1(9.745). (B) Same method used to analyze a naturally contaminated corn sample with AFB_2(7.565) and AFB_1(9.723).

the other two fractions is negligible. Both iodine and bromine generate fluorescent derivatives using an addition reaction, which introduces a halogen atom on the double bond of the dihydrofuran ring. The steric hindrance and electronic repulsion conferred by the halogens, which are spatially opposite to each other, render an aflatoxin molecule with far more torsion [−0.2812 (AFB_1) vs. 2.9320 (AFB_1I_2), using an MM2 energy minimization]. Hence, favoring a more rigid structure, a fluorescence prone one (**Figure 3**).

Machado Trombete and coworkers validated a fluorescence-based method and compared three methods of extraction; the authors found chloroform to be the most efficient solvent [86]. Pre-column derivatization with trifluoroacetic acid was used to increase sensitivity (reaching 0.6 μg kg^{-1} as the limit of detection). Although the authors intended this method for wheat projected for human consumption, this matrix is a widely used as a feed ingredient in some countries. Horizon technologies introduced to the market a thermostatically controlled heated water chamber that facilitates the pre-column derivatization of aflatoxins with trifluoroacetic acid (XcelVap®). Cortés and coworkers also used a fluorescence-based method without derivatization using 360 nm as an excitation wavelength and 418–700 nm as excitation [87]. The method included aflatoxicol (a reduced derivate from AFB1; cyclopentanone moiety is converted to cyclopentanol) and also assessed the recovery of aflatoxins and aflatoxicol in

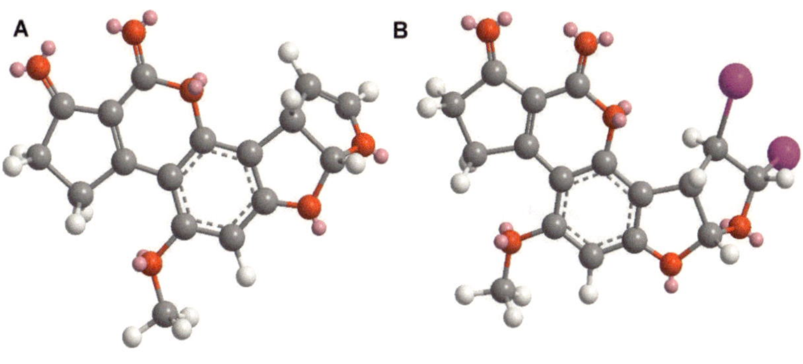

Figure 3. 3D structure minimized energy using MM2 calculations of (A) aflatoxin B_1 (total energy of 48.2584 kcal mol^{-1}) and (B) AFB_1 after iodine addition (total energy of 53.8536 kcal mol^{-1}); pink-colored beads represent non-bonding electron pairs.

poultry litter finding non-trivial levels of these contaminants. A relevant feature since poultry litter (urea, uric acid, and ammonium-rich by-products) has served on occasion as livestock feed. Interestingly, in a Waters Corporation application note, Benvenuti and coworkers used a fluorescence detector-based approach to quantitate aflatoxins B_1, B_2, G_1, G_2, and M_1 without derivatization [88]. The authors used 365 nm as excitation wavelength and emission wavelengths 429 and 455 nm for $AFM_1/AFB_2/AFB_1$ and AFG_1/AFG_2, respectively. Pickering Laboratories developed a fluorometric method using photochemical derivatization and λ_{ex} = 365 nm and λ_{em} = 455 nm for the detection of aflatoxins in dried distillers grains (DDGs). This process was conceived as a multiple toxin analysis using just a fluorescence detector.

Pirestani and coworkers measured aflatoxins both in dairy cattle feeds and milk samples from the province of Esfahan, Iran [89]. The authors compared results from HPLC (bromine post-column derivatization) and ELISA. It was concluded that there was no significant difference between the values obtained by the two procedures. However, sensitivity and specificity were determined to be superior to that of ELISA. Gomes Pereira and coworkers also did analyze dairy cattle feed and milk from the Lavras, Minas Gerais region of Brazil [90]. In the case of cattle feed samples, the authors state they used an AOAC method but failed to specified which.

5. Mass spectrometry coupled chromatography multiple toxin approaches (including aflatoxins)

With the advent of confirmatory and multi-analyte techniques such as tandem MS (mass spectrometry) coupled to LC, a whole new span of methods has been described which include the "classic" toxins and other not-so-known ones into feed vigilance schemes. In 2012, Warth and coworkers described a multiple-toxin method based on LC-MS/MS that included several metabolites, a total of 63 analytes were tested in corn, groundnut, sorghum, and feed produced in Burkina Faso and Mozambique [91]. De Souza and coworkers reported a LC-MS/MS using ESI^+ with a QTrap 4000 system, which was used to analyze 119 samples collected from poultry feed factory [92]. The researchers analyzed n = 74 whole corn samples, n = 36 chicken feed, and n = 9 feed mill residue. Limits of detection ranged from 0.5 (AFG_1) to 1 (AFG_2) μg kg^{-1}, and recoveries ranged from 71 to 87% for corn and 65 (AFB_2) to 77% (AFB_1) for feed. This analysis is relevant since most feed formulations recurred to corn products to supply energy and carbohydrates. Contamination in feed ingredients will concurrently have an adverse impact on feed safety.

Recently, Njumbe and coworkers reported a LC-MS/MS method that included 23 mycotoxins in different sorghum varieties, all analytes eluted under 14 minutes and stated a high sensitivity for all mycotoxins, specifically 2.5 and 5.0 μg kg^{-1} for AFB_1/AFB_2 and AFG_1/AFG_2, respectively [93]. Although sorghum, in some regions, has been substituted by other grains such as corn, it has seen a resurgence as a crop for feed in several parts, which is relevant since some grain production is not continuous throughout the year, and feed ingredient supplies are in high demand, and their availability is constrained. More interestingly, when the method was applied to a small subset of retailed samples from Belgium and Germany, 90% were positive for aflatoxin B_1. Regarding food and feed monitoring, a very comprehensive review was written by Zhang and coworkers [94]. This report is unique since it cites the techniques used by the US FDA to assess mycotoxins in different staple foods including LC-FLD (fluorescence detector), MS, tandem MS, and HRMS. For example, a LC-MS/MS method including 11 different mycotoxins (e.g. aflatoxins), using stable isotope dilution, has been developed and validated in various matrixes (including cat/dog food, corn, feeds, and wheat flour). Samples were fortified using ^{13}C-IS and prepared by solvent extraction. In general, the recoveries ranged from 70 to 120%, with RSDs < 20%. Limit of quantitation was calculated to be 0.005 μg kg^{-1} for AFB_1. The method above applied in our laboratory based on LC-FLD with post-column derivatization renders a similar acuteness for AFB_1 (limit of detection and quantification 0.005 and 0.15 μg kg^{-1}, respectively) nonetheless our method had to be modified to include a ca. 200-fold concentration step. Lattanzio and coworkers used a similar approach to analyze cereal-based foods using as a clean-up strategy SPE [95]. Zhang and coworkers opted for a LC/MS/MS approach to analyze mycotoxins in feed using isotope dilution and circumventing the clean-up step altogether [96]. For a thorough review of chromatographic and spectrometric techniques used for mycotoxin analysis, we suggest the paper wrote by Li and coworkers [97]. Ok and coworkers recently opted to include aflatoxins and sterigmatocystin in the same analysis using tandem MS for their assay in sorghum and rice [98].

DDGs is an essential matrix since the shortage and costs of other corn-based feed ingredients have pushed toward their extensive use [99, 100]. As this is a residue from ethanol production, any mycotoxins initially found in the raw material may be concentrated. On the other hand, Oplatowska-Stachowiak and coworkers developed a UPLC/MS/MS method capable of analyzing as much as 77 mycotoxins and other fungal metabolites [101]. The method analyzed 169 DDGs samples produced from wheat, corn, barley and other grains. Aflatoxin contamination was frequently encountered in corn DDGs. In contrast, wheat and mixed DDGs showed none or very few contaminated samples. In a very exhaustive analysis of European feedingstuffs, Zachariasova and coworkers used a UHPLC–QtrapMS/MS. The authors found that forages showed the lowest mycotoxin incidence while the most diversity of detected mycotoxins. In contrast, the highest concentrations, was quantified in DDGs [99]. For example, AFB_1 was found with a mean value and a maximum of 0.6 and 6.4 μg kg^{-1}, respectively.

Another important feed ingredient is palm kernel cake, which is used as a source of protein and energy for livestock and occasionally used as poultry feed supplement. Yibadatihan and coworkers developed a LC/MS/MS ESI^+ to analyze several toxins in palm kernel cake, including aflatoxins [102]. Recoveries ranged from 84 to 110, and the method sensitivity was calculated as 0.16 and 0.54 for AFB_2/AFG_2 and AFB_1/AFG_1, respectively. Twenty-five samples were analyzed using this approach, and a very high prevalence for aflatoxins (>85% samples tested positive for any of the fractions) was found. The lowest and highest concentrations found were 1.31 (for AFG_1) and 78.38 (for AFG_2) μg kg^{-1}. As with DDGs, any toxin found in palm kernel raw material will probably be concentrated as the palm kernel is mechanically pressed to extract vegetable oil.

Finally, in a provoking research, Escrivá and coworkers assessed mycotoxin (including aflatoxins) contamination of rat feed [100]. Twenty-seven commercial Spanish rat feed was analyzed using a liquid chromatography equipped with a 3200 QTrapVR mass spectrometry system with a Turbo electrospray ionization interface. Considerable mycotoxin burden was found in feeds. For example, concentrations of AFB_2 (21.61 μg/kg bw/day), and AFG_2 (15.09 μg/kg bw/day) were calculated for the assayed feeds. Since laboratory animals are used as models in other research, these contaminants' toxic effects may cause artifacts and confounding results. The authors detailed the feed composition listing each ingredient use during formulation, data that is usually overlooked during contaminant analysis. Feed composition plays a significant role in toxin pollution as the main ingredients may guide which contaminants will be more likely to be present [103]. McElhinney and coworkers developed a method for the determination of mycotoxins in grass silage [104]. In this case, they used a modified QuEChERS approach with almost no clean-up and an UHPLC/MS/MS technique. Polarity switch during the analysis permitted to assess both positive and negative ions. AFB_1 detection limit was calculated to be 3 μg kg^{-1} DM. This relatively low sensitivity is usually the cost of a swift sample preparation and avoiding thorough clean-up steps. A similar approach was used by Dzuman and coworkers for the analysis of cereals, complex compound feeds, extracted oil cakes, fermented silages, malt sprouts, or DDGs using U-HPLC-HRMS [105].

6. Novel approaches for aflatoxin determination in feed

An interesting earlier report made by Babu and Muriana stated that AF recovery was enabled by the use of primary polyclonal antibodies for AFB_1 [106]. Said antibodies, were covalently attached to 2.8 μm diameter magnetic beads using a cross-linking agent and a secondary antibody for the toxin covalently linked to DNA oligonucleotides based on the *luc* gene as a reporter DNA molecule which, in turn, was amplified using real-time immune quantitative polymerase chain reaction after aflatoxin capture if present. The authors prepared toxin suspensions in methanol:water solution. This mixture can also serve as an extraction solvent; the sensitivity of this method was calculated to be 0.1 μg kg^{-1}. The same authors [107], later on, applied a modified version of this approach and applied it to poultry, dairy and horse feed, and a whole kernel corn, corn gluten feed, and yellow corn meal. The authors conclude that the technique is useful for quantifying low natural aflatoxin levels in animal feed samples without the requirement of additional sample cleanup. However, samples artificially contaminated with high levels of aflatoxin (i.e. 200 μg kg^{-1}) exhibited recoveries of 60% which are considered poor.

Another novel approach for the extraction, preconcentration, and determination of aflatoxins in animal feedstuffs was carried on recently by Zhao and coworkers who developed a novel two-step extraction technique combining ionic liquid-based dispersive liquid-liquid microextraction with magnetic solid-phase before HPLC coupled with FLD [66]. The ionic liquid 1-octyl-3-methylimidazolium hexafluorophosphate was used as the toxin-retrieval agent, and hydrophobic pelargonic acid modified Fe_3O_4 magnetic nanoparticles as an active sorbent.

Ramesh and coworkers used a high-performance thin-layer chromatography method that uses a stationary phase based on silica gel 60G F254 and a mobile phase that consisted of acetone:chloroform (1: 9), n = 59 samples of feed were analyzed by this method that reported 0.5 μg kg^{-1} as the limit of detection [108]. A similar approach was used by Kotinagu and coworkers to assess AFB_1 in a total of 97 livestock feed samples. In this case, a chloroform:acetone:water mixture (28:4:0.06) was used to elute [109]. The toxin was revealed using a 366 nm wavelength. Finally, Zhang and coworkers developed an on-site analysis for aflatoxin B_1 in food and feed samples using a chromatographic time-resolved fluoroimmunoassay that offered a magnified positive signal and low signal-to-noise ratio [110]. Wang and coworkers developed a fluorescence based on europium nanospheres and monoclonal antibodies for the determination of total aflatoxin in the feed with a 0.16 μg kg^{-1} limit of detection [111]. Interestingly, the authors designed this method to use with a portable reader. A good association was found among the assay and HPLC results for corn, wheat bran, peanut meal, soybean meal, cottonseed meal, DDGs, alfalfa forage, silage, swine feed, and poultry feed.

Ren and coworkers used an immunochromatographic assay based on CdSe/Zn quantum dot beads, reaching values as low as 0.42 pg mL^{-1} AFB_1 [112]. Quantum dots were prepared using poly(methyl methacrylate), poly(maleicanhydride-*alt*-1-octadecene), and *N*-(3-(dimethylamino)propyl)-*N*′-ethylcarbodiimide hydrochloride. The method results were compared with ELISA and LC-MS/MS, exhibiting great association with both. Feed ingredients that included corn, soybean meal, rapeseed meal, cottonseed meal, distillers dried grain, and wheat, were tested. Setlem and coworkers generated high-affinity single-stranded DNA aptamers that specifically bind to AFB_1 by a modified Systemic Evolution of Ligands by Exponential Enrichment procedure [113]. The two aptamers with lower Gibbs energy threw 20–40 ng mL^{-1} sensitivity. Coupled with HPLC, the aptamers were able to recover and quantify 82.2–96.21%. He and coworkers constructed AFB_1-BSA conjugated "nanobody" from immunized alpacas [114]. The most interesting part about the outcome is that the authors describe the resulting nanobody to be thermally and organic solvent resistant. Recovery from spiked peanut, rice, corn, and feedstuff ranged from 80 to 115%. Xiong and coworkers reported an improved magnetic bead-based immunoaffinity extraction method, for the highly efficient purification of AFB_1 from corn samples, that circumvents common inherent disadvantages of this approach [115]. The method involves the expression of anti-AFB_1 nanobodies, with degeneration resistance, to replace conventional antibodies. Magnetic beads, carrying poly(acrylic acid) "brushes", expand significantly adsorption capacity (i.e. 623 μg g^{-1}) and reusability (10x without obvious loss of the capture efficiency for AFB_1). The reliability of the proposed method for AFB_1 extraction was further evaluated using AFB_1-spiked corn samples.

Finally, new efforts to quantitate and detect mycotoxins should include emerging analytes, such as other *Aspergillus* metabolites, such as STE and emodin, neither analyte routinely screened for in feed nor regulated by legislation [116].

7. Bearing of aflatoxigenic molds isolation from feed

Isolation and identification of fungi, especially those with aflatoxigenic capabilities, is an analytical feature, during aflatoxin determination, which is seldom considered. These data may easily be contrasted with concentrations obtained by any of the analysis methods aforementioned. However, a few papers have indeed tackled the issue. Suganthi and coworkers used both ELISA (based on urea peroxide and the chromogen tetramethylbenzidine) and thin-layer chromatography (using a mixture of chloroform and acetone) to detect aflatoxins in animal feed and also isolated molds using Czapek Dox Agar medium [117]. In this case, the authors concluded that 80% *Aspergillus* strains were *A. flavus* and none of the strains were toxigenic. Finally, they also showed an antifungal effect of *Lactobacillus* species. Chandra and coworkers assessed AFB_1 in corn from Indian markets using competitive ELISA [118]. The authors also determined mold count that ranged from 1.0×10^2 to 3.6×10^6 CFU/g. Relevant data since isolation of aflatoxigenic fungi from contaminated samples are seldom done. Although no association was found between microbial analysis and contamination, other more selective media can be used to isolate more specific *Aspergillus* species (e.g. AFPA Base Oxoid™), this may help to determine which species are responsible for the contamination. For example, Queiroz and coworkers used ELISA to assess aflatoxin contamination in three quality type of bird feeds from Brazil. Furthermore, they used Dichloran Rose Bengal Chloramphenicol agar, as a general medium used to estimate mold counts and dichloran glycerol 18% agar, a low a_w medium that facilitates the growth of xerophilic fungi [119]. The authors found that *Aspergillus* (82%), *Cladosporium* (50%), and *Penicillium* (42%) were the predominantly isolated genera. *Aspergillus niger* aggregate (35%), *Aspergillus fumigatus* (28%) and *A. flavus* (18%) had the highest relative densities. Finally, aflatoxins have rarely been detected in feeds and foods in Hungary. However, Sebök and coworkers analyzed several corn fields alongside Hungary between the years 2013 and 2014 and found the presence of aflatoxigenic fungi in corn and soil samples with isolation ratios of 26.9 to 16.1% and 42.3 to 34.7%, respectively, on both accounts [120]. The authors evaluated on the isolates the presence of partial calmodulin gene demonstrating the identity of the strains to be *A. flavus* (n = 110/114) and *A. parasitcus* (n = 4/114). Based on the strain genotoxic response, 45.5% of the 110 *A. flavus* strains were toxin producers. Carvalho and coworkers not only found a prevalence of 77.7% for aflatoxin in tropical corn silages (Minas Gerais, Brazil) but also identified *A. fumigatus* in all silages that presented growth of molds [121]. Yeast species including those from the *Candida* genera were isolated. As an additional example, Kaya-Celiker, in an interesting paper, successfully used Fourier transform mid-infrared and photoacoustic spectroscopy to identify and separate infected peanuts based on spectral characteristics [122]. Ibrahim and coworkers recently screened 102 feed samples (including poultry feed, cotton seed meal, and corn) for the presence of aflatoxin biosynthetic pathway genes *ver*-1, *apa*-2, and *omt*-1 using PCR assay, and thin-layer chromatography was performed to confirm the synthesis of aflatoxin in PCR-positive strains [123]. Nine samples exhibited the simultaneous presence of the three genes and all were capable of producing AFB_1 and AFB_2.

8. Conclusions and perspectives

As stated before, research based on animal feeds is somewhat lacking and usually, the importance of this matrix within the food chain sometimes omitted. Although this chapter is devoted to aflatoxins and regulatory standards, till this day, target only specific toxins, evidence suggests that other mycotoxins, contaminants, residues, and xenobiotics interact with aflatoxins sometimes even enhancing their carcinogenic potential. We consider that aflatoxin monitoring programs should be implemented and reinforced to minimize the impact of aflatoxins on animals and humans. On the other hand, policy makers and officials should concentrate efforts and prioritize the incorporation of country-wise feed monitoring systems where currently there are none. Based on the data recollected here-in, sampling and surveying should focus especially on corn and corn products and pet food and full ruminant rations including balanced feed. From the zootechnical standpoint, evidence indicates that general nutritional formulations can be modified to minimize fungal and toxin contamination and hence animal health impacts and still cover the traditional nutritional needs. On the other hand, farmers may equivocally attribute productivity loss to toxin presence where none is found. As occasionally farm feed practices are based on the exploitation of residue from other agricultural activities (e.g. fruit processing wastes, poultry litter, alcohol product by-products [DDGs]), strict control of this type of samples should be kept. Aflatoxin toxicity occurs at very low concentrations. Therefore, sensitive and reliable methods for their detection are required. Sampling and analysis of aflatoxins are paramount. Failure to achieve a verifiable analysis can lead to erroneous conclusions or judgments; contaminated feedlots being accepted or satisfactory batches unnecessarily rejected. Thanks to technological advancements method for aflatoxin detection are continuously improving in sensitivity, repeatability, accuracy, efficiency, and with less and less waste. Data herein demonstrate that even in countries where expensive technology (e.g. LC-MS/MS) is scarce or not readily available, feed monitoring is possible. Efforts have been made to provide proficiency testing for laboratories (e.g. American Association of Feed Control Officials [AAFCO] and Laboratory of Government Chemist [LGC Standards]), which improve method accuracy bias and reliability. However, feed-based certified materials available are still few. Finally, considering the relevance of feed in the food chain safety, countries should implement and improve monitoring programs for aflatoxin in foodstuffs; these programs should contemplate risk management, One Health or "MyToolBox" approaches, and farm-to-fork models that include all stakeholders to mitigate the economic and health burden that aflatoxin contamination generates.

Acknowledgements

The author would like to thank Vicerrectoría de Investigación for the financial support.

Author details

Fabio Granados-Chinchilla

Address all correspondence to: fabio.granados@ucr.ac.cr

Centro de Investigación en Nutrición Animal (CINA), Universidad de Costa Rica, San José, Costa Rica

References

[1] [IARC] International Agency for Research on Cancer. IARC Monographs on the Evaluation of Carcinogenic Risks to Humans. Some Organophosphate Insecticides and Herbicides, Volume 112 [Internet]. 2015. Available from: http://monographs.iarc.fr/ENG/Monographs/vol112/mono112.pdf [Accessed: 03/05/2017]

[2] Zhang D, Cui Y, Shen H, Xing L, Cui J, Wang J, Zhang X. Sterigmatocystin-induced DNA damage triggers G_2 arrest via an ATM/p53-related pathway in human gastric epithelium GES-1 cells *in vitro*. PLoS ONE. 2013;**8**(5):e65044. DOI: 10.1371/journal.pone.0065044

[3] Bbosa GS, Kitya D, Odda J, Ogwal-Okeng J. Aflatoxins metabolism, effects on epigenetic mechanisms and their role in carcinogenesis. Health. 2013;**5**(10A):14-34. DOI: 10.4236/health.2013.510A1003

[4] Dohnal V, Wu Q, Kuča K. Metabolism of aflatoxins: Key enzymes and interindividual as well as interspecies differences. Archives on Toxicology. 2014;**88**(9):1635-1644. DOI: 10.1007/s00204-014-1312-9

[5] Varga J, Frisvad JC, Samson RA. Two new aflatoxin producing species, and an overview of *Aspergillus* section *Flavi*. Studies in Mycology. 2011;**69**:57-80. DOI: 10.3114/sim.2011.69.05

[6] Baranyi N, Kocsubé S, Vágvölgyi C, Varga J. Current trends in aflatoxin research. Acta Biologica Szegediensis. 2013;**57**(2):95-107

[7] Samson RA, Visagie CM, Houbraken J, Hong S-B, Hubka V, Klaassen CHW, Perrone G, Seifert KA, Susca A, Tanney JB, Varga J, Kocsubé S, Szigeti G, Yaguchi T, Frisvad JC. Phylogeny, identification and nomenclature of the genus *Aspergillus*. Studies in Mycology. 2014;**78**:141-173. DOI: 10.1016/j.simyco.2014.07.004

[8] Yu J. Current understanding on aflatoxin biosynthesis and future perspective in reducing aflatoxin contamination. Toxins. 2012;**4**(11):1024-1057. DOI: 10.3390/toxins4111024

[9] Godet M, Munaut F. Molecular strategy for identification in *Aspergillus* section *Flavi*. FEMS Microbiology Letters. 2010;**304**:157-168. DOI: 10.1111/j.1574-6968.2009.01890.x

[10] Ranjbar S, Nazari R, Noori M. Isolation and molecular identification of *Aspergillus* species of Cattle. Veterinary Research. 2011;**7**(1):11-17

[11] Iheanacho HE, Njobeh PB, Dutton FM, Steenkamp PA, Steenkamp L, Mthombeni JQ, Daru BH, Makun AH. Morphological and molecular identification of filamentous *Aspergillus flavus* and *A. parasiticus* isolated from compound feeds in South Africa. Food Microbiology. 2014;**44**:180-184. DOI: 10.1016/j.fm.2014.05.019

[12] Habib MA, Abdu P, Kwanashie CN, Kabir J, Negedu A. Isolation and identification of *Aspergillus* species from poultry feeds in Kaduna State, Nigeria. Microbiology Research International. 2015;**3**(2):27-32

[13] da Silva JL, Aparecido CC, Hansen D, Pereira TAM, D'arc-Felicio J, Gonçalez E. Identification of toxigenic *Aspergillus* species from diet dairy goat using a polyphasic approach. Ciência Rural, Santa Maria, Online. 2015;**45**(8):1466-1471. DOI: 10.1590/0103-8478cr2014058

[14] Guchi E. Implication of aflatoxin contamination in agricultural products. American Journal of Food and Nutrition. 2015;**3**(1):12-20. DOI: 10.12691/ajfn-3-1-3

[15] Kana JR, Gnonlonfin BGJ, Harvey J, Wainaina J, Wanjuki I, Skilton RA, Teguia A. Assessment of aflatoxin contamination of maize, peanut meal and poultry feed mixtures from different agroecological zones in Cameroon. Toxins. 2013;**5**:884-894. DOI: 10.3390/toxins5050884

[16] Zain ME. Impact of mycotoxins on humans and animals. Journal of Saudi Chemical Society. 2011;**15**(2):129-144. DOI: 10.1016/j.jscs.2010.06.006

[17] Becker-Algeri TA, Castagnaro D, de Bortoli K, de Souza C, Drunkler DA, Badiale-Furlong E. Mycotoxins in bovine milk and dairy products: A review. Concise Reviews in Food Science. 2016;**81**(3):R544-R552. DOI: 10.1111/1750-3841.13204

[18] Smith M-C, Madec S, Coton E, Hymery N. Natural co-occurrence of mycotoxins in foods and feeds and their *in vitro* combined toxicological effects. Toxins (Basel). 2016;**8**(4):94. DOI: 10.3390/toxins8040094

[19] [AAFCO] Association of American Feed Control Officials. Official Publication. Atlanta, United States: Association of American Feed Control Officials; 2017. 325 p.

[20] Alonso VA, Pereyra CM, Keller LAM, Dalcero AM, Rosa CAR, Chiacchiera SM. Fungi and mycotoxins in silage: An overview. Journal of Applied Microbiology. 2013;**115**(3):637-643. DOI: 10.1111/jam.12178

[21] Hashimoto EH, do Santos MA, Ono EYS, Hayashi C, Bracarense APFRL, Hirooka EY. Bromatology and fumonisin and aflatoxin contamination in aquaculture feed of the region of Londrina, State of Paraná, Brazil. Ciências Agrárias, Londrina. 2003;**24**(1):123-132

[22] Prabakaran JJ, Dhanapal S. Analysis of proximate composition and aflatoxins of some poultry feeds. Asian Journal of Biotechnology. 2009;**1**(3):104-110

[23] Gupta SK, Venkitasubramanian TA. Production of aflatoxin on soybeans. Applied Microbiology. 1975;**29**(6):834-836

[24] Gupta SK, Maggon KK, Venkitasubramanian TA. Effect of zinc on tricarboxylic acid cycle intermediates and enzymes in relation to aflatoxin biosynthesis. Journal of General Microbiology. 1977;**99**(1):43-48. DOI: 10.1099/00221287-99-1-43

[25] Liu J, Sun L, Zhang N, Zhang J, Guo J, Li C, Rajput SA, Qi D. Effects of nutrients in substrates of different grains on aflatoxin B_1 production by *Aspergillus flavus*. BioMed Research International. 2016;**2016**(1-10):7232858. DOI: 10.1155/2016/7232858

[26] Herzallah SM. Aflatoxin B_1 residues in eggs and flesh of laying hens fed aflatoxin B_1 contaminated diet. American Journal of Agricultural and Biological Sciences. 2013;**8**(2):156-161. DOI: 10.3844/ajabssp.2013.156.161

[27] van der Fels-Klerx HJ, Camenzuli L. Effects of milk yield, feed composition, and feed contamination with aflatoxin B_1 on the aflatoxin M_1 concentration in dairy cows' milk investigated using Monte Carlo simulation modelling. Toxins. 2016;**8**:290-301. DOI: 10.3390/toxins8100290

[28] van der Fels-Klerx HJ, Bouzembrak Y. Modelling approach to limit aflatoxin B_1 contamination in dairy cattle compound feed. World Mycotoxin Journal. 2016;**9**(3):455-464. DOI: 10.3920/WMJ2015.1957

[29] Rabinowitz PM, Kock R, Kachani M, Kunkel R, Thomas J, Gilbert J, Wallace R, Blackmore C, Wong D, Karesh W, Natterson B, Dugas R, Rubin C. Toward proof of concept of a one health approach to disease prediction and control. Emerging Infectious Diseases. 2013;**19**(12):e130265. DOI: 10.3201/eid1912.130265

[30] Shenge KC, LeJeune JT. One Health: A focus on interdisciplinary collaboration. Advances in Plants & Agriculture Research. 2014;**1**(3):00018. DOI: 10.15406/apar.2014.01.00018

[31] Sirma AJ, Lindahl J, Kiama TN, Senerwa DM, Waithanji EM, Kang'ethe EK, Grace D. Systematic review of mycotoxins with a One Health perspective. In: 1st African Regional Conference of the International Association on Ecology and Health (Ecohealth); October 2013; Abidjan, Côte d'Ivoire

[32] Magnussen A, Parsi MA. Aflatoxins, hepatocellular carcinoma and public health. World Journal of Gastroenterology. 2013;**19**(10):1508-1512. DOI: 10.3748/wjg.v19.i10.1508

[33] Frazzoli C, Gherardi P, Saxena N, Belluzzi G, Mantovani A. The hotspot for (global) One Health in primary food production: Aflatoxin M_1 in dairy products. Frontiers in Public Health. 2017;**4**(1-11):294. DOI: 10.3389/fpubh.2016.00294

[34] Ogodo AC, Ugbogu OC. Public health significance of aflatoxin in food industry—A review. European Journal of Clinical and Biomedical Sciences. 2016;**2**(5):51-58. DOI: 10.11648/j.ejcbs.20160205.14

[35] Krska R, de Nijs M, McNerney O, Pichler M, Gilbert J, Edwards S, Suman M, Magan N, Rossi V, van der Fels-Klerx HJ, Bagi F, Poschmaier B, Sulyok M, Berthiller F, van Egmond HP. Safe food and feed through an integrated toolbox for mycotoxin management: The MyToolBox approach. World Mycotoxin Journal. 2016;**9**(4):487-495. DOI: 10.3920/WMJ2016.2136

[36] Cigić IK, Prosen H. An overview of conventional and emerging analytical methods for the determination of mycotoxins. International Journal of Molecular Science. 2009;**10**(1):62-115

[37] Stark AA. Molecular Mechanism of Detection of Aflatoxins and Other Mycotoxins. In: Rai M, Varma A, editors. Mycotoxins in Food, Feed and Bioweapons. 1st ed. Basel, Switzerland: Springer-Verlag Berlin Heidelberg; 2010. pp. 21-37. DOI: 10.1007/978-3-642-00725-5

[38] Yao H, Hruska Z, Di Mavungu D. Developments in detection and determination of aflatoxins. World Mycotoxin Journal. 2015;**8**(2):181-191. DOI: 10.3920/WMJ2014.1797

[39] Herzallah SM. Determination of aflatoxins in eggs, milk, meat and meat products using HPLC fluorescent and UV detectors. Food Chemistry. 2009;**114**(3):1141-1146. DOI: 10.1016/j.foodchem.2008.10.077

[40] Turner NW, Subrahmanyam S, Piletsky SA. Analytical methods for determination of mycotoxins: A review. Analytica Chimica Acta. 2009;**632**(2):168-180. DOI: 10.1016/j.aca.2008.11.010

[41] Wacoo AP, Wendiro D, Vuzi PC, Hawumba JF. Methods for detection of aflatoxins in agricultural food crops. Journal of Applied Chemistry. 2014;**2014**(1-15):706291. DOI: 10.1155/2014/706291

[42] Paterson RM, Lima N. How will climate change affect mycotoxins in food?. Food Research International. 2010;**43**:1902-1914. DOI: 10.1016/j.foodres.2009.07.010

[43] Kuiper HA, Paoletti C. Food and feed safety assessment: The importance of proper sampling. Journal of AOAC International. 2015;**98**(2):252-258. DOI: 10.5740/jaoacint.15-007

[44] Mallmann AO, Marchioro A, Oliveira MS, Rauber RH, Dilkin P, Mallmann CA. Comparison of the efficiency between two sampling plans for aflatoxins analysis in maize. Brazilian Journal of Microbiology. 2014;**45**(1):35-42

[45] Herrman TJ, Lee K-M, Jones B, McCormick C. Aflatoxin sampling and testing proficiency in the Texas grain industry. Journal of Regulatory Science. 2014;**2**(1):7-13

[46] [AAFCO] Association of American Feed Control Officials. Official Publication. Feed Inspector's Manual 6th Edition [Internet]. 20/2017. Available from: http://www.aafco.org/Portals/0/SiteContent/Publications/AAFCO_Feed_Inspectors_Manual_6th_ed.pdf [Accessed: 06/03/2017]

[47] European Commission. Commission Regulation (EC) No 401/2006 [Internet]. 23/02/2006. Available from: http://eur-lex.europa.eu/legal-content/EN/TXT/PDF/?uri=CELEX:32006R0401&from=EN [Accessed: 06/03/2017]

[48] Berthiller F, Brera C, Iha MH, Krska R, Lattanzio VMT, MacDonald S, Malone RJ, Maragos C, Solfrizzo M, Stranska-Zachariasova M, Stroka J, Tittlemier SA. Developments in mycotoxin analysis: An update for 2015-2016. World Mycotoxin Journal. 2017;**10**(1):5-29. DOI: 10.3920/WMJ2016.2138

[49] Lee K-M, Herrman TJ, Dai SY. Application and validation of a statistically derived risk-based sampling plan to improve efficiency of inspection and enforcement. Food Control. 2016;**64**:135-141. DOI: 10.1016/j.foodcont.2015.12.033

[50] Iheanacho HE. A study to evaluate aflatoxin contamination in food from Gauteng Province. Annals of Chromatography and Separation Techniques. 2015;**1**(1):1002

[51] Maniglia-Merida GC, Ascanio-Evanoff E, Riera JG, Colina-Rivero J, Briceño-Fereira E del C, Flores-Chona S, Ascanio-Evan A, Arrieta-Mendoza D Determinación de Aflatoxinas en Alimentos Balanceados para Cerdos en Granjas de los Estados Aragua y Carabobo, Venezuela. Revista Científica, FCV-LUZ. 2015;**25**(3):200-207

[52] Castegnaro M, Tozlovanu M, Wild C, Molinié A, Sylla A, Pfohl-Leszkowicz A. Advantages and drawbacks of immunoaffinity columns in analysis of mycotoxins in food. Molecular Nutrition and Food Research. 2006;**50**(6):480-487. DOI: 10.1002/mnfr.200500264

[53] Chavarría G, Granados-Chinchilla F, Alfaro-Cascante M, Molina A. Detection of aflatoxin M_1 in milk, cheese and sour cream samples from Costa Rica using enzyme-assisted extraction and HPLC. Food Additives & Contaminants: Part B. 2015;**8**(2):128-135

[54] Garon D, Richard E, Sage L, Bouchart V, Pottier D, Lebailly P. Mycoflora and multi-mycotoxin detection in corn silage: Experimental study. Journal Agricultural and Food Chemistry. 2006;**54**(9):3479-3484. DOI: 10.1021/jf060179i

[55] Soleimany F, Jinap S, Rahmani A, Khatib A. Simultaneous detection of 12 mycotoxins in cereals using RP-HPLC-PDA-FLD with PHRED and a post-column derivatization system. Food Additives & Contaminants. Part A, Chemistry, Analysis, Control, Exposure & Risk Assessment. 2011;**28**(4):494-501. DOI: 10.1080/19440049.2010.551547

[56] Khayoon WS, Saad B, Yan CB, Hashim NH, Ali ASM, Salleh MI, Salleh B. Determination of aflatoxins in animal feeds by HPLC with multifunctional column clean-up. Food Chemistry. 2010;**118**(3):882-886. DOI: 10.1016/j.foodchem.2009.05.082

[57] Berthiller F, Brera C, Crews C, Iha MH, Krska R, Lattanzio VMT, MacDonald S, Malone RJ, Maragos C, Solfrizzo M, Stroka J, Whitaker TB. Developments in mycotoxin analysis: An update for 2013-2014. World Mycotoxin Journal. 2015;**8**(1):5-36. DOI: 10.3920/WMJ2014.1840

[58] Dzuman Z, Stranska-Zachariasova M, Vaclavikova M, Tomaniova M, Veprikova Z, Slavikova P, Hajslova J. Fate of free and conjugated mycotoxins within the production of distiller's dried grains with solubles (DDGS). Journal of Agricultural and Food Chemistry. 2016;**64**(24):5085-5092. DOI: 10.1021/acs.jafc.6b00304

[59] León, N Pastor A, Yusà V Target analysis and retrospective screening of veterinary drugs, ergot alkaloids, plant toxins and other undesirable substances in feed using liquid chromatography-high resolution mass spectrometry. Talanta. 2016;**149**:43-52. DOI: 10.1016/j.talanta.2015.11.032

[60] Sirhan A, Tan GH, Al-Shunnaq A, Abdulra'uf L, Wong RCS. QuEChERS-HPLC method for aflatoxin detection of domestic and imported food in Jordan. Journal of Liquid Chromatography & Related Technologies. 2014;**37**(3):321-342. DOI: 10.1080/10826076.2012.745138

[61] Campone L, Piccinelli AL, Celano R, Rastrelli L. Application of dispersive liquid–liquid microextraction for the determination of aflatoxins B_1, B_2, G_1 and G_2 in cereal products. Journal of Chromatography A. 2011;**1218**:7648-7654. DOI: 10.1016/j.chroma.2011.05.028

[62] Afzali D, Ghanbarian M, Mostafavi A, Shamspur T, Ghaseminezhad S. A novel method for high preconcentration of ultra trace amounts of B_1, B_2, G_1 and G_2 aflatoxins in edible oils by dispersive liquid–liquid microextraction after immunoaffinity column clean-up. Journal of Chromatography A. 2012;**1247**:35-41. DOI: 10.1016/j.chroma.2012.05.051

[63] Lai XW, Sun DL, Ruan CQ, Zhang H, Liu CL. Rapid analysis of aflatoxins B_1, B_2, and ochratoxin A in rice samples using dispersive liquid-liquid microextraction combined with HPLC. Journal of Separation Science. 2014;**37**(1-2):92-98. DOI: 10.1002/jssc.201300970

[64] Amirkhizi B, Arefhosseini SR, Ansarin M, Nemati M. Aflatoxin B_1 in eggs and chicken livers by dispersive liquid-liquid microextraction and HPLC. Food Additives & Contaminants: Part B: Surveillance. 2015;**8**(4):245-249. DOI: 10.1080/19393210.2015.1067649

[65] Spietelun A, Marcinkowski Ł, de la Guardia M, Namieśnik J. Green aspects, developments and perspectives of liquid phase microextraction techniques. Talanta. 2014;**119**:34-45. DOI: 10.1016/j.talanta.2013.10.050

[66] Zhao J, Zhu Y, Jiao Y, Ning J, Yang Y. Ionic-liquid-based dispersive liquid-liquid microextraction combined with magnetic solid-phase extraction for the determination of aflatoxins B_1, B_2, G_1, and G_2 in animal feeds by high-performance liquid chromatography with fluorescence detection. Journal of Separation Science. 2016;**39**(19):3789-3797. DOI: 10.1002/jssc.201600671

[67] Shephard GS, Burger H-M, Gambacorta L, Krska R, Powers SP, Rheeder JP, Solfrizzo M, Sulyok M, Visconti A, Warth B, van der Westhuizen L. Mycological analysis and multimycotoxins in maize from rural subsistence farmers in the former Transkei, South Africa. Journal of Agricultural and Food Chemistry. 2013;**61**(4):8232-8240. DOI: 10.1021/jf4021762

[68] Arroyo-Manzanares N, Huertas-Pérez JF, García-Campaña AM, Gámiz-Gracia L. Aflatoxins in animal feeds: A straightforward and cost-effective analytical method. Food Control. 2015;**54**:74-78. DOI: 10.1016/j.foodcont.2015.01.027

[69] Ates E, Godula M, Stroka J, Senyuva H. Screening of plant and fungal metabolites in wheat, maize and animal feed using automated on-line clean-up coupled to high resolution mass spectrometry. Food Control. 2014;**142**:276-284. DOI: 10.1016/j.foodchem.2013.07.054

[70] Fabregat-Cabello N, Zomer P, Sancho JV, Roig-Navarro AF, Mol HGJ. Comparison of approaches to deal with matrix effects in LC-MS/MS based determinations of mycotoxins in food and feed. World Mycotoxin Journal. 2015;**9**(2):149-161. DOI: 10.3920/WMJ2014.1872

[71] Hu X, Hu R, Zhang Z, Li P, Zhang Q, Wang M. Development of a multiple immunoaffinity column for simultaneous determination of multiple mycotoxins in feeds using UPLC–MS/MS. Analytical and Bioanalytical Chemistry. 2016;**408**(22):6027-6036. DOI: 10.1007/s00216-016-9626-5

[72] Zhao Z, Liu N, Yang L, Deng Y, Wang J, Song S, Lin S, Wu A, Zhou Z, Hou J. Multimycotoxin analysis of animal feed and animal-derived food using LC-MS/MS system

with timed and highly selective reaction monitoring. Analytical and Bioanalytical Chemistry. 2015;**407**(24):7359-7368. DOI: 10.1007/s00216-015-8898-5

[73] Arroyo-Manzanares N, Huertas-Pérez JF, García-Campaña AM, Gámiz-Gracia L. Mycotoxin analysis: New proposals for sample treatment. Advances in Chemistry. 2014;**2014**(1-12):547506. DOI: 10.1155/2014/547506

[74] Lee S, Kim G, Moon J. Performance improvement of the one-dot lateral flow immunoassay for aflatoxin B_1 by using a smartphone-based reading system. Sensors. 2013;**13**(4): 5109-5116. DOI: 10.3390/s130405109

[75] Zhang D, Li P, Yang Y, Zhang Q, Zhang W, Xiao Z, Ding X. A high selective immunochromatographic assay for rapid detection of aflatoxin B_1. Talanta. 2011;**85**(1):736-742. DOI: 10.1016/j.talanta.2011.04.061

[76] Banerjee A, Shetty HS. Microbial load in poultry feed and detection of aflatoxin B_1 using monoclonal antibody-based enzyme linked immunosorbent assay. Letters in Applied Microbiology. 1992;**15**(3):89-91. DOI: 10.1111/j.1472-765X.1992.tb00733.x

[77] Rossi CN, Takabayashi CR, Ono MA, Saito GH, Itano EN, Kawamura O, Hirooka EY, Ono EYS. Immunoassay based on monoclonal antibody for aflatoxin detection in poultry feed. Food Chemistry. 2012;**132**(4):2211-2216. DOI: 10.1016/j.foodchem.2011.12.067

[78] Masinde LA, Sheng W, Xu X, Zhang Y, Yuan M, Kennedy IR, Wang S. Colloidal gold based immunochromatographic strip for the simple and sensitive determination of aflatoxin B_1 and B_2 in corn and rice. Microchimica Acta. 2013;**180**(9-10):921-928. DOI: 10.1007/s00604-013-1008-5

[79] Sun Q, Zhu Z, Deng Q-M, Liuab J-M, Shi G-Q. A "green" method to detect aflatoxin B_1 residue in plant oil based on a colloidal gold immunochromatographic assay. Analytical Methods. 2016;**8**:564-569. DOI: 10.1039/C5AY02779A

[80] Kok WT. Derivatization reactions for the determination of aflatoxins by liquid chromatography with fluorescence detection. Journal of Chromatography B Biomedical Applications. 1994;**659**(1-2):127-137

[81] Kok, WT, van Neer TC, Traag WA, Tuinstra LG. Determination of aflatoxins in cattle feed by liquid chromatography and post-column derivatization with electrochemically generated bromine. Journal of Chromatography. 1986;**367**(1):231-236

[82] Manetta AC, Di Giuseppe L, Giammarco M, Fusaro I, Simonella A, Gramenzi A, Formigoni A. High-performance liquid chromatography with post-column derivatisation and fluorescence detection for sensitive determination of aflatoxin M_1 in milk and cheese. Journal of Chromatography A. 2005;**1083**(1-2):219-222

[83] Woodman M, Zweigenbaum J. Agilent Technologies. Aflatoxin Analysis by HPLC with Aflatoxin Analysis by HPLC with Fluorescence Detection using Pos Fluorescence Detection using Post-Column Derivatization and Pre Column Derivatization and Pre-column Immunoaffinity Cleanup column Immunoaffinity [Internet]. 2008. Available

from: http://www.agilent.com/cs/library/posters/Public/Aflatoxin%20Analysis%20by%20HPLC%20with%20Fluorescence%20Detection.pdf [Accessed: 07/03/2017]

[84] Ramírez-Galicia G, Garduño-Juárez R, Vargas MG. Effect of water molecules on the fluorescence enhancement of Aflatoxin B_1 mediated by Aflatoxin B_1:β-cyclodextrin complexes. A theoretical study. Photochemical & Photobiological Sciences. 2007;**6**:110-118. DOI: 10.1039/B614107B

[85] Granados-Chinchilla F, Molina A, Chavarría G, Alfaro-Cascante M, Bogantes-Ledezma D, Murillo-Williams A. Aflatoxins occurrence trough the food chain in Costa Rica: Applying the One Health Approach to mycotoxins surveillance. Forthcoming

[86] Trombete FM, Santos TB, Direito GM, Fraga ME, Saldanha T. In-house validation of a method for determining aflatoxins B_1, B_2, G_1 and G_2 in wheat and wheat by-products. Pesquisa Agropecuária Tropical. 2014;**44**(3):255-262. DOI: 10.1590/S1983-40632014000300003

[87] Cortés G, Carvajal M, Méndez-Ramírez I, Ávila-González E, Chilpa-Galván N, Castillo-Urueta P, Flores CM. Identification and quantification of aflatoxins and aflatoxicol from poultry feed and their recovery in poultry litter. Processing, Products and Food Safety. 2010;**89**:993-1001. DOI: 10.3382/ps.2009-00019

[88] Benvenuti ME, Di Gioia AJ. Waters Corporation. Rapid Analysis of Aflatoxins Without Derivatization Using Ultra Performance Liquid Chromatography and Fluorescence Detection [Internet]. 2009. Available from: http://www.waters.com/webassets/cms/library/docs/720002644en.pdf [Accessed: 07/03/2017]

[89] Pirestani A, Tabatabaei SN, Fazeli MH, Antikchi M, Baabaei M. Comparison of HPLC and Elisa for determination of aflatoxin concentration in the milk and feeds of dairy cattle. Journal of Research in Agricultural Science. 2011;**7**(1):71-78

[90] Pereira MMG, de Carvalho EP, Prado G, da Rocha Rosa CA, Veloso T, de Souza LAF, Ribeiro JMM. Detection of aflatoxins in dairy cattle feed and milk in Lavras, Minas Gerais - Brazil. Ciência e Agrotecnologia. 2005;**29**(1):106-112. DOI: 10.1590/S1413-70542005000100013

[91] Warth B, Parich A, Atehnkeng J, Bandyopadhyay R, Schuhmacher R, Sulyok M, Krska R. Quantitation of mycotoxins in food and feed from Burkina Faso and Mozambique using a modern LC-MS/MS multitoxin method. Journal of Agricultural and Food Chemistry. 2012;**60**(36):9352-9363. DOI: 10.1021/jf302003n

[92] de Souza MLM, Sulyok M, Freitas-Silva O, Costa SS, Brabet C, Machinski MJ, Sekiyama BL, Vargas EA, Krska R, Schuhmacher R. Cooccurrence of mycotoxins in maize and poultry feeds from Brazil by liquid chromatography/tandem mass spectrometry. Scientific World Journal. 2013;**2013**(1-9):427369. DOI: 10.1155/2013/427369

[93] Njumbe EE, Van Poucke C, De Saeger S. A multi-analyte LC-MS/MS method for the analysis of 23 mycotoxins in different sorghum varieties: The forgotten sample matrix. Food Chemistry. 2015;**177**:397-404. DOI: 10.1016/j.foodchem.2015.01.060

[94] Zhang K, Wong JW, Krynitsky AJ, Trucksess MW. Perspective on advancing FDA regulatory monitoring for mycotoxins in foods using liquid chromatography and mass spectrometry (review). Journal of AOAC International. 2016;**99**(4):890-894(5). DOI: 10.5740/jaoacint.16-0116

[95] Lattanzio VM, Gatta SD, Suman M, Visconti A. Development and in-house validation of a robust and sensitive solid-phase extraction liquid chromatography/tandem mass spectrometry method for the quantitative determination of aflatoxins B_1, B_2, G_1, G_2, ochratoxin A, deoxynivalenol, zearalenone, T-2 and HT-2 toxins in cereal-based foods. Rapid Communications in Mass Spectrometry. 2011;**25**(13):1869-1880. DOI: 10.1002/rcm.5047

[96] Zhang K, Wong JW, Krynitsky AJ, Trucksess MW. Determining mycotoxins in baby foods and animal feeds using stable isotope dilution and liquid chromatography tandem mass spectrometry. Journal of Agricultural and Food Chemistry. 2014;**62**(36):8935-8943. DOI: 10.1021/jf503943r

[97] Li P, Zhang Z, Hu X, Zhang Q. Advanced hyphenated chromatographic-mass spectrometry in mycotoxin determination: Current status and prospects. Mass Spectrometry Reviews. 2013;**32**(6):420-445. DOI: 10.1002/mas.21377

[98] Ok HE, Tian F, Hong EY, Paek O, Kim S-H, Kim D, Chun HS. Harmonized collaborative validation of aflatoxins and sterigmatocystin in white rice and sorghum by liquid chromatography coupled to tandem mass spectrometry. Toxins. 2016;**8**(12):371. DOI: 10.3390/toxins8120371

[99] Zachariasova M, Dzuman Z, Veprikova Z, Hajkova K, Jiru M, Vaclavikova M, Zachariasova A, Pospichalova M, Florian M, Hajslova J. Occurrence of multiple mycotoxins in European feedingstuffs, assessment of dietary intake by farm animals. Animal Feed Science and Technology. 2014;**193**:124-140. DOI: 10.1016/j.anifeedsci.2014.02.007

[100] Escrivá L, Font G, Berrada H, Manyes L. Mycotoxin contamination in laboratory rat feeds and their implications in animal research. Toxicology Mechanisms and Methods. 2016;**26**(7):529-537. DOI: 10.1080/15376516.2016.1206163

[101] Oplatowska-Stachowiak M, Haughey SA, Chevallier OP, Galvin-King P, Campbell K, Magowan E, Adam G, Berthiller F, Krska R, Elliott CT. Determination of the mycotoxin content in distiller's dried grain with solubles using a multianalyte UHPLC–MS/MS method. Journal of Agricultural and Food Chemistry. 2015;**63**(43):9441-9451. DOI: 10.1021/acs.jafc.5b03844

[102] Yibadatihan S, Jinap S, Mahyudin NA. Simultaneous determination of multi-mycotoxins in palm kernel cake (PKC) using liquid chromatography-tandem mass spectrometry (LC-MS/MS). Food Additives & Contaminants. Part A, Chemistry, Analysis, Control, Exposure & Risk Assessment. 2014;**31**(12):2071-2079. DOI: 10.1080/19440049.2014.978396

[103] Pettersson H. Mycotoxin contamination of animal feed. In: N. Magan, M. Olsen, editors. Animal Feed Contamination Effects on Livestock and Food Safety. 1st ed. Philadelphia, PA, USA: Woodhead Publishing Limited; 2012. pp. 233-285

[104] Mcelhinney C, O'Kiely P, Elliott C, Danaher M. Development and validation of an UHPLC-MS/MS method for the determination of mycotoxins in grass silages. Food Additives & Contaminants. Part A, Chemistry, Analysis, Control, Exposure & Risk Assessment. 2015;32(12):2101-2112. DOI: 10.1080/19440049.2015.1094830

[105] Dzuman Z, Zachariasova M, Lacina O, Veprikova Z, Slavikova P, Hajslova J. A rugged high-throughput analytical approach for the determination and quantification of multiple mycotoxins in complex feed matrices. Talanta. 2014;**121**:263-272. DOI: 10.1016/j.talanta.2013.12.064

[106] Babu D, Muriana PM. Immunomagnetic bead-based recovery and real time quantitative PCR (RT iq-PCR) for sensitive quantification of aflatoxin B(1). Journal of Microbiological Methods. 2011;**86**(2):188-194. DOI: 10.1016/j.mimet.2011.05.002

[107] Babu D, Muriana PM. Sensitive quantification of aflatoxin B_1 in animal feeds, corn feed grain, and yellow corn meal using immunomagnetic bead-based recovery and real-time immunoquantitative-PCR. Toxins. 2014;**6**: 3223-3237. DOI: 10.3390/toxins6123223

[108] Ramesh J, Sarathchandra G, Sureshkumar V. Analysis of feed samples for aflatoxin B_1 contamination by HPTLC—A validated method. International Journal of Current Microbiology and Applied Science. 2013;**2**(5):373-377

[109] Kotinagu K, Mohanamba T, Rathna Kumari L. Assessment of aflatoxin B_1 in livestock feed and feed ingredients by high-performance thin layer chromatography. Veterinary World. 2015;**8**(12):1396-1399. DOI: 10.14202/vetworld.2015.1396-1399

[110] Zhang Z, Tang X, Wang D, Zhang Q, Li P, Ding X. Rapid on-site sensing aflatoxin B_1 in food and feed via a chromatographic time-resolved fluoroimmunoassay. PLoS ONE. 2015;**10**(4):e0123266. DOI: 10.1371/journal.pone.0123266

[111] Wang D, Zhang Z, Li P, Zhang Q, Ding X, Zhang W. Europium nanospheres-based time-resolved fluorescence for rapid and ultrasensitive determination of total aflatoxin in feed. Journal of Agricultural and Food Chemistry. 2015;**63**(47): 10313-10318. DOI: 10.1021/acs.jafc.5b03746

[112] Ren M, Xu H, Huang X, Kuang M, Xiong Y, Xu H, Xu Y, Chen H, Wang A. Immunochromatographic assay for ultrasensitive detection of aflatoxin B_1 in maize by highly luminescent quantum dot beads. ACS Applied Material Interfaces. 2014;**6** (16):14215-14222. DOI: 10.1021/am503517s

[113] Setlem K, Mondal B, Ramlal S, Kingston J. Immunoaffinity SELEX for simple, rapid, and cost-effective aptamer enrichment and identification against aflatoxin B_1. Frontiers in Microbiology. 2016;**7**:1909. DOI: 10.3389/fmicb.2016.01909

[114] He T, Wang Y, Li P, Zhang Q, Lei J, Zhang Z, Ding X, Zhou H, Zhang W. Nanobody-based enzyme immunoassay for aflatoxin in agro-products with high tolerance to cosolvent methanol. Analytical Chemistry. 2014;**86**(17):8873-8880. DOI: 10.1021/ac502390c

[115] Xiong Y, Tu Z, Huang X, Xie B, Xiong Y, Xu Y. Magnetic beads carrying poly(acrylic acid) brushes as "nanobody containers" for immunoaffinity purification of aflatoxin B_1 from corn samples. RSC Advances. 2015;**5**:77380-77387. DOI: 10.1039/C5RA15843E

[116] Gruber-Dorninger C, Novak B, Nagl V, Berthiller F. Emerging Mycotoxins: Beyond Traditionally Determined Food Contaminants. Journal of Agricultural and Food Chemistry. 2016. DOI: 10.1021/acs.jafc.6b03413 [Epub ahead of print]

[117] Suganthi F, Benazir JF, Pooja MP, Swathee C, Namitha R, Balaji KS. Detection of aflatoxins in animal feeds. Environmental Science. 2009;**4**(2):51-54

[118] Chandra H, Bahuguna J, Singh A. Detection of aflatoxin in *Zea mays* L. from Indian Markets by Competitive ELISA. Octa Journal of Biosciences. 2013;**1**(1):62-68

[119] Queiroz B, Pereyra CM, Keller KM, Almeida T, Cavaglieri LR, Magnoli CE, da Rocha Rosa CA. Fungal contamination and determination of fumonisins and aflatoxins in commercial feeds intended for ornamental birds in Rio de Janeiro, Brazil, Letters in Applied Microbiology. 2013;**57**(5):405-411. DOI: 10.1111/lam.12127

[120] Sebők F, Dobolyi C, Zágoni D, Risa A, Krifaton C, Hartman M, Cserháti M, Szoboszlay S, Kriszt B. Aflatoxigenic *Aspergillus Flavus* and *Aspergillus Parasiticus* strains in hungarian maize fields. Acta Microbiologica et Immunologica Hungarica. 2016;**63**(4):491-502. DOI: 10.1556/030.63.2016.012

[121] Carvalho BF, Avila CLS, Krempser PM, Batista LR, Pereira MN, Schwan RF. Occurrence of mycotoxins and yeasts and moulds identification in corn silages in tropical climate. Journal of Applied Microbiology. 2015;**120**:1181-1192. DOI: 10.1111/jam.13057

[122] Kaya-Celiker H, Mallikarjunan PK, Kaaya A. Characterization of invasion of genus *Aspergillus* on peanut seeds using FTIR-PAS. Food Analytical Methods. 2015;**9**(1):105-113. DOI: 10.1007/s12161-015-0159-x

[123] Ibrahim F, Jalal H, Khan AB, Asghar MA, Iqbal J, Ahmed A, Nadeem G. Prevalence of aflatoxigenic *Aspergillus* in food and feed samples from Karachi, Pakistan. Journal of Infection and Molecular Biology. 2016;**4**(1):1-8. DOI: 10.14737/journal.jimb/2016/4.1.1.8

[124] Perrone G, Gallo A. *Aspergillus* species and their associated mycotoxins. In: Moretti A, Susca A, editors. Mycotoxigenic Fungi: Methods and Protocols. 1st ed. Humana Press; U.S.A 2017. pp. 33-49. DOI: 10.1007/978-1-4939-6707-0